Mariano Giaquinta
Giuseppe Modica

Mathematical Analysis

An Introduction to Functions of Several Variables

Birkhäuser
Boston • Basel • Berlin

Mariano Giaquinta
Scuola Normale Superiore
Piazza dei Cavalieri, 7
I-56100 Pisa, Italy
giaquinta@sns.it

Giuseppe Modica
Dipartimento di Matematica Applicata
Università di Firenze
Via S. Marta, 3
I-50139 Firenze, Italy
giuseppe.modica@unifi.it

Library of Congress Control Number: 2009922164

ISBN 978-0-8176-4509-0 (hardcover)   e-ISBN 978-0-8176-4612-7
ISBN 978-0-8176-4507-6 (softcover)
DOI 10.1007/978-0-8176-4612-7

Mathematics Subject Classification (2000): 00A35, 32A10, 42B05, 49J40, 34D99

Cover design by Alex Gerasev.

Printed on acid-free paper.

www.birkhauser.com

Preface

This book[1] introduces the main ideas and fundamental methods of differential and integral calculus for functions of several variables.

In Chapter 1 we discuss differential calculus for functions of several variables with a short excursion into differential calculus in Banach spaces.

In Chapter 2 we present some of the most relevant results of the Lebesgue integration theory, including the limit and approximation theorems, Fubini's theorem, the area and coarea theorems, and Gauss–Green formulas. The aim is to provide the reader with all that is needed to use the power of Lebesgue integration. For this reason some details as well as some proofs concerning the formulation of the theory are skipped, as we think they are more appropriate in the general context of measure theory.

In Chapter 3 we deal with potentials and integration of differential 1-forms, focusing on solenoidal and irrotational fields.

Chapter 4 provides a sufficiently wide introduction to the theory of holomorphic functions of one complex variable. We present the fundamental theorems and discuss singularities and residues as well as Riemann's theorem on conformal representation and the related Schwarz and Poisson formulas and Hilbert's transform.

In Chapter 5, we discuss the notions of immersed and embedded surface in $\mathbb{R}^n$, and we present the implicit function theorem and some of its applications to vector fields, constrained minimization, and functional dependence. The chapter ends with the study of some notions of the local theory of curves and surfaces, such as of curvature, first variation of area, the Laplace–Beltrami operator, and distance function.

In Chapter 6, after a few preliminaries about systems of linear ordinary differential equations, we discuss a few results concerning the stability of nonlinear systems and the Poincaré–Bendixson theorem in order to show that dynamical systems with one degree of freedom do not present chaos, in contrast with the one-dimensional discrete dynamics or the higher-dimensional continuous dynamics.

---

[1] This book is a translated and revised edition of M. Giaquinta, G. Modica, *Analisi Matematica, IV. Funzioni di più variabili*, Pitagora Ed., Bologna, 2005.

The study of this volume requires a stronger effort compared to that of [GM1],[GM2],and [GM3][2] both because of intrinsic difficulties and broad scope of the themes we present. We think, in fact, that it is useful for the reader to have a wide spectrum of contexts in which these ideas play an important role and wherein even the technical and formal aspects play a role. However, we have tried to keep the same spirit, always providing examples, illustrations, and exercises to clarify the main presentation, omitting several technicalities or developments that we thought to be too advanced.

We are greatly indebted to Cecilia Conti for her help in polishing our first draft and we warmly thank her. We would like to thank also Paolo Acquistapace, Timoteo Carletti, Giulio Ciraolo, Roberto Conti, Giovanni Cupini, Matteo Focardi, Pietro Majer, and Stefano Marmi for their comments and their invaluable help in catching errors and misprints and Stefan Hildebrandt for his comments and suggestions concerning especially the choice of illustrations. Our special thanks also go to all members of the editorial and technical staff of Birkhäuser for the excellent quality of their work and especially to Rebecca Biega and the executive editor Ann Kostant.

Note: We have tried to avoid misprints and errors. But, as most authors, we are imperfect. We will be very grateful to anybody who wants to inform us about errors or just misprints, or wants to express criticism or other comments. Our e-mail addresses are

giaquinta@sns.it   giuseppe.modica@unifi.it

We shall try to maintain any errata and corrigenda at the following web pages:

http://www.sns.it/~giaquinta

http://www.dma.unifi.it/~modica

Mariano Giaquinta
Giuseppe Modica
Pisa and Firenze
July 2007

---

[2] We shall refer to the following sources as [GM1], [GM2], and [GM3], respectively: [GM1]: M. Giaquinta, G. Modica, *Mathematical Analysis, Functions of One Variable*, Birkhäuser, Boston, 2003; [GM2]: M. Giaquinta, G. Modica, *Mathematical Analysis, Approximation and Discrete Processes*, Birkhäuser, Boston, 2004; [GM3]: M. Giaquinta, G. Modica, *Mathematical Analysis, Linear and Metric Structures and Continuity*, Birkhäuser, Boston, 2007.

Contents

1. Differential Calculus

In this chapter we discuss the basic notions of differential calculus of functions of several variables.

1.1 Differential Calculus of Scalar Functions

1.1.1 Directional and partial derivatives, and the differential

a. Directional derivatives

Given $x_0 \in \mathbb{R}^n$ and a direction $v \in \mathbb{R}^n$, the map $r(t) := x_0 + vt$, $t \in \mathbb{R}$, is the parameterization of the line through x_0 and $x_0 + tv$, called the *parametric equation of the line through x_0 with direction v*. It represents the motion of a point that at time $t = 0$ is at 0 and moves with constant velocity v.

Let $A \subset \mathbb{R}^n$ be an open set, $x_0 \in A$, and suppose that the ball $B(x_0, \epsilon_0)$ of center x_0 and radius ϵ_0 is contained in A. For each $v \in \mathbb{R}^n$, then $r(t) := x_0 + tv$ belongs to A for $|t| \leq \epsilon_0/|v|$ if $v \neq 0$ or for all $t \in \mathbb{R}$ if $v = 0$. Consequently, given $f : A \to \mathbb{R}$, the composite function

$$\phi_v(t) := f(x_0 + tv), \qquad t \in r^{-1}(A), \tag{1.1}$$

called the *restriction of f to the line through x with direction v* is well defined in the interval $|t| < \epsilon_0/|v|$ ($\mathbb{R}$ if $v = 0$).

1.1 Definition. *We say that f has a* directional derivative *at x_0 in the direction v if the following limit exists and is finite*

$$\lim_{t \to 0} \frac{f(x_0 + tv) - f(x_0)}{t} = \lim_{t \to 0} \frac{\phi_v(t) - \phi_v(0)}{t}, \tag{1.2}$$

i.e., if ϕ_v is diferentiable at 0. The number $\phi'_v(0)$ is called the derivative *of f at x_0 in the direction v and is denoted by one of the following symbols*

M. Giaquinta and G. Modica, *Mathematical Analysis: An Introduction to Functions of Several Variables*, DOI: 10.1007/978-0-8176-4612-7_1,
© Birkhäuser Boston, a part of Springer Science + Business Media, LLC 2010

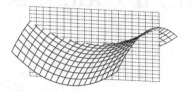

Figure 1.1. The restriction of the graph of f to a line.

$$\frac{\partial f}{\partial v}(x_0), \qquad D_v f(x_0), \qquad or \qquad f_v(x_0).$$

Notice that $\frac{\partial f}{\partial v}(x_0) = 0$ if $v = 0$.

Let $(e_1, e_2, \ldots, e_n)$ be a basis of $\mathbb{R}^n$ and let $(x^1, x^2, \ldots, x^n)$ be the corresponding system of coordinates. For $i = 1, \ldots, n$ the *partial derivative* of f in the direction x^i is defined as the derivative of f in the direction of the corresponding direction e_i of the basis,

$$\frac{\partial f}{\partial x^i}(x_0) := \frac{\partial f}{\partial e_i}(x_0),$$

if this directional derivatives exists. The partial derivative $\frac{\partial f}{\partial x^i}(x_0)$ is also denoted by

$$D_i f(x_0) \qquad or \qquad f_{x^i}(x_0).$$

1.2 ¶ Fermat's theorem. Suppose that $f : A \to \mathbb{R}$ has a maximum point or a minimum point at an interior point $x_0 \in A$ and that f has a derivative at x_0 in the direction v. Show that

$$\frac{\partial f}{\partial v}(x_0) = 0.$$

b. The differential

The mere existence of all directional derivatives has no further consequences such as continuity. The following example illustrates the situation and motivates the introduction of the stronger notion of *differentiability* of functions of several variables.

1.3 Example.   (i) The function $f : \mathbb{R}^2 \to \mathbb{R}$, which is defined to be zero at points in the coordinate axes and one outside, has zero derivatives in the directions of the axes and is not continuous at $(0,0)$.

(ii) The function

$$f(x,y) := \begin{cases} 1 & \text{if } y = x^2, \ x \neq 0, \\ 0 & \text{otherwise} \end{cases}$$

is discontinuous at $(0,0)$ even though all its directional derivatives vanish at $(0,0)$. Similarly, the function

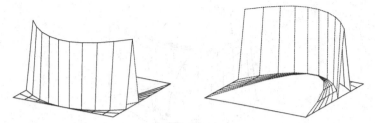

Figure 1.2. The graph of a function similar to the graph of the function in (ii) Example 1.3.

$$f(x,y) := \begin{cases} \left(\frac{x^2y}{x^4+y^2}\right)^2 & \text{if } (x,y) \neq (0,0), \\ 0 & \text{if } (x,y) = (0,0) \end{cases}$$

has vanishing directional derivatives at $(0,0)$ and is not continuous, see Figures 1.2 and 1.3.

(iii) The function

$$f(x,y) := \begin{cases} \frac{x^2y}{x^2+y^2} & \text{if } x \neq 0, \\ 0 & \text{if } x = 0 \end{cases}$$

is continuous at $(0,0)$, all its directional derivatives vanish at $(0,0)$, but the so-called *tangent map* $v \to \frac{\partial f}{\partial v}(x_0)$ is not linear since the partial derivatives of f are zero at $(0,0)$

$$0 = 1\frac{\partial f}{\partial(1,0)}(0,0) + 1\frac{\partial f}{\partial(0,1)}(0,0) \neq \frac{\partial f}{\partial(1,1)}(0,0) = 1/2,$$

see Figure 1.4.

(iv) It can happen that all directional derivatives of a function f vanish at $(0,0)$, the function f is continuous, and there exist two different paths through $(0,0)$ that are *tangent at* $(0,0)$ and, along those paths, we arrive at $(0,0)$ with different slopes. For instance, all directional derivatives at $(0,0)$ of the function

$$f(x,y) := \begin{cases} x & \text{if } y = x^2, \\ 0 & \text{otherwise} \end{cases}$$

vanish and f is continuous at $(0,0)$; however, along the curves $x \to (x,x^2)$ and $x \to (x,0)$ we find

$$\lim_{x\to 0} \frac{x-0}{x} = 1 \neq \frac{\partial f}{\partial(1,0)}(0,0) = 0.$$

In particular, there is no way of defining the slope of the graph of f at $(0,0)$ in the direction $(1,0)$.

1.4 Definition. *Let $f : A \to \mathbb{R}$ be a function with domain $A \subset \mathbb{R}^n$, and let x_0 be an interior point of A. We say that f is* differentiable *at x_0 if there exists a linear function $L : \mathbb{R}^n \to \mathbb{R}$ (depending on x_0) such that*

$$\lim_{h\to 0} \frac{f(x_0+h) - f(x_0) - L(h)}{|h|} = 0. \tag{1.3}$$

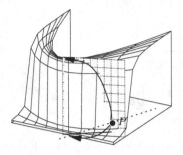

Figure 1.3. Moving along a straight line toward the precipice, we end up at the top. Moving instead along a sufficiently curved path, we end up at the bottom.

Here $|h|$ denotes the Euclidean norm of h.

The linear map L is unique provided it exists, by Proposition 1.5 below. L is called the *differential* or the linear *tangent map* of f at x_0 and will be denoted by df_{x_0} or $df(x_0)$. Whenever a domain A is given, we say that *f is differentiable in A if f is differentiable at every point of A.*

1.5 Proposition. *Let $f : A \to \mathbb{R}$, $A \subset \mathbb{R}^n$, and let x_0 be an interior point of A. If f is differentiable at x_0, then*

(i) *f is continuous at x_0,*
(ii) *f has derivatives at x_0 in every direction and*

$$\frac{\partial f}{\partial v}(x_0) = df_{x_0}(v) \qquad \forall v \in \mathbb{R}^n. \tag{1.4}$$

In particular,

○ *the tangent map $v \to \frac{\partial f}{\partial v}(x_0)$, $v \in \mathbb{R}^n$, is linear,*
○ *the differential is unique, if it exists.*

Proof. (i) Let $h \in \mathbb{R}^n$, $h \neq 0$. We have

$$\left| f(x_0 + h) - f(x_0) - L(h) \right| \leq |h| \left| \frac{f(x_0 + h) - f(x_0) - L(h)}{|h|} \right| \to 0 \cdot 0 = 0,$$

and, since $L(h) \to 0$ as $h \to 0$ (since linear maps of $\mathbb{R}^n$ are continuous), we conclude that $f(x_0 + h) - f(x_0) \to 0$.

(ii) Let L be a differential of f at x_0 and let $v \in \mathbb{R}^n$, $v \neq 0$. From the definition of differential and because of the theorems of the limits of composite function, we infer that

$$\left| \frac{f(x_0 + tv) - f(x_0)}{t} - L(v) \right| = \left| \frac{f(x_0 + tv) - f(x_0) - L(tv)}{t} \right| \to 0$$

as $t \to 0$. Therefore f has derivative at x_0 in the direction v; moreover,

$$\frac{\partial f}{\partial v}(x_0) = L(v),$$

and the uniqueness of the differential follows at once from the uniqueness of the directional derivatives. □

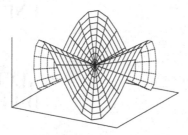

Figure 1.4. The restriction of the graph of $f(x,y) = (x^3 - 3xy^2)/(x^2 + y^2)$ to a line through the origin is a line; nevertheless, $\mathcal{G}_f$ has no *tangent plane* at $(0,0)$.

1.6 Remark. We can write (1.3) in Landau's notation as

$$f(x_0 + h) = f(x_0) + df_{x_0}(h) + o(|h|) \qquad |h| \to 0. \tag{1.5}$$

As we have just seen, if (1.5) holds, f is continuous. Therefore we can assume that $o(|h|)$ is defined and continuous at 0 by setting $o(0) := 0$.

For functions of one variable, differentiability is equivalent to existence of the derivative, see [GM1]. For functions of two or more variables, Proposition 1.5 shows that differentiability implies continuity of the function, existence of all directional derivatives, and linearity of the thangent map, i.e., (1.4). None of the opposite implications holds. This should not be surprising. In fact, differentiability is a property of approximation with a linear map expressed by a limit in several variables, whereas a directional derivative involves the restriction of f along a line through x_0 and a limit in one variable. Finally, we repeat, in general the behavior of a function along a curve γ through x_0 and along the tangent line to γ at x_0 can be very different. Indeed, the function in (ii) Example 1.3 has directional derivatives, but it is discontinuous; the function in (iii) Example 1.3 has directional derivatives but does not satisfy (1.4); the function in (iv) Example 1.3 is continuous at $(0,0)$, has vanishing directional derivatives, satisfies (1.4), but is not differentiable at $(0,0)$.

c. The gradient vector

Suppose that $\mathbb{R}^n$ is endowed with an inner product $\bullet$. By Riesz's theorem, see [GM3], to every linear map $L : \mathbb{R}^n \to \mathbb{R}$ we can associate a unique vector $x_L \in \mathbb{R}^n$ characterized by

$$L(h) = h \bullet x_L \qquad \forall h \in \mathbb{R}^n.$$

In particular, to the differential of f at x_0 we can associate a unique vector of $\mathbb{R}^n$, called the *gradient* of f at x_0 and denoted by $\nabla f(x_0)$ or $\operatorname{grad} f(x_0)$, such that

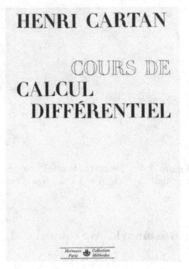

Figure 1.5. Frontispieces of two well-known treatises on the differential calculus of functions of several variables.

$$df_{x_0}(h) = h \bullet \nabla f(x_0) = h \bullet \operatorname{grad} f(x_0) \qquad \forall h \in \mathbb{R}^n. \qquad (1.6)$$

Notice that

$$\ker df_{x_0} = \operatorname{grad} f(x_0)^\perp.$$

d. Direction of steepest ascent

Let $f : A \subset \mathbb{R}^n \to \mathbb{R}$ be differentiable at an interior point x_0 of A. Cauchy's inequality yields

$$\left| \frac{\partial f}{\partial v}(x_0) \right| = |v \bullet \nabla f(x_0)| \le ||v|| \; ||\nabla f(x_0)||$$

and, if $\nabla f(x_0) \ne 0$, with equality

$$\frac{\partial f}{\partial v}(x_0) = ||v|| \; ||\nabla f(x_0)||$$

holding if and only if v is a positive multiple of $\nabla f(x_0)$; here $||x|| := \sqrt{x \bullet x}$ is the norm induced by the inner product. In other words, the maximum of the tangent map $v \to df_{x_0}(v)$, when v varies in $\{v \,|\, ||v|| = 1\}$ is $||\nabla f(x_0)||$ and, if $\nabla f(x_0) \ne 0$, the maximum is attained at

$$v = \frac{\nabla f(x_0)}{||\nabla f(x_0)||}. \qquad (1.7)$$

Let $\gamma :]-1, 1[\to A$ be a differentiable curve with $\gamma(0) = x_0$ and $\gamma'(0) = v$. It is easily seen that if f is differentiable at x_0, then the curve $t \to f(\gamma(t))$ is differentiable at 0 and

$$\frac{d(f \circ \gamma)}{dt}(0) = \nabla f(x_0) \bullet \gamma'(0), \tag{1.8}$$

see also Theorem 1.22. In particular, the growth of f at x_0 along every regular curve γ through x_0 depends on γ only through its velocity at zero, $\gamma'(0) = v$. Consequently, for a differentiable function f at x_0, we may introduce the notion of *slope at x_0 in the direction v* as the real number $\frac{\partial f}{\partial v}(x_0) = df_{x_0}(v)$, and state, compare Example 1.3 (iv), the following.

1.7 Proposition. *Let $f : A \subset \mathbb{R}^n \to \mathbb{R}$ be differentiable at an interior point x_0 of A and suppose that $\nabla f(x_0) \neq 0$. Then the direction of maximum slope of f at x_0 among all directions v such that $\|v\| = 1$ is $\nabla f(x_0)/\|\nabla f(x_0)\|$.*

1.1.2 Directional derivatives and differential in coordinates

a. Partial derivatives

Let $(e_1, e_2, \ldots, e_n)$ be a basis of $\mathbb{R}^n$ and let $(x^1, x^2, \ldots, x^n)$ be the corresponding coordinate system.

1.8 Definition. *The derivative of f at x_0 in the direction e_i is called the partial derivative of f with respect to x^i and will be denoted by one of the symbols*

$$\frac{\partial f}{\partial x^i}(x_0) = \frac{\partial}{\partial x^i} f(x_0) = D_i f(x_0) = f_{x^i}(x_0) := \frac{\partial f}{\partial e_i}(x_0).$$

By (1.2) the partial derivative of f with respect to x^i at $x_0 = (x_0^1, x_0^2, \ldots, x_0^n)$ is the derivative of the function of *one variable*

$$t \to f(x_0^1, \ldots, x_0^{i-1}, t, x_0^{i+1}, \ldots, x_0^n)$$

at $t = x_0^i$. In other words, the partial derivative with respect to x^i is computed by taking the remaining variables $(x^1, \ldots, x^{i-1}, x^{i+1}, x^n)$ as constant and differentiating with respect to x^i.

1.9 Example. If we want to compute $\frac{\partial f}{\partial x}(1, 1)$ where $f(x, y) := x^2 + y^2$, we consider $\varphi(x) := f(x, 1) = x^2 + 1$ and then $\frac{\partial f}{\partial x}(1, 1) = \frac{d\varphi}{dx}(1) = 2$.

b. Jacobian matrix

Recall that if $(e_1, e_2, \ldots, e_n)$ is a basis of $\mathbb{R}^n$, and $L : \mathbb{R}^n \to \mathbb{R}$ is a linear map, then $L(h)$, $h \in \mathbb{R}^n$, can be written as the product rows by columns of the $1 \times n$ matrix $\mathbf{L} := (L(e_1), L(e_2), \ldots, L(e_n))$ and the vector of the coordinates of h,

$$L(h) = L\left(\sum_{i=1}^{n} h^i e_i\right) = \sum_{i=1}^{n} L(e_i)h^i = \mathbf{L} \begin{pmatrix} h^1 \\ h^2 \\ \vdots \\ h^n \end{pmatrix}.$$

In particular, if $f : A \subset \mathbb{R}^n \to \mathbb{R}$ is differentiable at an interior point x_0 of A and if we introduce the $1 \times n$ matrix,

$$\mathbf{D}f(x_0) := \left(\frac{\partial f}{\partial x^1}(x_0), \frac{\partial f}{\partial x^2}(x_0), \ldots, \frac{\partial f}{\partial x^n}(x_0) \right),$$

called the *Jacobian matrix of f at x_0*, (1.4) can be rewritten as

$$df_{x_0}(h) = \sum_{i=1}^{n} df_{x_0}(e_i)h^i = \sum_{i=1}^{n} \frac{\partial f}{\partial x^i}(x_0)h^i = \mathbf{D}f(x_0)\, h \qquad (1.9)$$

for all $h = (h^1, h^2, \ldots, h^n)^T \in \mathbb{R}^n$.

1.10 ¶. Notice that $f : A \subset \mathbb{R}^2 \to \mathbb{R}$ is differentiable at an interior point x_0 of A if and only if there exist $a, b \in \mathbb{R}$ such that

$$f(u_0 + h, v_0 + k) - f(u_0, v_0) = ah + bk + o(\sqrt{h^2 + k^2}) \qquad \text{as } (h, k) \to (0, 0)$$

and moreover,

$$a = \frac{\partial f}{\partial u}(u_0, v_0), \qquad b = \frac{\partial f}{\partial v}(u_0, v_0).$$

c. The differential in the dual basis

Let $(e_1, e_2, \ldots, e_n)$ be a basis of $\mathbb{R}^n$ and let $x^i : \mathbb{R}^n \to \mathbb{R}$ be the linear maps that associate its ith component to each $h \in \mathbb{R}^n$. Since the map x^i is linear, it agrees with its differential (at any point),

$$dx^i_{x_0}(h) = x^i(h) = h^i, \qquad \forall h \in \mathbb{R}^n.$$

Therefore, for and any linear map $L : \mathbb{R}^n \to \mathbb{R}$ we can write

$$L(h) = \sum_{i=1}^{n} L(e_i)h^i = \sum_{i=1}^{n} L(e_i)\, dx^i(h)$$

or, equivalently, as maps

$$L = \sum_{i=1}^{n} L(e_i) dx^i. \tag{1.10}$$

In particular, if f is differentiable at x_0, then

$$df_{x_0} = \sum_{i=1}^{n} \frac{\partial f}{\partial x^i}(x_0) dx^i. \tag{1.11}$$

d. The gradient vector in coordinates

Writing (1.6) in coordinates with respect to a basis $(e_1, e_2, \ldots, e_n)$, we find the relation between the components of the gradient vector and the components of the Jacobian matrix. If we denote by $\mathbf{G} = (\mathbf{G}_{ij})$ the *metric tensor* defined by $\mathbf{G}_{ij} := e_i \bullet e_j$, see, e.g., [GM3], then we find

$$\nabla f(x_0) = \mathbf{G}^{-1} \mathbf{D} f(x_0)^T.$$

In particular, if $(e_1, e_2, \ldots, e_n)$ is an orthonormal basis, then $\mathbf{G} = \mathrm{Id}$, hence

$$\nabla f(x_0) = \left(\frac{\partial f}{\partial x^1}(x_0), \frac{\partial f}{\partial x^2}(x_0), \ldots, \frac{\partial f}{\partial x^n}(x_0) \right)^T. \tag{1.12}$$

e. The tangent plane

1.11 Graphs of linear maps. Recall that the *graph of $f : A \subset \mathbb{R}^n \to \mathbb{R}$* is the subset of $\mathbb{R}^{n+1}$ defined by

$$\mathcal{G}_f := \left\{ (x, y) \in \mathbb{R}^n \times \mathbb{R} \,\middle|\, x \in A, \ y = f(x) \right\},$$

and trivially

$$\mathcal{G}_f = \left\{ (x, f(x)) \in \mathbb{R}^n \times \mathbb{R} \,\middle|\, x \in A \right\} = \mathrm{Im}\,(\mathrm{Id} \times f)(A)$$

i.e., $\mathcal{G}_f$ is the image of the injective map $x \to \mathrm{Id} \times f(x) := (x, f(x))$, $x \in A$.

The graph $\mathcal{G}_L$ of a linear map $L : \mathbb{R}^n \to \mathbb{R}$, $\{(x, y) \,|\, y - L(x) = 0\}$ is therefore a linear subspace of $\mathbb{R}^n \times \mathbb{R}$ of dimension n, and, if $(e_1, e_2, \ldots, e_n)$ is a basis of $\mathbb{R}^n$, then the vectors

$$(e_1, L(e_1)), \ (e_2, L(e_2)), \ \ldots, \ (e_n, L(e_n))$$

of $\mathbb{R}^n \times \mathbb{R}$ form a basis of $\mathcal{G}_L$.

Let $f : A \subset \mathbb{R}^n \to \mathbb{R}$ be differentiable at an interior point $x_0 \in A$. The graph of the tangent map to f at x_0 is called the *tangent space to the graph of f at x_0* and denoted by

$$\mathrm{Tan}\,_{(x_0, f(x_0))}\mathcal{G}_f := \left\{ (x, y) \in \mathbb{R}^n \times \mathbb{R} \,\middle|\, y = df_{x_0}(x) \right\}.$$

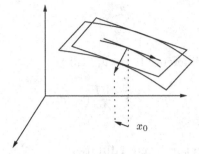

Figure 1.6. $(\nabla f(x_0), -1)$ is perpendicular to the tangent plane to the graph of f at $(x_0, f(x_0))$.

If we choose a basis $(e_1, e_2, \ldots, e_n)$ of $\mathbb{R}^n$, $\mathrm{Tan}_{(x_0, f(x_0))}\mathcal{G}_f$ is a linear space of dimension n since a basis is given by the column vectors of the $(n+1) \times n$ matrix

$$
\begin{pmatrix}
1 & 0 & \cdots & 0 \\
0 & 1 & \cdots & 0 \\
\vdots & \vdots & \ddots & \vdots \\
0 & 0 & \cdots & 1 \\
\frac{\partial f}{\partial x^1}(x_0) & \frac{\partial f}{\partial x^2}(x_0) & \cdots & \frac{\partial f}{\partial x^n}(x_0)
\end{pmatrix}.
\tag{1.13}
$$

The translate to $(x_0, f(x_0))$ of $\mathrm{Tan}_{(x_0, f(x_0))}\mathcal{G}_f$

$$
\left\{ (x, y) \in \mathbb{R}^n \times \mathbb{R} \,\middle|\, y = f(x_0) + \mathbf{D}f(x_0)(x - x_0) \right\}
\tag{1.14}
$$

is called the *tangent plane to the graph of f at x_0*.

f. The orthogonal to the tangent space

Let $\bullet$ denote the inner product in $\mathbb{R}^n$. Then

$$
g((x_1, y_1), (x_2, y_2)) = x_1 \bullet x_2 + y_1 y_2
$$

for all $(x_1, y_1), (x_2, y_2) \in \mathbb{R}^n \times \mathbb{R}$ defines an inner product $g(\ ,\ )$ in $\mathbb{R}^n \times \mathbb{R}$ for which the factors $\mathbb{R}^n$ and $\mathbb{R}$ are orthogonal. If $x \in \mathbb{R}^n$ is given and $L(h) := x \bullet h \ \forall h \in \mathbb{R}^n$, clearly the vector in $\mathbb{R}^{n+1}$

$$
\nu := (x, -1)
$$

is g-orthogonal to the graph of L, i.e., to all vectors of the form $(h, L(h)) \in \mathbb{R}^n \times \mathbb{R}$, $h \in \mathbb{R}^n$, since

$$
g(\nu, (h, L(h))) = x \bullet h - L(h) = L(h) - L(h) = 0.
$$

We therefore conclude the following.

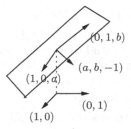

Figure 1.7. $(e_i, L(e_i))$, $i = 1, \ldots, n$, is a basis for $\mathcal{G}_L$.

1.12 Proposition. *Let $f : A \subset \mathbb{R}^n$ be differentiable at an interior point of A. Then the vector*

$$\nu_{x_0} := (\nabla f(x_0), -1)^T \in \mathbb{R}^{n+1}$$

is g-orthogonal to the tangent space to the graph of f at x_0.

g. The tangent map

Suppose that $f : A \subset \mathbb{R}^n \to \mathbb{R}$ possesses a directional derivative at an interior point x_0 of A in the direction v. From (1.2), we easily infer that for all $\lambda \neq 0$ we have

$$\frac{\partial f}{\partial (\lambda v)}(x_0) = \lim_{t \to 0} \frac{f(x_0 + t\lambda v) - f(x_0)}{t} = \lambda \frac{\partial f}{\partial v}(x_0),$$

i.e., f has directional derivative in the direction λv. In other words, the set S of directions with respect to which f has directional derivative is a *cone*, and the map

$$v \;\rightarrow\; \frac{\partial f}{\partial v}(x_0), \qquad v \in S,$$

called the *tangent map to f at x_0*, is defined on S and is homogeneous of degree one. Its graph

$$\left\{ (v, t) \in S \times \mathbb{R} \,\Big|\, t = \frac{\partial f}{\partial v}(x_0) \right\}$$

is a (piece of) cone $S \times \mathbb{R}$ and trivially reduces to a (piece of) plane if and only if the tangent map is linear. This does not always happen, see (iii) Example 1.3 and Figure 1.4.

When f has directional derivative in any direction and the tangent map is linear, we say that f is *Gâteaux-differentiable* and the tangent map is also called the *Gâteaux-differential* of f at x_0 and is still denoted by $df_{x_0}(v)$. There is no ambiguity in doing that since the Gâteaux-differential agrees with the ordinary differential if f is differentiable, see Proposition 1.5. Notice that the converse is instead false, see (iv) Example 1.3.

Of course, the Gâteaux-differential can be written in coordinates and, for a Gâteaux differentiable map, the Jacobian matrix, the gradient vector, and the tangent plane are well defined and we have

$$df_{x_0}(h) = \mathbf{D}f(x_0)h = \nabla f(x_0) \bullet h, \qquad h \in \mathbb{R}^n;$$

moreover, the column vectors of the matrix (1.13) form a basis of the graph of the Gâteaux-differential of f at x_0, and the vector $(\nabla f(x_0), -1)$ is orthogonal to the graph of the Gâteaux-differential of f at x_0.

h. Differentiability and blow-up

The difference between Gâteaux-differentiability and ordinary differentiability is not geometric but analytic: it has to do with the meaning we may attribute to the claim "the tangent plane is a good approximation of the graph of f". The directional derivatives are the result of a blow-up procedure, see [GM1]: we imagine looking at the graph of f through a microscope of higher and higher power centered at $(x_0, f(x_0))$. In the observer's coordinates,

$$X = \lambda(x - x_0), \qquad Y = \lambda(y - y_0),$$

where λ is the magnification factor, the graph $y = f(x)$ of f looks like the graph of the map

$$F_\lambda(X) := \lambda\left(f\left(x_0 + \frac{X}{\lambda}\right) - f(x_0)\right).$$

1.13 ¶. Show the following.
   (i) f has a directional derivative $\frac{\partial f}{\partial v}(x_0)$ in the direction v if and only if $F_\lambda(v) \to \frac{\partial f}{\partial v}(x_0)$ as $\lambda \to +\infty$. Therefore f *has directional derivative in all directions if and only if $F_\lambda(v) \to \frac{\partial f}{\partial v}(x_0)$ pointwise in $\mathbb{R}^n$ as $\lambda \to \infty$.*
   (ii) The limit map of $F_\lambda(v)$ for $\lambda \to \infty$, that is, by (i), the tangent map of f, is linear if and only if f is Gâteaux-differentiable at x_0.
   (iii) f is differentiable at x_0 if and only if the limit map as $\lambda \to \infty$ of $F_\lambda(v)$, i.e., $v \to \frac{\partial f}{\partial v}(x_0)$, is linear, and the maps $\{F_\lambda(v)\}_\lambda$ converge uniformly on compact sets of $\mathbb{R}^n$ to $v \to \frac{\partial f}{\partial v}(x_0)$ as $\lambda \to \infty$.

Therefore it is the way the blow-ups $F_\lambda(v)$ converge to $\frac{\partial f}{\partial v}(x_0)$ that distinguishes the Gâteaux-differentiabity from the differentiability. Finally, we notice that the notion of *slope* is meaningless for Gâteaux-differentiable functions, see also the discussion that precedes Proposition 1.7. As a consequence, the chain rule Theorem 1.22 does not hold in general for Gâteaux-differentiable functions.

1.2 Differential Calculus for Vector-valued Functions

The notions of calculus discussed in the previous section easily extend to maps from an open set $A \subset \mathbb{R}^n$ into $\mathbb{R}^m$, $n, m \geq 1$. Examples of vector-valued maps, $m \geq 2$, are

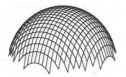

Figure 1.8. One sees a rectangular mesh pulled into the upper hemisphere: on the left, (i), by the graph map, i.e., $(x, y) \to (x, y, \sqrt{1 - x^2 - y^2})$, $x^2 + y^2 < 1$ and on the right, (ii), by the spherical coordinates, i.e., $(\theta, \varphi) \to (cos\theta \sin \varphi, \sin \theta \sin \varphi, \cos \varphi)$ where $\theta \in [0, 2\pi[$ and $\varphi \in [0, \pi/2[$.

(i) maps from an interval $I \subset \mathbb{R}$ into $\mathbb{R}^m$, $m > 1$, i.e., *curves in $\mathbb{R}^m$*,

(ii) maps $A \subset \mathbb{R}^2 \to \mathbb{R}^3$ that parameterize 2-dimensional surfaces in $\mathbb{R}^3$, as for instance,

$$(x, y) \to (x, y, \sqrt{1 - x^2 - y^2}), \qquad (x, y) \in B(0, 1) \subset \mathbb{R}^2,$$

which parameterize the unit upper hemisphere in $\mathbb{R}^3$ centered at the origin, or the map

$$(\theta, \varphi) \to (x, y, z), \qquad x = \cos \theta \sin \varphi, \quad y = \sin \theta \sin \varphi, \quad z = \cos \varphi,$$

which, when defined on $[0, 2\pi[\times [0, \pi]$, parameterizes the unit sphere of $\mathbb{R}^3$, θ having the meaning of *longitude* and φ of *latitude*, see Figure 1.8,

(iii) in general, transformations $\mathbb{R}^n \to \mathbb{R}^n$, $n > 1$, or nonlinear changes of coordinates as

$$(\rho, \theta) \to (x, y), \qquad x = \rho \cos \theta, \qquad y = \rho \sin \theta,$$

which, when defined in $[0, +\infty[\times [0, 2\pi[$, yield the *polar coordinates* in $\mathbb{R}^2$, see Figure 1.10, or as the map

$$(r, \theta, \varphi) \to (x, y, z), \qquad \begin{cases} x = r \cos \theta \sin \varphi, \\ y = r \sin \theta \sin \varphi, \\ z = \cos \varphi \end{cases}$$

which transforms the parallelepiped $[0, 1[\times [0, 2\pi[\times [0, \pi]$ into the unit ball $B(0, 1)$ of $\mathbb{R}^3$.

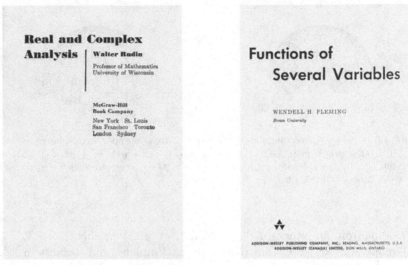

Figure 1.9. Frontispieces of two books about functions of several variables.

1.2.1 Differentiability

a. Jacobian matrix

1.14 Definition. *A map* $f : A \subset \mathbb{R}^n \to \mathbb{R}^m$, $n, m \geq 1$, *is said to be differentiable at an interior point* x_0 *of* A *if there exists a linear map* $L : \mathbb{R}^n \to \mathbb{R}^m$, *called the* tangent linear map *of* f *at* x_0 *such that*

$$\frac{f(x_0 + h) - f(x_0) - L(h)}{|h|} \to 0 \qquad as \ h \to 0. \tag{1.15}$$

When A *is open and* f *is differentiable at every point of* A, *we say that* f *is differentiable in* A.

If we fix an (ordered basis) in $\mathbb{R}^m$, (1.15) is in fact the system of m limits

$$\begin{cases} f^1(x_0 + h) - f^1(x_0) - L^1(h) = o(|h|) & as \ h \to 0, \\ f^2(x_0 + h) - f^2(x_0) - L^2(h) = o(|h|) & as \ h \to 0, \\ \vdots \\ f^m(x_0 + h) - f^m(x_0) - L^m(h) = o(|h|) & as \ h \to 0 \end{cases}$$

for the components $f^1(x), f^2(x), \ldots, f^m(x)$ of f and $L^1, L^2, \ldots, L^m$ of L. Therefore f is differentiable if and only if all components of f are differentiable at x_0, and, in this case, the tangent map to f is $L = (L^1, L^2, \ldots, L^m)^T$ where for every $i = 1, \ldots, m$, $L^i : \mathbb{R}^n \to \mathbb{R}$ is the differential of f^i at x_0. In particular, the differential is unique, if it exists, and for all $h = (h^1, h^2, \ldots, h^n)$ we have

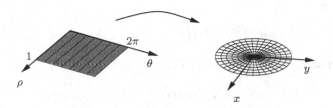

Figure 1.10. Polar coordinates in the unit disk $x^2 + y^2 \leq 1$.

$$L^1(h) = df^1_{x_0}(h) = \frac{\partial f^1}{\partial h}(x_0) = \sum_{j=1}^{n} \frac{\partial f^1}{\partial x^j}(x_0)h^j,$$

$$L^2(h) = df^2_{x_0}(h) = \frac{\partial f^2}{\partial h}(x_0) = \sum_{j=1}^{n} \frac{\partial f^2}{\partial x^j}(x_0)h^j,$$

$$\vdots \qquad\qquad (1.16)$$

$$L^m(h) = df^m_{x_0}(h) = \frac{\partial f^m}{\partial h}(x_0) = \sum_{j=1}^{n} \frac{\partial f^m}{\partial x^j}(x_0)h^j.$$

In terms of the *Jacobian matrix of f at x_0* defined by

$$\mathbf{D}f(x_0) := \left[\frac{\partial f^i}{\partial x^j}(x_0)\right] = \begin{pmatrix} \frac{\partial f^1}{\partial x^1}(x_0) & \frac{\partial f^1}{\partial x^2}(x_0) & \cdots & \frac{\partial f^1}{\partial x^n}(x_0) \\ \frac{\partial f^2}{\partial x^1}(x_0) & \frac{\partial f^2}{\partial x^2}(x_0) & \cdots & \frac{\partial f^2}{\partial x^n}(x_0) \\ \vdots & \vdots & \ddots & \vdots \\ \frac{\partial f^m}{\partial x^1}(x_0) & \frac{\partial f^m}{\partial x^2}(x_0) & \cdots & \frac{\partial f^m}{\partial x^n}(x_0) \end{pmatrix},$$

the system of equations (1.16) can be rewritten as

$$L(h) = \begin{pmatrix} L^1(h) \\ L^2(h) \\ \vdots \\ L^m(h) \end{pmatrix} = \mathbf{D}f(x_0)h.$$

Summarizing we have: *if $f : A \subset \mathbb{R}^n \to \mathbb{R}^m$ is differentiable at x_0, then the linear tangent map to f at x_0 defined by (1.15) and denoted by one of the symbols*

$$df_{x_0}, \qquad T_{x_0}f,$$

is unique; moreover,

$$df_{x_0}(h) = \mathbf{D}f(x_0)h, \qquad h \in \mathbb{R}^n. \tag{1.17}$$

If we choose an inner product in $\mathbb{R}^n$, so that

$$\mathbf{D}f^i(x_0)(h) = \nabla f^i(x_0) \bullet h \qquad \forall i = 1, \dots, m,$$

then we have

$$L(h) = \mathbf{D}f(x_0)h = \begin{pmatrix} \nabla f^1(x_0) \bullet h \\ \nabla f^2(x_0) \bullet h \\ \dots \\ \nabla f^m(x_0) \bullet h \end{pmatrix};$$

in particular,

$$\ker L = \ker \mathbf{D}f(x_0) = \mathrm{Span}\left\{\nabla f^1(x_0), \dots, \nabla f^m(x_0)\right\}^\perp.$$

Finally, in the case $n = m$, the determinant of the Jacobian matrix $\mathbf{D}f(x_0)$ is called the *Jacobian determinant* or simply the *Jacobian* of f at x_0 and is denoted by one of the symbols

$$J_f(x_0) = J(\mathbf{D}f(x_0)) = \frac{\partial(f^1, f^2, \dots, f^n)}{\partial(x^1, x^2, \dots, x^n)}(x_0) := \det \mathbf{D}f(x_0).$$

b. The tangent space

Let $f : A \subset \mathbb{R}^n \to \mathbb{R}^m$ be a differentiable map at an interior point x_0 of A. As in the scalar case, we call the graph of the linear tangent map to f at x_0 the *tangent space* to the graph of f at $(x_0, f(x_0))$

$$\mathrm{Tan}_{(x_0, f(x_0))}\mathcal{G}_f := \mathcal{G}_{df_{x_0}}$$

and its translate at $(x_0, f(x_0))$,

$$\left\{(x, y) \in \mathbb{R}^n \times \mathbb{R}^m \,\middle|\, y = f(x_0) + df_{x_0}(x - x_0)\right\},$$

the *tangent plane to the graph of f at x_0*.

Clearly, the tangent space to the graph of f at x_0 is the image of the injective linear map $x \to (x, df_{x_0}x)$, hence its dimension is n. With respect to bases respectively in $\mathbb{R}^n$ and $\mathbb{R}^m$ we have

$$\mathrm{Tan}_{(x_0, f(x_0))}\mathcal{G}_f = \left\{(x, y) \in \mathbb{R}^n \times \mathbb{R}^m \,\middle|\, y = \mathbf{D}f(x_0)x\right\},$$

and the n-tuple of column vectors of the $(n + m) \times n$ matrix

$$\mathbf{A} := \begin{pmatrix} \boxed{\text{Id}} \\ \boxed{\mathbf{D}f(x_0)} \end{pmatrix} \tag{1.18}$$

form a basis of the tangent space to the graph of f at $(x_0, f(x_0))$.

1.15 The normal space. Clearly

$$\mathrm{Tan}_{(x_0,f(x_0))}\mathcal{G}_f = \Big\{(x,y) \in \mathbb{R}^n \times \mathbb{R}^m \;\Big|\; y - \mathbf{D}f(x_0)x = 0\Big\} = \ker \mathbf{B}$$

where $\mathbf{B}$ is the $n \times (n+m)$ matrix

$$\mathbf{B} := \begin{pmatrix} \boxed{\mathbf{D}f(x_0)} & \boxed{-\text{Id}} \end{pmatrix}.$$

If the target $\mathbb{R}^m$ and the product space $\mathbb{R}^n \times \mathbb{R}^m$ are endowed with inner products, then by the alternative theorem,

$$\Big(\mathrm{Tan}_{(x_0,f(x_0))}\mathcal{G}_f\Big)^{\perp} = \ker \mathbf{B}^{\perp} = \mathrm{Im}\,\mathbf{B}^*,$$

i.e., the m column vectors of the $(n+m) \times n$ *adjoint matrix* $\mathbf{B}^*$ span the orthogonal subspace to $\mathrm{Tan}_{(x_0,f(x_0))}\mathcal{G}_f$.

1.16 ¶. Assuming that $\mathbb{R}^n$ and $\mathbb{R}^m$ are endowed with inner products respectively $g_1(\ ,\ )$ and $g_2(\ ,\ )$, then

$$g((x,z),(y,w)) := g_1(x,y) + g_2(z,w) \qquad \forall x,y \in \mathbb{R}^n,\ \forall z,w \in \mathbb{R}^m,$$

is an inner product on $\mathbb{R}^n \times \mathbb{R}^m$ for which the factors $\mathbb{R}^n$ and $\mathbb{R}^m$ are orthogonal. Choose now an orthonormal basis in the target $\mathbb{R}^m$ to compute the components of $f = (f^1, f^2, \ldots, f^m)$. Show that

$$\mathbf{B}^* = \begin{pmatrix} \boxed{\nabla f^1 \Big| \cdots \Big| \nabla f^m} \\ \boxed{-\text{Id}} \end{pmatrix}$$

where ∇f^i denotes the gradient of the component f^i of f with respect to metric g_1.

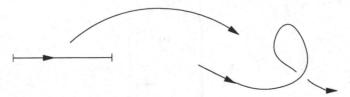

Figure 1.11. A curve in $\mathbb{R}^2$.

1.17 Example (Curves in $\mathbb{R}^m$). The maps $r : [a, b] \subset \mathbb{R} \to \mathbb{R}^m$, $m \geq 1$, $r(t) = (r^1(t), \ldots, r^m(t))$ in the standard basis are called *curves* in $\mathbb{R}^n$. The map r parameterizes the image or the *trajectory* of the curve r. The map r is differentiable at t_0 if and only if its components $r = (r^1, r^2, \ldots, r^m)$ in a given basis are differentiable at t_0, and in this case the Jacobian matrix is the $m \times 1$ matrix

$$\mathbf{D}r(t_0) = \begin{pmatrix} r^{1'}(t_0) \\ \vdots \\ r^{m'}(t_0) \end{pmatrix} =: r'(t_0),$$

i.e., $\mathbf{D}r(t_0)$ is the velocity vector at the time t_0. The linear tangent map is the map $t \to r'(t_0)t$ and yields the parametric equation of the tangent line to the curve $r(t)$ at each point $r(t_0)$ at which r is injective and $r'(t_0) \neq 0$.

The graph of r is the curve in $\mathbb{R} \times \mathbb{R}^m$ given by

$$t \to \begin{pmatrix} t \\ r(t) \end{pmatrix}$$

and its tangent space at t_0 is the line in $\mathbb{R} \times \mathbb{R}^m$ through the origin image of the map

$$t \to \begin{pmatrix} t \\ r'(t_0)t \end{pmatrix} = \begin{pmatrix} 1 \\ r'(t_0) \end{pmatrix} t$$

while the normal plane to the graph at $(t, r(t))$, assuming we are using an orthonormal basis in $\mathbb{R}^m$, is generated by the m row vectors of the matrix

$$\begin{pmatrix} r^{1'}(t) & -1 & 0 & \cdots & 0 \\ r^{2'}(t) & 0 & -1 & \cdots & 0 \\ \vdots & \vdots & \vdots & \ddots & \vdots \\ r^{m'}(t) & 0 & 0 & \cdots & -1 \end{pmatrix}.$$

1.18 Example (Immersed surfaces). A map $r : A \subset \mathbb{R}^2 \to \mathbb{R}^3$,

$$r(u, v) = (x(u, v), y(u, v), z(u, v))^T,$$

may be regarded as a parameterization of the surface $r(A) \subset \mathbb{R}^3$, see Figure 1.12.
If r is differentiable at (u_0, v_0), its Jacobian matrix is

$$\mathbf{D}r(u_0, v_0) = \begin{pmatrix} x_u & x_v \\ y_u & y_v \\ z_u & z_v \end{pmatrix}$$

where we shortened

$$x_u := \frac{\partial x}{\partial u}(u_0, v_0), \quad x_v := \frac{\partial x}{\partial v}(u_0, v_0), \quad \cdots .$$

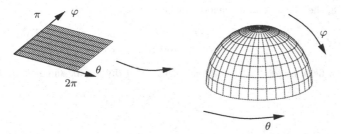

Figure 1.12. The hemisphere parameterized by $(u, v) \rightarrow (\cos u \sin v, \sin u \sin v, \cos v)$, $u \in [0, 2\pi]$, $v \in [0, \pi/2]$.

The column vectors of the Jacobian matrix $r_u := (x_u, y_u, z_u)^T$ and $r_v := (x_v, y_v, z_v)^T$ are the velocity vectors respectively, of the curves $u \rightarrow r(u, v_0)$ at $u = u_0$ and $v \rightarrow r(u_0, v)$ at $v = v_0$. In turn the curves $u \rightarrow r(u, v_0)$ and $v \rightarrow r(u_0, v)$ are the images respectively of the lines $u \rightarrow (u, v_0)$ and $v \rightarrow (u_0, v)$ di $\mathbb{R}^2$.

The tangent space to $\mathcal{G}_r$ at (u_0, v_0) is the 2-dimensional subspace of $\mathbb{R}^2 \times \mathbb{R}^3$ given by

$$\left\{ (u, v, x, y, z) \in \mathbb{R}^2 \times \mathbb{R}^3 \,\middle|\, \begin{pmatrix} x \\ y \\ z \end{pmatrix} = \begin{pmatrix} x_u & x_v \\ y_u & y_v \\ z_u & z_v \end{pmatrix} \begin{pmatrix} u \\ v \end{pmatrix} \right\},$$

and a basis of it is given by the two column vectors of the matrix

$$\begin{pmatrix} 1 & 0 \\ 0 & 1 \\ x_u & x_v \\ y_u & y_v \\ z_u & z_v \end{pmatrix}$$

while the three rows of the matrix

$$\begin{pmatrix} x_u & x_v & -1 & 0 & 0 \\ y_u & y_v & 0 & -1 & 0 \\ z_u & z_v & 0 & 0 & -1 \end{pmatrix}$$

form a basis of the normal space (assuming that we are using orthonormal bases).

1.19 Example. *Affine transformations*, i.e., maps of the form

$$f(x) := x_0 + L(x) \qquad L : \mathbb{R}^n \rightarrow \mathbb{R}^n \text{ linear}$$

provide simple examples of transformations from $\mathbb{R}^n$ into $\mathbb{R}^n$. They are differentiable with $df_{x_0} = L$.

1.20 Example (Vector fields). A *vector field* in $A \subset \mathbb{R}^n$ is the datum of a vector $f(x) \in \mathbb{R}^n$ at every point x of A; it can be regarded at first glance as a map $f : A \subset \mathbb{R}^n \rightarrow \mathbb{R}^n$, $f = (f^1, \ldots, f^n)$. The field of velocities of particles in a fluid, the electrostatic field, or the gravitational field are all examples of vector fields. Two operators acting on vector fields are particularly important: the *divergence* operator

$$\operatorname{div} f = \nabla \bullet f := \sum_{i=1}^{n} \frac{\partial f^i}{\partial x^i},$$

and, for $n = 3$, the *curl* operator, denoted by curl or rot

$$\operatorname{curl} f = \operatorname{rot} f = \nabla \times f := \left(\frac{\partial f^3}{\partial x^2} - \frac{\partial f^2}{\partial x^3}, \frac{\partial f^1}{\partial x^3} - \frac{\partial f^3}{\partial x^1}, \frac{\partial f^2}{\partial x^1} - \frac{\partial f^1}{\partial x^2} \right).$$

We say that a field $f : A \subset \mathbb{R}^n \to \mathbb{R}^n$ is *solenoidal* if $\operatorname{div} f = 0$, and *irrotational*, if $\operatorname{curl} f = 0$.

1.2.2 The calculus

The following is easily verified.

1.21 Proposition. *Let f and $g : A \to \mathbb{R}$ be two differentiable maps at an interior point x_0 of A and let $c \in \mathbb{R}$. Then $cf, f + g, fg,$ and f/g provided $g(x_0) \neq 0$, are also differentiable at x_0 and we have*

- $\mathbf{D}(cf)(x_0) = c\mathbf{D}f(x_0),$
- $\mathbf{D}(f + g)(x_0) = \mathbf{D}f(x_0) + \mathbf{D}g(x_0),$
- $\mathbf{D}(fg)(x_0) = g(x_0)\mathbf{D}f(x_0) + f(x_0)\mathbf{D}g(x_0),$
- $\mathbf{D}(f/g)(x_0) = \dfrac{g(x_0)\mathbf{D}f(x_0) - f(x_0)\mathbf{D}g(x_0)}{g(x_0)^2}.$

It is also easily seen that if f and $g : A \subset \mathbb{R}^n \to \mathbb{R}^m$ are differentiable, then for $i = 1, \ldots, m$ and $j = 1, \ldots, n$, we have

$$\frac{\partial(f + g)^i}{\partial x^j}(x) = \frac{\partial f^i}{\partial x^j}(x) + \frac{\partial g^i}{\partial x^j}(x),$$

or in matrix notation

$$\mathbf{D}(f + g)(x) = \mathbf{D}f(x) + \mathbf{D}g(x).$$

Similarly, if $f : A \subset \mathbb{R}^n \to \mathbb{R}^m$ and $\lambda : A \subset \mathbb{R}^n \to \mathbb{R}$ are differentiable, then for $i = 1, \ldots, m$ and $j = 1, \ldots, n$, we have

$$\frac{\partial(\lambda f)^i}{\partial x^j}(x) = \lambda(x)\frac{\partial f^i}{\partial x^j}(x) + f^i(x)\frac{\partial \lambda}{\partial x^j}(x),$$

or

$$\mathbf{D}(\lambda f)(x) = \lambda(x)\mathbf{D}f(x) + f(x)\mathbf{D}\lambda(x). \tag{1.19}$$

Notice that $f(x) \in \mathbb{R}^m$ is a column vector and $D\lambda(x_0)$ is a row vector with n entries, hence $f(x)\mathbf{D}\lambda(x)$ is an $m \times n$-matrix.

1.2.3 Differentiation of compositions

1.22 Theorem. *Let $U \subset \mathbb{R}^n$ and $V \subset \mathbb{R}^m$ be open sets, and let $f : U \to \mathbb{R}^m$ and $g : V \to \mathbb{R}^p$ be such that $f(U) \subset V$. Suppose that f and g are differentiable respectively, at $x_0 \in U$ and $f(x_0)$. Then $g \circ f$ is differentiable at x_0 and*

$$d(g \circ f)_{x_0} = dg_{f(x_0)} \circ df_{x_0} \qquad (1.20)$$

or, in terms of Jacobian matrices,

$$\mathbf{D}(g \circ f)(x_0) = \mathbf{D}g(f(x_0))\mathbf{D}f(x_0). \qquad (1.21)$$

Notice that (1.21) is the natural extension of the formula (1.8) for the calculus of the derivative of a function along a curve. Notice also that in order for (1.21) to hold, g needs in general to be differentiable, see Example 1.3 (iv).

Proof. Since $f(U) \subset V$, the map $g \circ f : U \to \mathbb{R}^p$ is well defined and, by assumption,

$$f(x) = f(x_0) + \mathbf{D}f(x_0)(x - x_0) + o(|x - x_0|) \qquad \text{as } x \to x_0,$$
$$g(y) = g(f(x_0)) + \mathbf{D}g(f(x_0))(y - f(x_0)) + o(|y - f(x_0)|) \qquad \text{as } y \to f(x_0).$$

We then infer, since f is continuous at x_0, that

$$g(f(x)) = g(f(x_0)) + \mathbf{D}g(f(x_0))\mathbf{D}f(x_0)(x - x_0) + \mathbf{D}g(f(x_0))o(|x - x_0|)$$
$$+ o(|\mathbf{D}f(x_0)(x - x_0) + o(|x - x_0|)|),$$

hence, noticing that $|\mathbf{B}h| = O(|h|)$ as $h \to 0$ for any matrix $\mathbf{B}$, we have

$$g(f(x)) = g(f(x_0) + \mathbf{D}g(f(x_0))\mathbf{D}f(x_0)(x - x_0) + o(|x - x_0|). \qquad \text{as } x \to x_0.$$

$\square$

1.23 Chain rule. Suppose that $y : A \subset \mathbb{R}^n \to \mathbb{R}^m$ and $g : \mathbb{R}^m \to \mathbb{R}$ are differentiable at, respectively, x_0 and $y(x_0)$. If $y(x) = (y^1(x), \dots, y^m(x))^T$, computing row by columns, (1.21) yields

$$\frac{\partial g \circ y}{\partial x^j}(x) = \frac{\partial}{\partial x^j}\left[g(y^1, y^2, \dots, y^m) \right] = \mathbf{D}g(y(x))\mathbf{D}y(x) \qquad (1.22)$$

$$= \sum_{i=1}^{m} \frac{\partial g}{\partial y^i}\frac{\partial y^i}{\partial x^j} = \frac{\partial g}{\partial y^1}\frac{\partial y^1}{\partial x^j} + \frac{\partial g}{\partial y^2}\frac{\partial y^2}{\partial x^j} + \cdots + \frac{\partial g}{\partial y^m}\frac{\partial y^m}{\partial x^j}$$

for $j = 1, \dots, m$, where, of course,

$$\frac{\partial g}{\partial y^i} := \frac{\partial g}{\partial y^i}(y(x)) \qquad \text{and} \qquad \frac{\partial y^i}{\partial x^j} := \frac{\partial y^i}{\partial x^j}(x).$$

Formula (1.22) is the so-called *chain rule* for the calculus of the derivatives of the composition.

1.24 Example. Let $r : [0,1] \to \mathbb{R}^m$ and $g : \mathbb{R}^m \to \mathbb{R}^p$. If $s(t) := g(r(t))$, then

$$s'(t) = \mathbf{D}g(r(t))\, r'(t).$$

Consequently, if $r'(t_0)$ and $s'(t_0)$ are nonzero, the Jacobian matrix of g maps the tangent line at t_0 to the curve $r(t)$ into the tangent line to the curve image $s(t)$ at t_0.

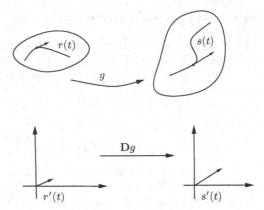

Figure 1.13. The map g and its tangent map.

1.2.4 Calculus for matrix-valued maps

Of course, the calculus for functions $t \to \mathbf{A}(t)$ with values $m \times n$ matrices can be subsumed to the calculus for curves in $\mathbb{R}^{mn}$. However, for its relevance it is convenient to state a few formulas explicitly. Operating on the entries, one can easily prove:

$$(\mathbf{A}(t)x(t))' = \mathbf{A}'(t)x(t) + \mathbf{A}(t)x'(t),$$
$$(\mathbf{A}(t)\mathbf{B}(t))' = \mathbf{A}'(t)\mathbf{B}(t) + \mathbf{A}(t)\mathbf{B}'(t), \qquad (1.23)$$
$$(\lambda(t)\mathbf{A}(t))' = \lambda'(t)\mathbf{A}(t) + \lambda(t)\mathbf{A}'(t),$$
$$(\operatorname{tr}\mathbf{A}(t))' = \operatorname{tr}\mathbf{A}'(t), \qquad (1.24)$$
$$\frac{d}{dt}\,x(t) \bullet y(t) = x'(t) \bullet y(t) + x(t) \bullet y'(t).$$

We also have the following.

(i) Starting from $\mathbf{A}(t)\mathbf{A}(t)^{-1} = \operatorname{Id}$, we infer, on account of (1.23), that $\mathbf{A}(t)^{-1}$ is differentiable if $\mathbf{A}(t)$ is differentiable, and

$$(\mathbf{A}(t)^{-1})' = -\mathbf{A}(t)^{-1}\mathbf{A}'(t)\mathbf{A}(t)^{-1}. \qquad (1.25)$$

(ii) By induction, from (1.23) and (1.24) we infer that

$$D(\operatorname{tr}\mathbf{A}(t)^2) = \operatorname{tr}(2\mathbf{A}(t)\mathbf{A}'(t)),$$
$$D(\operatorname{tr}\mathbf{A}(t)^3) = \operatorname{tr}(3\mathbf{A}(t)^2\mathbf{A}'(t)),$$

$$\vdots$$

$$D(\operatorname{tr}\mathbf{A}(t)^n) = \operatorname{tr}(n\mathbf{A}(t)^{n-1}\mathbf{A}'(t))$$

$$\vdots$$

Therefore, if p is a polynomial, then

$$\frac{d}{dt}\Big(\operatorname{tr} p(\mathbf{A}(t))\Big) = \operatorname{tr}\Big(p'(\mathbf{A}(t))\,\mathbf{A}'(t)\Big). \tag{1.26}$$

(iii) Again from (1.23) and (1.24) we infer that, if $\mathbf{A}(t)$ and $\mathbf{A}'(t)$ commute, we have

$$D(\mathbf{A}(t)^n) = n\mathbf{A}(t)^{n-1}\mathbf{A}'(t) \qquad \forall n;$$

hence for any polynomial p we have

$$D(p(\mathbf{A}(t))) = p'(\mathbf{A}(t))\,\mathbf{A}'(t) \tag{1.27}$$

provided $\mathbf{A}(t)$ and $\mathbf{A}'(t)$ commute. Notice that the chain rule does not hold in general; in fact, $(\mathbf{A}(t)^2)' = \mathbf{A}(t)\mathbf{A}'(t) + \mathbf{A}'(t)\mathbf{A}(t)$. Hence $(\mathbf{A}(t)^2)' = 2\mathbf{A}(t)\mathbf{A}'(t)$ if and only if $\mathbf{A}(t)$ and $\mathbf{A}'(t)$ commute.

(iv) Since the determinant of a matrix $\mathbf{A}$ is multilinear in the columns of $\mathbf{A}$, we infer for $\mathbf{A}(t) = [A_1(t)\,|\,A_2(t)\,|\,\ldots\,|\,A_n(t)] \in M_{n,n}$ that

$$\frac{d}{dt}\det \mathbf{A}(t) = \det[A_1'\,|\,A_2\,|\,\ldots\,|\,A_n]$$
$$+ \det[A_1\,|\,A_2'\,|\,\ldots\,|\,A_n] + \cdots + \det[A_1\,|\,A_2\,|\,\ldots\,|\,A_n'].$$

Thus, if $\mathbf{A}(0) = \mathrm{Id}$, we have

$$\det[A_1'(0)\,|\,A_2(0)\,|\,\ldots\,|\,A_n(0)] = (A_1^1)'(0),$$
$$\vdots$$
$$\det[A_1(0)\,|\,A_2(0)\,|\,\ldots\,|\,A_n'(0)] = (A_n^n)'(0),$$

from which

$$\frac{d\det \mathbf{A}(t)}{dt}(0) = \operatorname{tr} \mathbf{A}'(0) \qquad \text{if} \qquad \mathbf{A}(0) = \mathrm{Id}. \tag{1.28}$$

More generally, if $\mathbf{Y}(s) \in M_{n,n}$ is invertible at $s = t$, then (1.28) yields

$$\frac{1}{\det \mathbf{Y}(t)}\frac{d\det \mathbf{Y}(s)}{ds}(t) = \frac{d\det(\mathbf{Y}(t)^{-1}\,\mathbf{Y}(s))}{ds}(t) = \operatorname{tr}\Big(\mathbf{Y}(t)^{-1}\,\mathbf{Y}'(t)\Big). \tag{1.29}$$

1.25 ¶. Some typical facts relative to real-valued functions extend to matrix-valued functions. For instance, we have:

(i) If $\mathbf{A}(t)$ is self-adjoint, then $\mathbf{A}'(t)$ is self-adjoint.

(ii) If $\mathbf{A}'(t) > 0$, then $\mathbf{A}(s) < \mathbf{A}(t)$ if $s < t$.

(iii) If $\mathbf{R}$ and $\mathbf{S}$ are self-adjoint, and $0 < \mathbf{R} < \mathbf{S}$, then $\mathbf{R}^{-1} > \mathbf{S}^{-1}$ and $\sqrt{\mathbf{R}} < \sqrt{\mathbf{S}}$.

(iv) If $\mathbf{A}$ and $\mathbf{B}$ are self-adjoint, $\mathbf{A} > 0$ and $\mathbf{AB} + \mathbf{BA} > 0$, then $\mathbf{B} > 0$; notice however that $\mathbf{A}, \mathbf{B} > 0$ does not imply $\mathbf{AB} + \mathbf{BA} > 0$.

[*Hint:* To prove (iii), apply (ii) to $\mathbf{A}^{-1}(t)$, where $\mathbf{A}(t) := \mathbf{R} + t(\mathbf{S} - \mathbf{R})$, $t \in [0,1]$. To prove (iv), consider $\mathbf{B}(t) := \mathbf{B} + t\mathbf{A}$ and $\mathbf{S}(t) := \mathbf{AB}(t) + \mathbf{B}(t)\mathbf{A}$, and apply (iii).]

1.3 Theorems of Differential Calculus

1.3.1 Maps with continuous derivatives

a. Functions of class $C^1(A)$

As we have seen, the existence of partial derivatives does not imply differentiability. We shall see later that the existence of partial derivatives in conjunction with convexity does imply differentiability, see Section 2, Chap. 2 of Vol. V. Here we state a general theorem.

1.26 Theorem (of total derivative). *Let* $f : B(x_0, r) \subset \mathbb{R}^n \to \mathbb{R}$, $r > 0$, *be a map. Suppose that all partial derivatives of* f *exist at every point of* $B(x_0, r)$ *and are continuous at* x_0. *Then* f *is differentiable at* x_0.

Proof. We shall deal with the case $n = 2$, leaving to the reader the task of convincing himself that the argument extends to any dimension. It is convenient to slighty change notation: we assume that f is defined in $B(P_0, r)$ where $P_0 := (x_0, y_0)$ and we let $P := (x, y) \in B(P_0, r)$. We have

$$f(P) - f(P_0) = f(x, y) - f(x_0, y_0) = f(x, y) - f(x, y_0) + f(x, y_0) - f(x_0, y_0). \quad (1.30)$$

Since by assumption the function $g_1(t) := f(t, y_0)$ is differentiable in the closed interval with extremal points x_0 and x, Lagrange's theorem yields a point $\xi = \xi(x)$ with $0 < |\xi - x_0| < |x - x_0|$ such that

$$f(x, y_0) - f(x_0, y_0) = f_x(\xi, y_0)(x - x_0) \quad (1.31)$$
$$= f_x(x_0, y_0)(x - x_0) + [f_x(\xi, y_0) - f_x(x_0, y_0)](x - x_0).$$

Similarly, for any x the function $g_2(t) = f(x, t)$ is differentiable in the interval of extremal points y_0 and y, and again Lagrange's theorem yields $\eta = \eta(x, y)$ with $0 < |\eta - y_0| \leq |y - y_0|$ such that

$$f(x, y) - f(x, y_0) = f_y(x, \eta)(y - y_0) = f_y(x_0, y_0)(y - y_0) + [f_y(x, \eta) - f_y(x_0, y_0)](y - y_0).$$

Of course the distance of the points (ξ, y_0) and (x, η) from $P_0 = (x_0, y_0)$ is less than $|P - P_0|$, hence

$$(\xi, y_0) = (\xi(x), y_0) \to (x_0, y_0), \qquad (x, \eta) = (x, \eta(x, y)) \to (x_0, y_0), \qquad \text{as } P \to P_0,$$

and, on account of the continuity of the partial derivatives at (x_0, y_0),

$$\left| f_x(\xi, y_0) - f_x(x_0, y_0) \right| |x - x_0| + \left| f_y(x, \eta) - f_y(x_0, y_0) \right| |y - y_0| = o(|P - P_0|)$$

as $P \to P_0$. We therefore conclude from (1.30), (1.31) that

$$f(P) - f(P_0) = f_x(P_0)(x - x_0) + [f_x(\xi, y_0) - f_x(x_0, y_0)](x - x_0)$$
$$+ f_y(P_0)(y - y_0) + [f_y(x, \eta) - f_y(x_0, y_0)](y - y_0)$$
$$= f_x(P_0)(x - x_0) + f_y(P_0)(y - y_0) + o(|P - P_0|) \qquad \text{as } P \to P_0.$$

$\square$

1.27 Remark. Notice that in Theorem 1.26 the function f as well as the derivatives of f may not be continuous in any neighborhood of x_0. However, the partial derivatives $\frac{\partial f}{\partial x^i}(x^1, x^2, \ldots, x^n)$ are assumed to be continuous at x_0 not only as functions of the variable of differentiation x^i, but as *functions of several variables.*

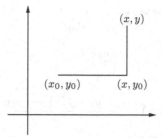

Figure 1.14. Illustration of the proof of the theorem of total derivative.

1.28 ¶. Let x_0 be an interior point of the domain A of a function $f : A \subset \mathbb{R}^n \to \mathbb{R}$. Suppose that f is continuous at x_0, all partial derivatives of f exist in $A \setminus \{x_0\}$, and, for all $i = 1, \ldots, n$, we have $f_{x^i}(x) \to a_i \in \mathbb{R}$ as $x \to x^0$. Show that f is differentiable at x_0 and $df_{x_0}(v) = \sum_{i=1}^n a_i v^i$.

1.29 ¶. Show that
 (i) polynomials in several variables are everywhere differentiable,
 (ii) if $L : \mathbb{R}^n \to \mathbb{R}$ is linear, then $dL_{x_0} = L$,
 (iii) if $\mathbf{A} = (A_{ij})$ is an $n \times n$-matrix, then the *quadratic form* $\phi(x) := \mathbf{A}x \bullet x = \sum_{i,j} A_{ij} x^i x^j$, $x \in \mathbb{R}^n$, is a homogeneous polynomial of degree 2 and $\mathbf{D}\phi(x) := (\mathbf{A} + \mathbf{A}^T)x$.

1.30 Definition. *We say that $f : A \subset \mathbb{R}^n \to \mathbb{R}^m$ is of class C^1 in the open set A and we write $f \in C^1(A, \mathbb{R}^m)$, or $f \in C^1(A)$, if all partial derivatives of f exist and are continuous in A.*

Every map $f \in C^1(A)$ is differentiable in A by Theorem 1.26, hence continuous by Proposition 1.5, i.e.,

$$C^1(A) \subset C^0(A);$$

moreover, by Proposition 1.21, $C^1(A)$ is a vector space over $\mathbb{R}$.

b. Functions of class $C^1(\overline{A})$

1.31 Definition. *Let A be an open subset of $\mathbb{R}^n$. We say that $f : \overline{A} \to \mathbb{R}$ is of class $C^1(\overline{A})$ if there exists an open set $U \supset A$ and an extension $F : U \to \mathbb{R}$ of f to U, $F = f$ on $\overline{A}$, of class $C^1(U)$.*

The differential of f in $x \in \overline{A}$ is defined as the differential at x of one of its C^1-extensions, as the differentials of two different C^1-extensions necessarily agree at x.

Hassler Whitney (1907–1989) has given a definition of function of class C^1 that is less naive than Definition 1.31 and that is worth mentioning.

1.32 Definition (Whitney). *Let E be a closed set in $\mathbb{R}^n$ without isolated points. We say, according to Whitney, that $f : E \to \mathbb{R}$ is of class $C^1(E)$ if*

(i) f *is differentiable at every point of* E, *i.e., for every* $x \in E$ *there is a linear map* $L_x : \mathbb{R}^n \to \mathbb{R}$ *such that*

$$\lim_{\substack{y \to x \\ y \in E}} R(y; x) = 0$$

where

$$R(y; x) := \frac{|f(y) - f(x) - L_x(y - x)|}{|y - x|},$$

(ii) *the gradient vector* $\nabla f(x)$ *associated to* L_x *is continuous in* E,
(iii) $R(y; x) \to 0$ *as* $y \to x$ *uniformly on the compact sets of* E.

It is easily seen that if A is an open set and $f : \overline{A} \to \mathbb{R}$ has a C^1 extension in a neighborhood U of A, then f is of class $C^1(\overline{A})$ according to Whitney's definition. A celebrated theorem of Whitney's claims that the converse also holds. We state it without proof.

1.33 Theorem (Whitney). *Let* $E \subset \mathbb{R}^n$ *be a closed set without isolated points. Suppose that* $f : E \to \mathbb{R}$ *is of class* $C^1(E)$ *in the sense of Whitney. Then there exists* $F : \mathbb{R}^n \to \mathbb{R}$ *of class* $C^1(\mathbb{R}^n)$ *that extends* f, *i.e.,* $F = f$ *in* E. *Moreover, the extension has the following properties:*

(i) *if* L_x *is the Whitney differential of* f *at* $x \in E$, *then* $DF(x) = L_x$,
(ii) *we have*

$$\|F\|_{\infty, \mathbb{R}^n} \le C \|f\|_{\infty, E},$$

$$\|DF\|_{\infty, \mathbb{R}^N} \le C \max \left(\|L_x\|_{\infty, E}, \max_{x, y \in E} \frac{|f(x) - f(y)|}{|x - y|} \right)$$

where C *is a constant depending only on the dimension* n.

c. Functions of class $C^2(A)$

Suppose that $f : A \to \mathbb{R}$, where $A \subset \mathbb{R}^n$ is open, has first derivatives in a neighborhood of $x_0 \in A$, and that the first derivatives have partial derivatives at x_0, then we say that f has second derivatives at x_0. If $(x^1, x^2, \ldots, x^n)$ are the coordinates in $\mathbb{R}^n$, the partial derivative of f first with respect to x^j and then to x^i is denoted by one of the symbols:

$$\frac{\partial^2 f}{\partial x^i \partial x^j}(x_0), \qquad D_i D_j f(x_0), \qquad \text{or} \qquad D_{ij} f(x_0).$$

The $n \times n$ matrix of the second derivatives of f,

$$\mathbf{H}f(x_0) := [D_i D_j f(x_0)],$$

is called the *Hessian matrix* of f at x_0.
    In general it may happen that

$$D_i D_j f \ne D_j D_i f \qquad \text{for } i \ne j,$$

as for instance with the function

$$f(x,y) = \begin{cases} y^2 \arctan \frac{x}{y} & \text{if } y \neq 0, \\ 0 & \text{if } y = 0 \end{cases}$$

for which $\frac{\partial^2 f}{\partial x \partial y}(0,0) = 0$ and $\frac{\partial^2 f}{\partial y \partial x}(0,0) = 1$. However, the following holds.

1.34 Theorem (Schwarz). *Let $f : B(x_0, r) \subset \mathbb{R}^n \to \mathbb{R}$, $r > 0$. Suppose that for $i \neq j$ the mixed derivatives*

$$\frac{\partial^2 f}{\partial x^i \partial x^j}(x) \quad and \quad \frac{\partial^2 f}{\partial x^j \partial x^i}(x)$$

exist in $B(x_0, r)$ and are continuous at x_0, then they agree at x_0.

Proof. We deal here with functions of two variables, leaving to the reader the task of convincing himself that the proof extends to functions of more than two variables.

Let $P_0 := (x_0, y_0)$ and $P = (x, y) \in B(P_0, r)$. Consider the so-called *second differential quotient*

$$A(t) := \frac{f(x_0 + t, y_0 + t) - f(x_0 + t, y_0) - f(x_0, y_0 + t) + f(x_0, y_0)}{t^2}$$

that is well defined for $0 < |t| < r$. Now introduce

$$g(x) := f(x, y_0 + t) - f(x, y_0), \qquad h(y) := f(x_0 + t, y) - f(x_0, y),$$

so that

$$A(t) = t^{-2}(g(x_0 + t) - g(x_0)) = t^{-2}(h(y_0 + t) - h(y_0)).$$

As a consequence of the mean value theorem for functions in one variable, we can write

$$A(t) = t^{-1}g'(\xi) = t^{-1}(f_x(\xi, y_0 + t) - f_x(\xi, y_0))$$

for some ξ between x_0 and $x_0 + t$; again the mean value theorem then yields

$$A(t) = \frac{\partial^2 f}{\partial y \partial x}(\xi, \eta)$$

where η is between y_0 and $y_0 + t$, and, similarly we can write

$$A(t) = t^{-1}h'(\beta) = \frac{\partial^2 f}{\partial x \partial y}(\alpha, \beta)$$

where α is between x_0 and $x_0 + t$, and β is between y_0 and $y_0 + t$.

When $t \to 0$, both points (α, β) and (ξ, η) (that depend on t) tend to (x_0, y_0). On account of the continuity of the mixed derivatives at (x_0, y_0), we conclude when $t \to 0$ that

$$\frac{\partial^2 f}{\partial y \partial x}(x_0, y_0) = \frac{\partial^2 f}{\partial x \partial y}(x_0, y_0).$$

$\square$

1.35 Definition. *If all second derivatives of a function f exist and are continuous in an open set A, we say that f is of class $C^2(A)$ and we write $f \in C^2(A)$.*

1.36 ¶. Let $|x| := \sqrt{\sum_{i=1}^{n} x_i^2}$. Compute for $x \neq 0$ the first and second derivatives of $|x|$, and, more generally, of $|x|^{\alpha}$, $\alpha \in \mathbb{R}$.

1.37 ¶. Let $\mathbf{A} \in M_{n,n}$. Compute the first and second derivatives of the linear map $x \to \mathbf{A}x$ and of the quadratic form $x \to \mathbf{A}x \bullet x$.

A trivial consequence of Schwarz's theorem is the following.

1.38 Corollary. *Every function $f \in C^2(A)$, A open in $\mathbb{R}^n$, has equal mixed second derivatives, $D_{ij}f(x) = D_{ji}f(x) \; \forall x \in A, \; \forall i, j = 1, \ldots, n$. In other words, the Hessian matrix $\mathbf{H}f(x)$ is symmetric $\forall x \in A$.*

d. Functions of classes $C^k(A)$ and $C^{\infty}(A)$

By induction we now define the partial derivatives of order k. We say that a function $f : A \to \mathbb{R}$ has *partial derivatives of order k*, $k \geq 2$, at an interior point x_0 of A if f has first partial derivatives in a neighborhood of x_0 that in turn have partial derivatives of order $k-1$ at x_0. The partial derivatives of order k of f are defined as the derivatives of order $k-1$ of the first-order partial derivatives of f. By taking into account at each step of the induction the theorems of total differentiation and of Schwarz, we easily infer if f has derivatives of order k at x_0 that

○ f has all partial derivatives of order less than k in a neighborhood of x_0, and these derivatives are continuous at x_0,
○ the derivatives of order h with $2 \leq h \leq k-1$ do not depend on the order of differentiation.

1.39 Definition. *If all derivatives of order k of a function exist and are continuous in an open set A, we say that f is of class $C^k(A)$ and we write $f \in C^k(A)$. If f has continuous derivatives of any order in an open set A, we say that f is of class $C^{\infty}(A)$ and we write $f \in C^{\infty}(A)$.*

Clearly,

$$C^{\infty}(A) \subset C^k(A) \subset C^{k-1}(A) \subset \cdots \subset C^2(A) \subset C^1(A) \subset C^0(A),$$

and again by Schwarz's theorem we have:

1.40 Corollary. *Let A be an open set of $\mathbb{R}^n$ and let $f \in C^k(A)$. Then the derivatives of f of order less than or equal to k do not depend on the order they are taken.*

Consequently, in order to specify a derivative of order k of a function of class C^k, it suffices to specify the number of derivatives we take in each variable; for instance, if $f \in C^6(\mathbb{R}^3)$, its sixth derivative 3 times with respect to x, 2 times with respect to y, and one time with respect to z at (x_0, y_0, z_0) is denoted by

$$\frac{\partial^6 f}{\partial x^3 \partial y^2 \partial z}(x_0, y_0, z_0).$$

1.3.2 Mean value theorem

a. Scalar functions

1.41 Theorem. *Let $A \subset \mathbb{R}^n$ be open and let $f : A \to \mathbb{R}$ be a function that has directional derivatives at all points of A. Suppose that x_0, x are two points of A such that the line segment joining x_0 to x is contained in A, and let $h := x - x_0$. Then the function $g(t) := f(x_0 + th)$, $t \in [0,1]$, is well defined and differentiable in $[0,1]$ and*

$$g'(t) = \frac{\partial f}{\partial h}(x_0 + th) \qquad \forall t \in [0,1]. \tag{1.32}$$

Moreover we have:

(i) (MEAN VALUE THEOREM) *There exists $s \in]0,1[$ such that*

$$f(x_0 + h) - f(x_0) = g(1) - g(0) = \frac{\partial f}{\partial h}(x_0 + sh), \tag{1.33}$$

(ii) (INTEGRAL MEAN VALUE THEOREM) *If $g'(t)$, $t \in [0,1]$, is continuous, then*

$$f(x_0 + h) - f(x_0) = \int_0^1 \frac{\partial f}{\partial h}(x_0 + th)\, dt. \tag{1.34}$$

Proof. (i) If $z := x_0 + th$ we have

$$g(t + \tau) - g(t) = f(x_0 + (t + \tau)h) - f(x_0 + th) = f(z + \tau h) - f(z);$$

hence

$$\frac{g(t + \tau) - g(t)}{\tau} = \frac{f(z + \tau h) - f(z)}{\tau} \to \frac{\partial f}{\partial h}(z) \qquad \text{as } \tau \to 0.$$

Therefore g is differentiable and (1.32) holds.

(i) and (ii) follow respectively from Lagrange's theorem and the fundamental theorem of calculus, see [GM1], applied to $g(t)$, $t \in [0,1]$. □

Theorem 1.41 applies of course to functions that are Gâteaux-differentiable. In this case

$$\frac{\partial f}{\partial h}(x) = \mathbf{D}f(x)h = \sum_{i=1}^{n} \frac{\partial f}{\partial x^i}(x)h^i$$

for all $h \in \mathbb{R}^n$. For future use we restate it as follows.

1.42 Corollary. *Let $f : A \to \mathbb{R}$ be Gâteaux-differentiable in an open set $A \subset \mathbb{R}^n$ and let $x_0 \in A$. Then for all $x \in B(x_0, r)$ we have*

(i) $f(x) - f(x_0) = \mathbf{D}f(x_0 + s(x - x_0))(x - x_0)$ *for some $s \in]0,1[$,*

(ii) *if the functions* $s \to \frac{\partial f}{\partial x^i}(x_0 + sh)$, $s \in [0,1]$, $i = 1, \ldots, n$, *are contin-uous, then*

$$f(x_0 + h) - f(x_0) = \int_0^1 \mathbf{D}f(x_0 + s(x - x_0))(x - x_0)\, ds \qquad (1.35)$$

$$= \sum_{i=1}^n \left(\int_0^1 \frac{\partial f}{\partial x^i}(x_0 + s(x - x_0))\, ds \right)(x - x_0)^i.$$

A trivial consequence, see also Corollary 1.51, is the following.

1.43 Proposition. *Let* $f : A \to \mathbb{R}$, *where* $A \subset \mathbb{R}^n$ *is open, be a function that has directional derivatives at every point of* A *and let*

$$M := \sup_{x \in A, |v| \le 1} \left| \frac{\partial f}{\partial v}(x) \right| < +\infty.$$

Then f *is continuous in* A. *Moreover, if* A *is convex, then* f *is Lipschitz-continuous in* A *and*

$$|f(y) - f(x)| \le M\, |y - x| \qquad \forall x, y \in A. \qquad (1.36)$$

1.44 ¶. Show that (1.36) does not hold in general if A is not convex. Show instead the following.

Corollary. *Let* $f \in C^1(\Omega)$, $\Omega \subset \mathbb{R}^n$ *open. Then* f *is Lipschitz-continuous in every compact subset* K *of* Ω.

1.45 Corollary. *Let* $f : A \to \mathbb{R}$, *where* $A \subset \mathbb{R}^n$ *is open, be a function that has directional derivatives at every point of* A. *If* A *is connected and* $\frac{\partial f}{\partial v}(x) = 0$ $\forall v \in \mathbb{R}^n$ *and* $\forall x \in A$, *then* f *is constant in* A.

Proof. Let $x_0 \in \Omega$ and let

$$B := \left\{ x \in \Omega \,\middle|\, f(x) = f(x_0) \right\}.$$

By Theorem 1.41 B is open, while Proposition 1.43 yields that f is continuous, hence B is closed. Hence $B = A$ since A is connected. $\qquad \square$

1.46 Remark. Of course, in Corollary 1.45 the assumption A connected is essential. We notice also that the assumption A open is needed. In fact one can show, though this is not trivial, the existence of a nonconstant function in a connected set with zero differential.[1]

---

[1] H. Whitney, *A function not constant on a connected set of critical points*, Duke Math. J. **1** (1935), 514–517.

For all $x, y \in A$, where $A \subset \mathbb{R}^n$ is open, let $\gamma : [0, 1] \to A$ be a curve of class C^1 such that $\gamma(0) = x$ and $\gamma(1) = y$, and assume that $f : A \to \mathbb{R}$ is as above. Then

$$f(y) - f(x) = \int_0^1 \frac{d}{dt} f(\gamma(t)) \, dt = \int_0^1 \mathbf{D}f(\gamma(t))\gamma'(t) \, dt,$$

hence

$$|f(y) - f(x)| \leq \int_0^1 |\mathbf{D}f(\gamma(t))| \, |\gamma'(t)| \, dt \leq M \, L(\gamma)$$

where $M := \sup_{x \in A} |\mathbf{D}f(x)|$ and $L(\gamma)$ is the length of γ. If $\delta_A(x, y)$ denotes the infimum of the lengths of the curves in A joining x to y, i.e., the *minimal connection* of x to y in A, we then have

$$|f(y) - f(x)| \leq M \, \delta_A(x, y).$$

Consequently, we infer the following.

1.47 Proposition. *Let $f \in C^1(A)$ with $|\mathbf{D}f(x)| \leq M \; \forall x \in A$. If there exists $C > 0$ such that*

$$\delta_A(x, y) \leq C \, |x - y| \qquad \forall x, y \in A,$$

then f is Lipschitz-continuous in A.

If $A = B(x_0, r)$ or A is convex, then clearly $\delta_A(x, y) = |x - y| \; \forall x, y \in A$.

1.48 ¶. Show that for any compact $F \subset A$ there exists C_F such that $\delta_A(x, y) \leq C_F \, |x - y| \; \forall x, y$. In particular, every function of class $C^1(A)$ is Lipschitz-continuous on the compact subsets of A.

b. Vector-valued functions

The mean value theorem in the form (1.33) does not hold for vector-valued maps $f : A \to \mathbb{R}^m$, $m > 1$.

1.49 Example. For instance, if $f(t) = (\cos t, \sin t)$, $t \in [0, 2\pi]$, we have $0 = f(2\pi) - f(0)$ but $f'(s) \neq 0 \; \forall s \in [0, 2\pi]$ since $|f'(s)| = 1$.

It instead holds in the integral form (1.34). In fact, recalling that for $f \in C^0([a, b], \mathbb{R}^m)$, $f := (f^1, f^2, \ldots, f^m)^T$, we have

$$\int_a^b f(s) \, ds := \left(\int_a^b f^1(s) \, ds, \int_a^b f^2(s) \, ds, \ldots, \int_a^b f^m(s) \, ds \right)^T,$$

we easily infer the following.

1.50 Theorem (Mean value). *Let $f : B(x_0, r) \subset \mathbb{R}^n \to \mathbb{R}^m$ be a map of class C^1. Then for all $x, y \in B(x_0, r)$ the following integral mean formula holds*

$$f(x) - f(y) = \int_0^1 \mathbf{D}f(y + t(x - y))(x - y) \, dt.$$

This allows us to estimate finite increments of f in terms of the Jacobian matrix of f. For $\mathbf{A} \in M_{m,n}(\mathbb{R})$ recall that

$$\|\mathbf{A}\| := \sup\left\{ \frac{|\mathbf{A}(h)|}{|h|} \,\middle|\, h \neq 0 \right\}$$

is a norm in $M_{m,n}(\mathbb{R})$ and that $|\mathbf{A}h| \leq \|\mathbf{A}\|\,|h| \; \forall h \in \mathbb{R}^n$.

1.51 Corollary. *Let $f : B(x_0, r) \subset \mathbb{R}^n \to \mathbb{R}^m$ be a map that has directional derivatives at every point of $B(x_0, r)$ and let*

$$K := \sup\left\{ \|\mathbf{D}f(z)\| \,\middle|\, z \in B(x_0, r) \right\}.$$

Then

$$|f(x) - f(y)| \leq K\,|x - y|. \tag{1.37}$$

Proof. Of course,

$$|\mathbf{D}f(z)(h)| \leq \|\mathbf{D}f(z)\|\,|h| \leq K\,|h| \qquad \forall h \in \mathbb{R}^n.$$

Thus

$$|f(x) - f(y)| \leq \left| \int_0^1 \mathbf{D}f(y + t(x-y))(x-y)\,dt \right| \leq \int_0^1 |\mathbf{D}f(y + t(x-y))(x-y)|\,dt$$

$$\leq \int_0^1 \|\mathbf{D}f(y + t(x-y))\|\,dt\,|x-y| \leq K\,|x-y|.$$

$\square$

1.3.3 Taylor's formula

Let A be an open set in $\mathbb{R}^n$ and let $f : A \to \mathbb{R}$ be a function of class C^k, $k \geq 1$. Suppose that the segment joining two points x_0, x is contained in A and let $h := x - x_0$. The function

$$F(t) := f(x^0 + th)$$

is well defined for $t \in [0, 1]$ and $F(0) = f(x_0)$ e $F(1) = f(x)$. Moreover $F \in C^k([0, 1])$, and we may compute for $t \in [0, 1]$

$$F'(t) = \sum_{i=1}^n \frac{\partial f}{\partial x_i}(x_0 + th)h_i = \sum_{i=1}^n D_i f(x_0 + th)h_i$$

$$F''(t) = \sum_{i=1}^n [D_i f(x_0 + th)]' h_i = \sum_{i,j=1}^n D_j D_i f(x_0 + th)h_i h_j$$

$$F'''(t) = \sum_{i,j,k=1}^n D_k D_j D_i f(x_0 + th)h_i h_j h_k \tag{1.38}$$

$$\cdots$$

$$F^{(k)}(t) = \sum_{i_1,i_2,\ldots,i_k=1}^n D_{i_k} D_{i_{k-1}} \cdots D_{i_1} f(x_0 + th)h_{i_1} h_{i_2} \ldots h_{i_k}.$$

Here, as in the rest of this section, we denote the components $(h_1, \ldots, h_n)$ of h with lower indices.

Notice that $F^{(j)}(0)$ is a homogeneous polynomial of degree j in the components $h_1, \ldots, h_n$ of h.

a. Taylor's formula of second order

We can write F' and F'' in (1.38) as

$$F'(t) = \nabla f(x_0 + th) \bullet h, \quad \text{and} \quad F''(t) = \mathbf{H}f(x_0 + th)h \bullet h, \quad (1.39)$$

where ∇f and $\mathbf{H}f$ are respectively the gradient and the Hessian of the function f, and $x \bullet y$ is the standard inner product in $\mathbb{R}^n$.

1.52 Theorem (Taylor's formula). *Let* $f \in C^2(B(x_0, r))$, $r > 0$. *For* $h \in \mathbb{R}^n$, $|h| < r$, *we have*

(i) (TAYLOR'S FORMULA WITH INTEGRAL REMAINDER)

$$f(x_0 + h) = f(x_0) + \nabla f(x_0) \bullet h + \int_0^1 (1 - s) \Big[\mathbf{H}f(x_0 + sh)h \bullet h \Big] ds$$

$$= f(x_0) + \nabla f(x_0) \bullet h$$

$$+ \sum_{i,j=1}^n \Big(\int_0^1 (1 - s) \frac{\partial^2 f}{\partial x_i \partial x_j}(x_0 + sh) \, ds \Big) h_i h_j,$$

(ii) (TAYLOR'S FORMULA WITH LAGRANGE'S REMAINDER)

$$f(x_0 + h) = f(x_0) + \nabla f(x_0) \bullet h + \frac{1}{2} \mathbf{H}f(x_0 + sh)h \bullet h$$

$$= f(x_0) + \nabla f(x_0) \bullet h + \sum_{i,j=1}^n \frac{\partial^2 f}{\partial x_i \partial x_j}(x_0 + sh)h_i h_j$$

for some $s \in]0, 1[$,

(iii) (TAYLOR'S FORMULA WITH PEANO'S REMAINDER)

$$f(x_0 + h) - f(x_0) - \nabla f(x_0) \bullet h - \frac{1}{2} \mathbf{H}f(x_0)h \bullet h = o(|h|^2) \quad \text{as } h \to 0.$$

Proof. (i) and (ii) are Taylor's formulas for $F(t) := f(x_0 + t(x - x_0))$, $t \in [0, 1]$, see [GM1], taking into account (1.39). Then (iii) follows at once from (ii). □

b. Taylor formulas of higher order

It is convenient to rewrite (1.38) in a more convenient form. An n-tuple of nonnegative integers is called a *multiindex*

$$\alpha = (\alpha_1, \alpha_2, \ldots, \alpha_n), \qquad \alpha_i \in \mathbb{N}.$$

The *length* of α is the number

$$|\alpha| := \alpha_1 + \alpha_2 + \cdots + \alpha_n,$$

and it is convenient to define

$$\alpha! := \alpha_1! \, \alpha_2! \ldots \alpha_n!,$$

and, for $x = (x_1, x_2, \ldots, x_n) \in \mathbb{R}^n$,

$$x^\alpha := x_1^{\alpha_1} x_2^{\alpha_2} \cdots x_n^{\alpha_n}.$$

Notice that $|x^\alpha| \leq |x|^{|\alpha|}$. Finally, if $\alpha = (\alpha_1, \alpha_2, \ldots, \alpha_n)$ is a multiindex with n elements, the derivative of f of order α_1 times with respect to x_1, α_2 times with respect to x_2, $\ldots$, α_n times with respect to x_n is denoted by

$$D^\alpha f \qquad \text{or} \qquad \frac{\partial^{|\alpha|} f}{\partial x_1^{\alpha_1} \partial x_2^{\alpha_2} \ldots \partial x_n^{\alpha_n}}.$$

Grouping in each of the equations in (1.38) the terms containing the derivatives of order α, $|\alpha| < j$, we may rewrite the equation in (1.38) as

$$F^{(j)}(t) = \sum_{|\alpha|=j} C_\alpha D^\alpha f(x_0 + th) \, h^\alpha, \qquad j = 1, \ldots, k, \qquad (1.40)$$

where C_α is the number of lists of $|\alpha|$ differentiations, α_1 times with respect to x_1, α_2 times with respect to x_2, $\ldots$, α_n times with respect to x_n.

Computing C_α is now a combinatorial problem, see, e.g., [GM2, 3.2.4]. There are $\binom{n}{\alpha_1}$ ways of disposing α_1 objects of type 1 in a list of n elements; hence $\binom{n}{\alpha_1}\binom{n-\alpha_1}{\alpha_2}$ ways of disposing α_1 objects of type 1 and α_2 objects of type 2 in a list of n, $\ldots$. Thus

$$C_\alpha = \binom{|\alpha|}{\alpha_1}\binom{|\alpha|-\alpha_1}{\alpha_2}\binom{|\alpha|-\alpha_1-\alpha_2}{\alpha_3}\cdots\binom{\alpha_n}{\alpha_n} = \frac{|\alpha|!}{\alpha_1!\alpha_2!\ldots\alpha_n!} = \frac{|\alpha|!}{\alpha!},$$

and (1.40) becomes

$$\frac{1}{j!}F^{(j)}(t) = \sum_{|\alpha|=j} \frac{D^\alpha f(x_0 + t(x - x_0))}{\alpha!}(x - x_0)^\alpha, \qquad j = 1, \ldots, k. \quad (1.41)$$

1.53 Definition. *Let* $f \in C^k(B(x_0, r))$, $r > 0$. *Taylor's polynomial of* f *centered at* x_0 *of order* k *is the polynomial in* $\mathbb{R}^n$ *of order at most* k

$$P_k(x; x_0) := \sum_{|\alpha| \leq k} \frac{D^\alpha f(x_0)}{\alpha!} (x - x_0)^\alpha.$$

Taylor formulas for $F(t) := f(x_0 + t(x - x_0))$, $t \in [0, 1]$, give rise, on account of (1.41), to Taylor formulas for f.

1.54 Theorem (Taylor's formula with Lagrange's remainder). *Let* f *be a function in* $C^k(B(x_0, r))$, $r > 0$. *Then*

$$f(x) = P_{k-1}(x; x_0) + \sum_{|\alpha|=k} \frac{D^\alpha f(x_0 + s(x - x_0))}{\alpha!} (x - x_0)^\alpha$$

for some $s \in]0, 1[$. *Moreover, if we set*

$$R_{k-1}(x; x_0) := f(x) - P_{k-1}(x; x_0),$$
$$M_k := \sup_{|\alpha|=k} \sup_{z \in B(x_0, r)} |D^\alpha f(z)|,$$

we have

$$\left| R_{k-1}(x; x_0) \right| \leq \frac{n^k M_k}{k!} |x - x_0|^k. \tag{1.42}$$

Proof. Taylor's formula for $F(t) := f(x_0 + t(x - x_0))$, $t \in [0, 1]$, yields

$$F(1) = \sum_{j=1}^{k-1} \frac{F^{(j)}(0)}{j!} + \frac{1}{k!} F^{(k)}(s)$$

for some $s \in]0, 1[$. This yields the result computing $F^{(j)}(0)$ with (1.41). From $F^{(k)}(s)$ in (1.38), we infer for $t \in [0, 1]$

$$|F^{(k)}(t)| \leq M_k |h|^k \left(\sum_{i_1, i_2, \ldots, i_k = 1}^{n} 1 \right) = M_k |h|^k n^k.$$

□

1.55 Corollary (Taylor's formula with Peano's remainder). *Let* f *be a function in* $C^k(B(x_0, r))$, $r > 0$. *Then*

$$f(x) = \sum_{|\alpha| \leq k} \frac{D^\alpha f(x_0)}{\alpha!} (x - x_0)^\alpha + o(|x - x_0|^k) \qquad \text{as } x \to x_0.$$

Proof. In fact,

$$f(x) = \sum_{|\alpha| \leq k-1} \frac{D^\alpha f(x_0)}{\alpha!} (x - x_0)^\alpha + \sum_{|\alpha|=k} \frac{D^\alpha f(x_0 + s(x - x_0))}{\alpha!} (x - x_0)^\alpha$$

$$= \sum_{|\alpha|=k} \frac{D^\alpha f(x_0)}{\alpha!} (x - x_0)^\alpha + \sum_{|\alpha|=k} \frac{D^\alpha f(x_0 + s(x - x_0)) - D^\alpha f(x_0)}{\alpha!} (x - x_0)^\alpha$$

for some $s = s(x)$ with $|s - x_0| < |x - x_0|$. When $x \to x_0$, also $x_0 + s(x)(x - x_0) \to x_0$, hence

$$|D^\alpha f(x_0 + s(x - x_0)) - D^\alpha f(x_0)| \to 0.$$

Since $|h^\alpha| \le |h|^{|\alpha|}$, this yields

$$\sum_{|\alpha|=k} \frac{D^\alpha f(x_0 + s(x - x_0)) - D^\alpha f(x_0)}{\alpha!}(x - x_0)^\alpha = o(|x - x_0|^k) \qquad \text{as } x \to x_0.$$

$\square$

1.56 ¶. Show that *Taylor's polynomial* $P_k(x; x_0)$ of degree k centered at x_0 of f is the unique polynomial $Q(x)$ of degree less than or equal to k such that

$$f(x) - Q(x) = o(|x - x_0|^k) \qquad \text{as } x \to x_0.$$

1.57 ¶. Show the following.

Proposition (Taylor's formula with integral remainder). *Let f be a function in* $C^{k+1}(B(x_0, r))$, $r > 0$. *Then*

$$f(x) = P_k(x; x_0) + (k+1) \sum_{|\alpha|=k+1} \left(\int_0^1 (1-t)^k D^\alpha f(x_0 + t(x - x_0)) \, dt \right) \frac{(x - x_0)^\alpha}{\alpha!}. \quad (1.43)$$

1.58 ¶. Applying Taylor's formula with integral remainder with $k = 0$ and $k = 1$, show the following.

Lemma (Hadamard). *If f is of class C^∞ near $0 \in \mathbb{R}^n$, then there exist functions* $g_i(x)$ *and* $g_{ij}(x)$ *of class C^∞ such that*

$$f(x) - f(0) = \sum_{i=1}^n g_i(x) x^i, \qquad f(x) - f(0) = \sum_{i=1}^n f_{x^i}(0) x^i + \sum_{i,j=1}^n g_{ij}(x) x^i x^j \quad (1.44)$$

in a neighborhood of zero.

c. Real analytic functions

Let f be a function of class C^∞ in an open set A of $\mathbb{R}^n$. Then f has Taylor polynomials of any order at every point x_0 of A. The sequence of Taylor polynomials $P_k(x; x_0)$ of f, equivalently the series of functions

$$\sum_{k=0}^\infty \left(\sum_{|\alpha|=k} \frac{D^\alpha f(x_0)}{\alpha!} (x - x_0)^\alpha \right),$$

is called the *Taylor's series* of f centered at $x_0 \in A$.

1.59 Definition. *We say that $f \in C^\infty(A)$ is an* analytic function *if every point $x_0 \in A$ has a neighborhood in which f agrees with the sum of its Taylor's series with center x_0.*

There exist functions of class C^∞ in an open set that are not analytic. For instance, compare [GM1], the function

$$f(x) := \begin{cases} e^{-1/x^2} & \text{if } x > 0, \\ 0 & \text{if } x \le 0 \end{cases}$$

is of class C^∞ and nonzero, but all its derivatives exist and vanish at zero, consequently the sum of its Taylor's series is zero. However, the following holds.

1.60 Proposition. *Let A be an open set of $\mathbb{R}^n$ and $f \in C^\infty(A)$. Suppose that for every $x_0 \in A$ there exist C and $r > 0$ such that $B(x_0, r) \subset A$ and*

$$\sup_{|\alpha|=k} \sup_{B(x_0,r)} |D^\alpha f(x)| \le C^k k! \qquad \forall k \in \mathbb{N},$$

then $f(x)$ is analytic in A.

1.61 ¶. Prove Proposition 1.60. [*Hint:* Use (1.42).]

d. A converse of Taylor's theorem

The following theorem may be read as a converse of Corollary 1.55.

1.62 Theorem (Marcinkiewicz–Zygmund). *Let Ω be an open set of $\mathbb{R}^n$ and let $f(x)$ and $a_\alpha(x)$, $|\alpha| \le k$, be continuous functions in Ω. Suppose that, for $x \in \Omega$ and $|h| < \text{dist}\,(x, \partial\Omega)$, we have*

$$f(x+h) = \sum_{|\alpha|\le k} \frac{a_\alpha(x)}{\alpha!} h^\alpha + g(x, h) \qquad (1.45)$$

where $g(x,h)/|h|^k \to 0$ as $h \to 0$ uniformly with respect to x on the compact subsets of Ω. Then f is of class $C^k(\Omega)$.

Proof. We convolute the two sides of (1.45) by means of a family of mollifiers, see Section 2.3 Chapter 2, to get

$$f_\epsilon(x+h) = \sum_{|\alpha|\le k} \frac{(a_\alpha)_\epsilon(x)}{\alpha!} h^\alpha + g_\epsilon(x, h).$$

Since $f_\epsilon \in C^k(\widetilde{\Omega})$ for all $\widetilde{\Omega} \subset\subset \Omega$ and

$$\|g_\epsilon(x,h)\|_{\infty,\widetilde{\Omega}} \le \|g(x,h)\|_{\infty,\Omega},$$

we infer from Exercise 1.56

$$D_\alpha f_\epsilon(x) = (a_\alpha)_\epsilon(x) \qquad \forall x \in \widetilde{\Omega},\ \forall \alpha,\ |\alpha| \le k$$

and letting $\epsilon \to 0$, see Section 2.3 Chapter 2, we conclude $D^\alpha f(x) = a_\alpha(x)\ \forall x \in \widetilde{\Omega}$. This concludes the proof as $\widetilde{\Omega}$ is arbitrary. □

1.3.4 Critical points

Let $A \subset \mathbb{R}^n$ and let $f : A \to \mathbb{R}$ be a function. A point $x_0 \in A$ such that

$$f(x_0) \leq f(x) \qquad \forall\, x \in A \qquad (1.46)$$

is called a *minimum point* or an *absolute minimizer* of f in A, the value of f at a minimum point

$$f(x_0) = \min\Big\{ f(x) \,\Big|\, x \in A \Big\}$$

is called the *minimum (value)* of f in A. In case the inequality in (1.46) is strict (except for $x = x_0$) we say that x_0 is a *strict* absolute minimizer. As we know, functions may or may not have minimizers: for a given $f : A \to \mathbb{R}$, they exist if A is bounded and closed and f is lower semicontinuous in A, see, e.g., [GM2, Chapter 2] and [GM3, Chapter 4].

We say that x_0 is a *local* (or *relative*) *minimizer* for $f : A \to \mathbb{R}$ if there is a neighborhood U_{x_0} of x_0 in which it is a minimizer for f, i.e.,

$$f(x_0) \leq f(x) \qquad \forall\, x \in U_{x_0} \cap A.$$

If the previous inequality is strict for $x \neq x_0$, we say that x_0 is an *isolated* local minimizer for f. Similar definitions of course hold for maximum points. Local minimizers and local maximizers of f are both called *extremal points* and the values of f at extremal points are called *extremal values*.

Sometimes the research of extremal points reduces to a simple inspection. For instance, clearly $f(x) := |x|$, $x \in \mathbb{R}^n$, has an absolute minimizer at 0, as well as the function $\log(1 + x^2 + 2y^2)$: since both functions are nonnegative and vanish if and only if $(x, y) = (0, 0)$. In other cases it is easy to conclude by looking at the *level lines* of the function. However, it is useful to develop some general remarks.

Suppose x_0 is an extremal for f. Then x_0 is also an extremal for the restriction of f to any line through x_0. It follows from Fermat's theorem: Suppose that x_0 is interior to A and that f has directional derivative in the direction v at x_0. Then $D_v f(x_0) = 0$. Of course, both assumptions x_0 interior to A and f has directional derivative in the direction v are essential as shown by the function $f(x) = |x|$.

When A is an open set of $\mathbb{R}^n$ and $f \in C^1(A)$, we can state more.

1.63 Proposition. *Let $f : A \to \mathbb{R}$ be a function that is differentiable at an interior point x_0 to A. Then x_0 is an extremal point for f if*

$$df_{x_0} = 0 \qquad equivalently \qquad \nabla f(x_0) = 0. \qquad (1.47)$$

1.64 Definition. *Let A be an open set of $\mathbb{R}^n$ and let $f : A \to \mathbb{R}$ be differentiable in A. The points $x \in A$ such that $df_x = 0$ (equivalently $\nabla f(x) = 0$) are called* critical points *of f in A.*

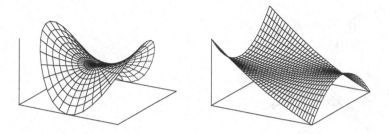

Figure 1.15. Two saddle points.

As for functions of one variable, whereas all interior extremal points are critical points, not every critical point is an extremal point. For instance, $(0,0)$ is a critical point of the function $f(x,y) = x^2 - y^2$, $(x,y) \in \mathbb{R}^2$, but it is not a local minimizer or a local maximizer for f. Looking at level lines of f, one readily infers that $(0,0)$ is a *saddle* point for f.

At this point we should warn the reader that the intuition relative to critical points for functions of several variables is not as reliable as for functions of one variable. The following example may be useful.

1.65 Example. The function $f(x,y) = y(y - x^2)$, $(x,y) \in \mathbb{R}^2$, has a critical point at $(0,0)$. The point $(0,0)$ is a minimizer for the restriction of f to any straight line through the origin, but it is a maximum point for the restriction of f to the parabola, see Figure 1.16.

1.66 Example. The function

$$f(x,y) := x^3 - 3x + (e^y - x)^2, \quad (x,y) \in \mathbb{R}^2,$$

has a unique critical point at $(1,0)$ that is a local minimizer, see Figure 1.17, moreover

$$\inf_{\mathbb{R}^2} f(x,y) = -\infty,$$

though f has no relative maximum point.

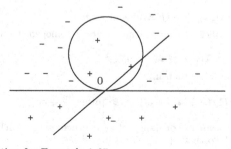

Figure 1.16. Illustration for Example 1.65.

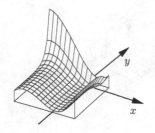 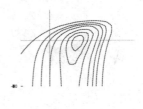

Figure 1.17. $f(x, y) = x^3 - 3x + (e^y - x)^2$ has a local minimizer at $(1, 0)$, tends to $-\infty$ along the curve $y = x^3$ as $x \to -\infty$, and has no relative maximizers.

Let $f \in C^2(A)$, with A open in $\mathbb{R}^n$, and let $x_0 \in A$ be a critical point of f. Taylor's formula with Peano's remainders tells us that

$$f(x_0 + h) = f(x_0) + \frac{1}{2} \mathbf{H}f(x_0)h \bullet h + o(|h|^2) \qquad \text{as } |h| \to 0, \qquad (1.48)$$

$\mathbf{H}f(x_0) = [D_{ij}f(x_0)]$ being the Hessian matrix of f at x_0.

1.67 Proposition. *Let $f \in C^2(A)$, with A open in $\mathbb{R}^n$, and let $x_0 \in A$ be a critical point of f. Then*

(i) *if x_0 is a local minimizer, then $\mathbf{H}f(x_0)\xi \bullet \xi \geq 0 \ \forall \xi \in \mathbb{R}^n$,*
(ii) *if $\mathbf{H}f(x_0)\xi \bullet \xi > 0 \ \forall \xi \neq 0$, then x_0 is an isolated local minimizer.*

Proof. (i) From (1.48) we have

$$0 \leq f(x + \lambda\xi) - f(x_0) = \frac{1}{2}\Big[\mathbf{H}f(x_0)(\lambda\xi) \bullet (\lambda\xi) \Big] + o(\lambda^2)$$

as $\lambda \to 0$, $\lambda \in \mathbb{R}$, and, dividing by λ^2 we get

$$\mathbf{D}f(x_0)\xi \bullet \xi + o(1) \geq 0 \qquad \text{as } \lambda \to 0,$$

i.e., the claim.

(ii) Let $\phi(\xi) := \mathbf{H}f(x_0)\xi \bullet \xi$. Since $\phi(\xi)$ is a homogeneous polynomial of degree two in the components of ξ, the restriction of $\xi \to \mathbf{H}f(x_0)\xi \bullet \xi$ to the unit sphere $S^{n-1} := \{\xi \,|\, |\xi| = 1\} \subset \mathbb{R}^n$ is continuous. Weierstrass's theorem then yields a point $\xi_0 \in S^{n-1}$ such that

$$\mathbf{H}f(x_0)\xi \bullet \xi \geq \mathbf{H}f(x_0)\xi_0 \bullet \xi_0 =: m_0 \qquad \forall \xi \in S^{n-1}; \qquad (1.49)$$

while the assumption implies that $m_0 > 0$ and, using 2-homogeneity of $\xi \to \mathbf{H}f(x_0)\xi \bullet \xi$, we get the estimate

$$\mathbf{H}f(x_0)\xi \bullet \xi \geq m_0|\xi|^2 \qquad \forall \xi \in \mathbb{R}^n.$$

From Taylor's formula we then infer

$$f(x) - f(x_0) = \frac{1}{2} \mathbf{H}f(x_0)h \bullet h + o(|h|^2) \geq |h|^2\Big(\frac{m_0}{2} + o(1)\Big), \qquad h := x - x_0.$$

Since $m_0 > 0$, the theorem of constancy of sign provides us with a ball $B(x_0, \delta)$ on which $m_0/2 + o(1) > 0$, so that $f(x) > f(x_0)$ for all $x \in B(x_0, \delta)$, $x \neq x_0$. $\qquad \square$

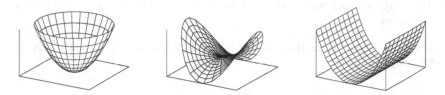

Figure 1.18. The Hessian matrix of the paraboloid on the left has two positive eigenvalues, the Hessian matrix of the saddle has eigenvalues of opposite sign, and the Hessian matrix of the cylinder has a positive and a vanishing eigenvalue.

1.68 Quadratic forms. We recall that the positivity of a quadratic form $\phi(h) = \mathbf{A}h \bullet h$ associated to a symmetric matrix $\mathbf{A} \in M_{n,n}$ can be checked in terms of the *signature* of the metric $(h, k) \to \mathbf{A}h \bullet k$, see [GM2, Chapter 2], that is, writing ϕ as a sum of squares

$$\mathbf{A}h \bullet h = \sum_{i=1}^{n} m_i w_i^2.$$

Several methods are available to do this: for instance by computing the eigenvalues of $\mathbf{A}$. Since $\mathbf{A}$ is symmetric, the spectral theorem tells us that there exist an orthonormal basis $(u_1, u_2, \ldots, u_n)$ of $\mathbb{R}^n$ and real numbers $\lambda_1, \lambda_2, \ldots, \lambda_n$ such that

$$\mathbf{A}h = \sum_{i=1}^{n} \lambda_i \, h \bullet u_i \, u_i \qquad \forall h,$$

which says in particular that $u_1, u_2, \ldots, u_n$ are eigenvectors of $\mathbf{A}$ and, for every $i = 1, \ldots, n$, λ_i is the eigenvalue of $\mathbf{A}$ corresponding to u_i. Consequently,

$$\mathbf{A}h \bullet h = \sum_{i=1}^{m} \lambda_i \, h \bullet u_i \, u_i \qquad \forall h.$$

Since $|h|^2 = \sum_{i=1}^{n} |h \bullet u_i|^2$, we get

$$\lambda_m |h|^2 \leq \mathbf{A}h \bullet h \leq \lambda_M |h|^2 \qquad \forall h \in \mathbb{R}^n$$

where $\lambda_m := \min_i(\lambda_i)$ and $\lambda_M = \max_i(\lambda_i)$. We can therefore restate Proposition 1.67 as follows.

1.69 Proposition. *Let $f \in C^2(A)$, with A open in $\mathbb{R}^n$ and let $x_0 \in A$ be a critical point of f and let $\mathbf{H}f(x_0)$ be the Hessian matrix of f at x_0. Then we have the following.*

(i) *If x_0 is a local minimizer for f, then the eigenvalues of $\mathbf{H}f(x_0)$ are nonnegative,*

(ii) *If the eigenvalues of* $\mathbf{H}f(x_0)$ *are positive, then* x_0 *is an isolated local minimizer.*

1.70 ¶. Show that, in $\mathbb{R}^2$, the quadratic form $\mathbf{A}h \bullet h$ is positive if and only if $\operatorname{tr} \mathbf{A} > 0$ and $\det \mathbf{A} > 0$.

1.3.5 Some classical partial differential equations

1.71 The Laplace operator. The differential operator

$$\Delta u := \operatorname{div} \operatorname{grad} u = \sum_{i=1}^{n} D_i D_i u$$

is called the *Laplace operator* or *Laplacian*. The functions $u \in C^2(\Omega)$ for which $\Delta u = 0$ in Ω are said to be *harmonic* in Ω.

1.72 ¶. Show that, in dimension two, and in polar coordinates (ρ, θ), Δu writes as

$$\Delta u = u_{\rho\rho} + \frac{1}{\rho} u_\rho + \frac{1}{\rho^2} u_{\theta\theta} = \frac{1}{\rho}\left[\frac{\partial}{\partial\rho}(\rho f_\rho) + \frac{\partial}{\partial\theta}\left(\frac{1}{\rho}f_\theta\right)\right]$$

whereas, in dimension $n = 3$ and in spherical coordinates, Δu writes as

$$\Delta u = \frac{1}{\rho \sin^2 \varphi}\left\{\frac{\partial}{\partial\rho}(\rho^2 f_\rho \sin\varphi) + \frac{\partial}{\partial\theta}\left(\frac{1}{\sin\varphi}f_\theta\right) + \frac{\partial}{\partial\varphi}(\sin\varphi f_\varphi)\right\}.$$

1.73 ¶ Laplacian and gravitational forces. The gravitational force acting on a unit mass at the point (x, y, z) due to the interaction with a mass placed at the origin is given, according to Newton's gravitational law, by

$$F = -g\frac{M}{r^2}\frac{r}{|r|}, \qquad g > 0, \ r := (x, y, z).$$

If

$$V = \frac{g}{2}\frac{M}{r}$$

denotes the gravitational potential, observe that $F = -\nabla V$. Moreover, show that

(i) V is harmonic in $\mathbb{R}^3 \setminus \{(0,0,0)\}$, i.e., $\Delta V = 0$ in $\mathbb{R}^3 \setminus \{(0,0,0)\}$.
(ii) The unique spherical symmetric harmonic function in $\mathbb{R}^n \setminus \{0\}$, i.e., the unique harmonic functions of the type $u(x) = \varphi(|x|)$, are

$$A\gamma(|x|) + B \qquad \text{where } A, B \in \mathbb{R} \ \text{ and } \ \gamma(r) := \begin{cases} \log r & \text{if } n = 2, \\ \dfrac{1}{r^{n-2}} & \text{if } n \geq 3. \end{cases}$$

(iii) The functions $e^{kx}\cos ky$ and $e^{kx}\sin ky$ are harmonic in $\mathbb{R}^2$.
(iv) The function $e^{3x+4y}\sin 5y$ is harmonic in $\mathbb{R}^3$.
(v) If $f(x, y)$ is harmonic in $\mathbb{R}^2$, then also $f\left(\frac{x}{x^2+y^2}, \frac{y}{x^2+y^2}\right)$ is harmonic.

1.74 ¶. Show the following theorem.

Figure 1.19. Frontispieces of two works respectively by Constantin Carathéodory (1873–1950) and Richard Courant (1888–1972).

Theorem (Maximum principle). *Let* $u \in C^0(\overline{\Omega}) \cap C^2(\Omega)$ *be harmonic in* Ω. *Then*

$$\sup_{\Omega} |u| = \sup_{\partial\Omega} |u|.$$

[*Hint:* Let $x_0 \in \Omega$ and, for $\epsilon > 0$, let $u_\epsilon(x) := u(x) - \epsilon|x - x_0|^2$. We then have

$$\Delta(u - \epsilon|x - x_0|^2) = -2n\epsilon < 0;$$

hence x_0 cannot be a maximum point for u_ϵ since in this case $\mathbf{H}u_\epsilon(x_0) \geq 0.$]

1.75 ¶. Consider the following problem, called *Dirichlet's problem*: find $u \in C^2(\Omega) \cap C^0(\overline{\Omega})$ such that

$$\begin{cases} \Delta u = f & \text{in } \Omega, \\ u = g & \text{on } \partial\Omega, \end{cases}$$

where g is a given continuous function on $\partial\Omega$. Infer from the maximum principle, see Exercise 1.74, that it has at most one solution.

1.76 ¶. Functions $u : \Omega \subset \mathbb{R}^n \to \mathbb{R}$ such that $-\Delta u \leq 0$ in Ω, are called *subharmonic*, whereas functions u such that $-\Delta u \geq 0$ in Ω are called *superharmonic*. Show that

(i) subharmonic functions in Ω have no (interior) maximum point in Ω,
(ii) superharmonic functions in Ω have no (interior) minimum point in Ω,
(iii) if u is subharmonic in Ω, and v is superharmonic in Ω and $u \leq v$ on $\partial\Omega$, then $u \leq v$ in Ω.

Finally, show that, if u is harmonic and $f \in C^2(\mathbb{R})$ satisfies $f''(t) \geq 0$, then $f(u)$ is subharmonic.

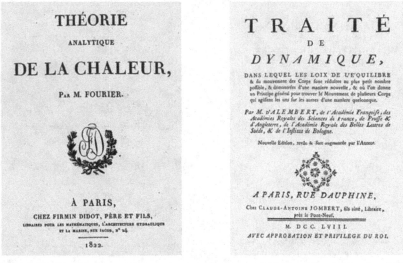

Figure 1.20. Frontispieces of two monographs respectively by Joseph Fourier (1768–1830) and Jean d'Alembert (1717–1783).

1.77 The wave equation. If f and g are two functions of class C^2, the function
$$u(t, x) := f(x - ct) + g(x + ct) \tag{1.50}$$
satisfies the 1-*dimensional wave equation*
$$\frac{\partial^2 u}{\partial t^2} = c^2 \frac{\partial^2 u}{\partial x^2}.$$

This equation is supposed to describe the vibrations of a string pulled tight between fixed ends: $u(x, t)$ represents the height at time t and it is assumed that the longitudinal displacement is negligible. Notice that $f(x - ct)$ and $g(x + ct)$ represent waves that propagate respectively, to the left and to the right with velocity c.

1.78 ¶. Show that by the change of variables $r = x + ct$, $s = x - ct$, the wave equation transforms into
$$\frac{\partial^2 u}{\partial r \partial s} = 0.$$
Infer from this that (1.50) represents a general solution of the 1-dimensional wave equation.

1.79 ¶. Show that the initial value problem
$$\begin{cases} u_{tt}(x, t) = c^2 u_{xx}(x, t), \\ u(x, 0) = p(x), \\ u_t(x, 0) = q(x) \end{cases}$$
admits as solution the function

Figure 1.21. Olga Ladyzhenskaya (1922–2004) and the first page of a celebrated paper by Jean Leray (1906–1998) on Navier–Stokes equations.

$$u(x,t) = \frac{1}{2}\Big(p(x-ct) + p(x+ct)\Big) + \frac{1}{2c}\int\limits_{x-ct}^{x+ct} q(s)\,ds.$$

1.80 The heat equation. The heat diffusion in a bar is described according to Joseph Fourier (1768–1830), by the *heat* or *diffusion equation*

$$u_t = u_{xx}.$$

Here $u(t,x)$ is the temperature in x at time t in suitable units. One checks that the function $t^{-1/2}e^{-x^2/4t}$ is a solution for $t > 0$ and $x \in \mathbb{R}$.

1.81 Schrödinger's equation. A complex variant of the heat equation is *Schrödinger's equation* in quantum mechanics

$$\frac{i\hbar}{2m}\psi_t = -\Delta\psi + V(x)\psi,$$

where $V(x)$ is a potential, $\hbar$ is Planck's constant, and $\psi = \psi(x,t)$, $x \in \mathbb{R}^3$, $t \geq 0$, is a "wave function", i.e., $\psi \in C^2(\Omega \times \mathbb{R}, \mathbb{C})$ with $\int |\psi(x,t)|^2\,dx = 1$ for all $t \geq 0$.

1.82 Euler's equation. The velocity field of a perfect fluid solves *Euler's equation*

$$v_t + (v \bullet \nabla)v = f - \frac{1}{\rho}\nabla p$$

where p is the *pressure*, f the exterior force and ρ the density of the fluid. The density is transported by the velocity field according to the *continuity equation*

$$\rho_t + \operatorname{div}(\rho v) = 0.$$

In particular, for incompressible fluids, $\rho(t) =$cost, Euler's equation reduces to

$$\operatorname{div} v = 0.$$

1.83 The Navier–Stokes equations. If the fluid is viscous, Euler's equation modifies into *Navier–Stokes equations* in which the *diffusion term* $\nu\Delta v$ appears, ν being the *viscosity coefficient*

$$v_t + (v \bullet \nabla)v - \frac{\nu}{\rho}\Delta v = f - \frac{1}{\rho}\nabla p.$$

In both Euler's and Navier–Stokes equations the notation is $v = (v^1, v^2, v^3)$, $(v \bullet \nabla)v := (v \bullet \mathbf{D}v^i)_{i=1,2,3}$, $v \bullet \mathbf{D}v^i := \sum_{j=1}^{3} v^j D_j v^i$.

1.4 Invertibility of Maps $\mathbb{R}^n \to \mathbb{R}^n$

Let $f : \Omega \subset \mathbb{R}^n \to \mathbb{R}^n$ be a map. When f is linear, $f(x) = \mathbf{A}x$, $\mathbf{A} \in M_{n,n}(\mathbb{R})$, its *invertibility*, i.e., the possibility of solving in x the system $\mathbf{A}x = y$ for all $y \in \mathbb{R}^n$, is equivalent to the invertibility of the matrix $\mathbf{A}$, and, in turn, this is equivalent to $\det \mathbf{A} \neq 0$.

When $f : \mathbb{R} \to \mathbb{R}$ is a differentiable function of one variable, we know, see, e.g., [GM1], that the condition $f' > 0$ (or $f' < 0$) implies monotonicity, consequently invertibility of f, and also the differentiability of the inverse function. Actually, if f is of class C^1 and $f'(x_0) \neq 0$ at some point x_0, there exists an interval $I(x_0, r)$ in which $f'(x)$ has the same sign of $f'(x_0)$, consequently $(f_{|I})^{-1}$ is strictly monotone, continuous with differentiable inverse, and

$$(f^{-1})'(y) = \frac{1}{f'(f^{-1}(y))} \qquad \forall y \in f(I(x_0, r)).$$

In Section 1.4.2, we state and prove (another proof will be presented in Section 1.4.4) a similar *local invertibility theorem*, known as the *Inverse Function Theorem*, for mappings f of class C^1: the condition $f'(x_0) \neq 0$ will be replaced by the nondegeneracy condition $\det \mathbf{D}f(x_0) \neq 0$, which may be seen as an invertibility condition for the linear tangent map to f at x_0, or as the nondegeneracy of the first-order Taylor expansion of f at x_0

$$f(x) = f(x_0) + \nabla f(x_0) \bullet (x - x_0) + o(|x - x_0|) \qquad \text{as } x \to x_0.$$

1.4.1 Banach's fixed point theorem

For the reader's convenience we begin by stating a few facts about Banach's fixed point theorem, see, e.g., [GM3, 9.5.1].

Let X be a metric space with distance d. A map $T : X \to X$ is said to be a *contraction* or a *contractive map* if there is a constant k, $0 \le k < 1$, such that $d(T(x), T(y)) \le k\, d(x, y)$, $\forall x, y \in X$. In this case we also say that T is *k-contractive*. In other words a contraction is a Lipschitz-continuous map with Lipschitz constant strictly less than one. A point $x \in X$ for which $T(x) = x$ is called a *fixed point* for T.

1.84 Theorem (Banach's fixed point theorem). *Let X be a complete metric space and let $T : X \to X$ be k-contractive, $0 \le k < 1$. Then T has a unique fixed point $\overline{x}$. Moreover, given $x_0 \in X$, the sequence $\{x_n\}$, $n \ge 0$, defined recursively by $x_{n+1} := T(x_n)$, converges with an exponential rate to the fixed point $\overline{x}$, and the following estimates hold:*

$$d(x_n, x) \le \frac{k^n}{1 - k} d(x_1, x_0),$$

$$d(x_{n+1}, x) \le \frac{k}{1 - k} d(x_{n+1}, x_n),$$

$$d(x_{n+1}, x) \le k\, d(x_n, x).$$

Proof. (i) *(Uniqueness)* If x and y are two fixed points, from $d(x, y) = d(Tx, Ty) \le k\, d(x, y)$ we infer $d(x, y) = 0$ since $0 < k < 1$.

(ii) *(Existence)* Let $x_0 \in X$ and for $n \ge 0$ let $x_{n+1} := T(x_n)$. We have

$$d(x_{n+1}, x_n) \le kd(x_n, x_{n-1}) \le k^n d(x_1, x_0) = k^n d(T(x_0), x_0),$$

hence, for $p > n$

$$d(x_p, x_n) \le \sum_{j=n}^{p-1} d(x_{j+1}, x_j) \le \sum_{j=n}^{p-1} k^j d(x_1, x_0) \le \frac{k^n}{1 - k} d(x_1, x_0).$$

Therefore $d(x_p, x_n) \to 0$ as $n, p \to \infty$, i.e., $\{x_n\}$ is a Cauchy sequence, hence it has a limit $x \in X$ and x is a fixed point as it is easily seen passing to the limit in $x_{n+1} = T(x_n)$. We leave the proof of the convergence estimates to the reader. $\square$

Notice that the first estimate in Theorem 1.84 allows us to evaluate the number of iterations that are sufficient to reach a desired accuracy; the second estimate allows us to evaluate the accuracy of x_{n+1} as an approximate value of x in terms of $d(x_{n+1}, x_n)$.

1.85 Example. Let $\phi : X \to X$ be a (nonlinear) map, X being a Banach space. Given $y \in X$ we would like to solve

$$\phi(x) = y. \tag{1.51}$$

We may write this equation as $x = x - \phi(x) + y$ so that, setting $g(x) := x - \phi(x) + y$, (1.51) is equivalent to finding a fixed point of g. If g is a contraction, we infer from the Banach fixed point theorem the existence of a fixed point and exponential convergence of the sequence $\{x_n\}$ defined by

$$\begin{cases} x_0 \in X, \\ x_{n+1} = x_n - \phi(x_n) + y \end{cases} \tag{1.52}$$

to the fixed point of g, hence to the solution of (1.51).

In the special case of $X = \mathbb{R}^n$ and $\phi(x) = \mathbf{L}x$, i.e., ϕ is linear, we have $g(x) = (\mathrm{Id} - \mathbf{L})x + y$ and g is a contraction if and only if $\mathbf{L}$ is close to the identity, $\|\mathrm{Id} - \mathbf{L}\| < 1$; in fact,

$$\sup_{x_1, x_2 \in \mathbb{R}^n} \frac{|g(x_2) - g(x_1)|}{|x_2 - x_1|} = \sup_{x_1, x_2 \in \mathbb{R}^n} \frac{|(\mathrm{Id} - \mathbf{L})(x_2 - x_1)|}{|x_2 - x_1|} = \|\mathrm{Id} - \mathbf{L}\|.$$

Moreover, (1.52) can be written as

$$x_{n+1} = (\mathrm{Id} - \mathbf{L})^{n+1} x_0 = \sum_{k=0}^{n} (\mathrm{Id} - \mathbf{L})^k y;$$

hence, when $n \to \infty$, we have

$$\bar{x} = \sum_{k=0}^{\infty} (\mathrm{Id} - \mathbf{L})^k y.$$

1.86 Example. A slight variant of the previous remark is the following. Suppose as in Example 1.85 we want to solve $\phi(x) = y$ given $y \in X$. For any invertible map $M : X \to X$, the equation $\phi(x) = y$ can be written as $Mx = Mx - \phi(x) + y$; thus x solves $\phi(x) = y$ if and only if x is a fixed point for

$$T(x) := x - M^{-1}\phi(x) + M^{-1}y.$$

Assuming T a contraction on X, we then infer that $\phi(x) = y$ has a unique solution defined as the limit point of the sequence $\{x_n\}$ defined by

$$\begin{cases} x_0 \in X, \\ x_{n+1} = x_n + M^{-1}\phi(x_n) + M^{-1}y. \end{cases}$$

1.87 ¶. Let X be a Banach space and let $T : X \to X$ be a Lipschitz-continuous map. Show that, for μ sufficiently large, the equation

$$Tx + \mu x = y$$

has, for any $y \in X$, a unique solution.

1.4.2 Local invertibility

Let $f = (f^1, f^2, \ldots, f^n)$ be a map of class C^1 from an open set $\Omega \subset \mathbb{R}^n$, $n \geq 1$. We recall that the *Jacobian* of f at $x_0 \in \Omega$ is the determinant of the Jacobian matrix

$$\det \mathbf{D}f(x_0),$$

that $f_{|U}$ denotes the restriction of f to $U \subset \Omega$, and, finally, that $x_0 + U$ stands for $\{x \in \mathbb{R}^n \mid x - x_0 \in U\}$; for instance, $x_0 + B(0, r) = B(x_0, r)$.

1.88 Definition. *We say that $f : \Omega \subset \mathbb{R}^n \to \mathbb{R}^n$ is locally invertible if for every $x \in \Omega$ there is a neighborhood U of x such that $f_{|U}$ is injective.*

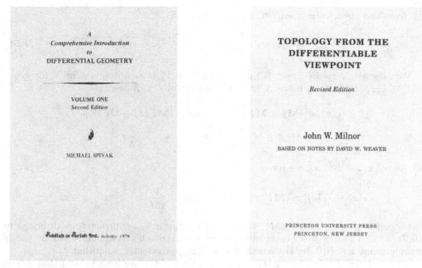

Figure 1.22. Frontispieces of two celebrated volumes.

1.89 Theorem (Inverse Function Theorem). *Let* $f : \Omega \to \mathbb{R}^n$ *be a map of class* C^k, $k \geq 1$, *from an open set* $\Omega \subset \mathbb{R}^n$ *into* $\mathbb{R}^n$, *let* $x_0 \in \Omega$ *and assume* $\det \mathbf{D}f(x_0) \neq 0$. *Then there exists an open neighborhood* U *of* x_0 *such that*

(i) $f_{|U}$ *is injective,*

(ii) $V := f(U)$ *is open,* $(f_{|U})^{-1} : V \to U$ *is continuous, i.e.,* $f_{|U}$ *is an open map,*

(iii) $(f_{|U})^{-1} : V \to U$ *is of class* C^k, *moreover,* $\forall\, y \in V$

$$\mathbf{D}(f_{|U})^{-1}(y) = \left[\mathbf{D}f(x)\right]^{-1} \qquad x := (f_{|U})^{-1}(y). \qquad (1.53)$$

Therefore a map $f : \Omega \to \mathbb{R}^n$ *from an open set of* $\mathbb{R}^n$ *into* $\mathbb{R}^n$ *of class* C^k, $k \geq 1$, *such that* $\det \mathbf{D}f(x) \neq 0$ $\forall x \in \Omega$ *is locally invertible, open with local inverses of class* C^k.

Proof. Without loss of generality we may assume $x_0 = 0$ and $f(x_0) = 0$.

Step 1. We set

$$\mathbf{M} := \mathbf{D}f(0)^{-1} \qquad M := ||\mathbf{M}|| \qquad \text{and} \qquad F(x) := f(x) - \mathbf{D}f(0)x,$$

and we write the equation $f(x) = y$ as

$$x = x + \mathbf{M}(-f(x) + y).$$

Therefore $f(x) = y$ if and only if x is a fixed point of the map

$$T_y(x) := x - \mathbf{M}f(x) + \mathbf{M}y = -\mathbf{M}F(x) + \mathbf{M}y.$$

Since $f \in C^1(\Omega, \mathbb{R}^n)$ by assumption, there exists $r > 0$ such that

$$\sup_{||z|| \leq r} ||\mathbf{D}f(z) - \mathbf{D}f(0)|| < \frac{1}{4\,M}$$

and, from the mean value theorem, see Corollary 1.51,

$$|F(x) - F(z)| \leq \frac{1}{4M}|x - z| \qquad \forall x, z \in \overline{B(0,r)}. \tag{1.54}$$

We now set $X := \overline{B(0,r)} \subset \mathbb{R}^n$ and prove that T_y is a contraction in X for every $y \in B(0, \frac{r}{2M})$. In fact, T_y maps X into X since for all $x \in X$ we have

$$|T_y(x)| = |\mathbf{M}y - \mathbf{M}F(x)| \leq |\mathbf{M}y| + |\mathbf{M}F(x) - \mathbf{D}F(0)|$$
$$\leq M|y| + M\frac{1}{4M}|x - 0| = \frac{r}{2} + \frac{r}{4} = \frac{3}{4}r$$

and, for all $x, z \in \overline{B(0,r)}$, we have

$$|T_y(x) - T_y(z)| = |\mathbf{M}F(x) - \mathbf{M}F(z)| \leq M\frac{1}{4M}|x - z| = \frac{1}{4}|x - z|.$$

Therefore for all $y \in B(0, r/(2M))$ the map $x \to T_y x$ is $1/4$-contractive with image in $B(0, 3r/4) \subset X$. Since X is a complete metric space, Banach's fixed point theorem yields a unique point $x \in B(0, 3r/4)$ such that $x = T_y x$, equivalently, such that $f(x) = y$.

Setting $U := f^{-1}(B(0, r/(2M)))$, $V := B(0, r/(2M))$, we have proved that $f_{|U}$ is invertible, thus (i).

Step 2. Let us show that $(f_{|U})^{-1} : V \to U$ is continuous. For $y, w \in V$ set $x := (f_{|U})^{-1}(y)$ and $z := (f_{|U})^{-1}(w)$. From (1.54) we infer

$$|\mathbf{D}f(0)(x - z)| = | - f(x) + \mathbf{D}f(0)x + f(z) - \mathbf{D}f(0)z + f(x) - f(z)|$$
$$\leq |F(x) - F(z)| + |f(x) - f(z)|$$
$$\leq \frac{1}{4M}|x - z| + |f(x) - f(z)|.$$

Hence

$$|x - z| \leq \frac{1}{4}|x - z| + M|f(x) - f(z)|,$$

i.e.,

$$|(f_{|U})^{-1}(y) - (f_{|U})^{-1}(w)| \leq \frac{4M}{3}|y - w|. \tag{1.55}$$

Step 3. It remains to prove that $g := (f_{|U})^{-1}$ is differentiable at every $y \in V = f(U)$. Without loss of generality we assume $y = 0$ and $g(0) = 0$. Setting $x = g(z)$, we have

$$g(z) - \mathbf{M}z = x - \mathbf{M}f(x) = -\mathbf{M}F(x)$$

and, on the other hand, by (1.55), $|f(x)| \geq \frac{3M}{4}|x|$ hence

$$\frac{|g(z) - \mathbf{M}z|}{|z|} = \frac{|\mathbf{M}(f(x) - \mathbf{D}f(0)x)|}{|x|}\frac{|x|}{|f(x)|}$$
$$\leq M\frac{o(|x|)}{|x|}\frac{|x|}{|f(x)|} \leq \frac{4}{3}M^2\frac{o(|x|)}{|x|}.$$

If $z \to 0$, then $x = g(z) \to 0$ since g is continuous. Consequently the right-hand side of the last inequality tends to zero as $z \to 0$. This yields that g is differentiable at 0 with $\mathbf{D}g(0) = \mathbf{M} = \mathbf{D}f(0)^{-1}$.

Step 4. Finally, if f is of class C^k, the formula (1.53) yields at once that the local inverse is of class C^k. □

Figure 1.23. From the left: (i) Polar coordinates in $\mathbb{R}^2$ and (ii) Spherical coordinates in $\mathbb{R}^3$.

1.90 Remark. With the notations of Theorem 1.89, we in fact have proved that for any $y \in B(0, r/(2M))$, the sequences defined by

$$\begin{cases} x_0 \in \overline{B(0, r)}, \\ x_{k+1} = x_k - \mathbf{M}f(x_k) + \mathbf{M}y, \end{cases}$$

exponentially converge to the unique solution of

$$\begin{cases} f(x) = y, \\ x \in \overline{B(0, r)}. \end{cases}$$

1.4.3 A few examples

1.91 Polar coordinates. The transformation $\phi(\rho, \theta) = (\rho \cos \theta, \rho \sin \theta)$, $(\rho, \theta) \in \mathbb{R}^2$, which yields the *polar coordinates* in $\mathbb{R}^2$, has Jacobian matrix and Jacobian

$$\mathbf{D}\phi(\rho, \theta) = \begin{pmatrix} \cos \theta & -\rho \sin \theta \\ \sin \theta & \rho \cos \theta \end{pmatrix}, \qquad \det \mathbf{D}\phi(\rho, \theta) = \rho.$$

By the inverse function theorem, Theorem 1.89, ϕ is locally invertible in $\mathbb{R}^2 \setminus \{0\}$, but ϕ is not (globally) invertible as $\phi(\rho, \theta + 2\pi) = \phi(\rho, \theta)$. However, the restriction of ϕ to the strip $S_a :=]0, +\infty[\times]a - \pi, a + \pi[$, $a \in \mathbb{R}$, is injective, hence globally invertible from S_a onto its image $\phi(S_a)$ (with inverse of class C^∞). Notice that, since

$$\mathbf{D}\phi(\rho, \theta) = \begin{pmatrix} \frac{x}{\rho} & -y \\ \frac{y}{\rho} & x \end{pmatrix},$$

we have

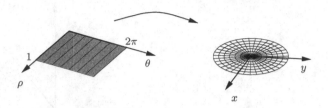

Figure 1.24. Polar coordinates in $\mathbb{R}^2$.

$$\mathbf{D}\phi^{-1}(x,y) = \frac{1}{\sqrt{x^2+y^2}} \begin{pmatrix} x & y \\ -y & x \end{pmatrix};$$

and $\phi(S_a)$ is $\mathbb{R}^2$ minus the half-line from the origin that forms an angle $a+\pi$ with the positive half-line of abscissa. The inverse of $\phi_{|S_a}$ can be written explicitly. For example, when $a = 0$, we can solve in $(\rho, \theta) \in]0, \infty[\times]-\pi, \pi[$ the system

$$\begin{cases} x = \rho\cos\theta \\ y = \rho\sin\theta \end{cases}, \qquad \rho > 0,\ \theta \in]a - \pi, a + \pi[$$

for all $(x, y) \in \phi(S_a)$, finding $\rho = \sqrt{x^2+y^2}$ and

$$\theta = \begin{cases} \frac{\pi}{2} + \arctan\frac{y}{x} & \text{if } x < 0, y > 0, \\ \frac{\pi}{2} & \text{if } x = 0, y > 0, \\ \arctan\frac{y}{x} & \text{if } x > 0, \\ -\frac{\pi}{2} & \text{if } x = 0, y < 0, \\ -\frac{\pi}{2} + \arctan\frac{y}{x} & \text{if } x < 0, y < 0, \end{cases}$$

i.e., the angle formed by the positive half-line of abscissa and the half-line from the origin through (x, y) measured in radians anticlockwise from $-\pi$ to π.

1.92 Cylindrical coordinates. Similar considerations may be developed for the transformation that yields *cylindrical coordinates* in $\mathbb{R}^3$

$$(x, y, z) = \phi(\rho, \theta, z), \qquad \begin{cases} x = \rho\cos\theta, \\ y = \rho\sin\theta, \\ z = z. \end{cases}$$

Its Jacobian matrix and its Jacobian are given by

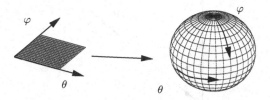

Figure 1.25. Spherical coordinates.

$$\mathbf{D}\phi(\rho, \theta, z) = \begin{pmatrix} \cos\theta & -\rho\sin\theta & 0 \\ \sin\theta & \rho\cos\theta & 0 \\ 0 & 0 & 1 \end{pmatrix}, \qquad \det \mathbf{D}\phi(\rho, \theta, z) = \rho.$$

Thus ϕ is locally invertible from $\mathbb{R}^3 \setminus \{\rho = 0\}$ into $\mathbb{R}^3 \setminus \{z = 0\}$ with local inverses of class C^∞. Moreover, the restriction of ϕ to $\Omega :=$ $]0, +\infty[\times]0, 2\pi[\times\mathbb{R}$ is injective, thus globally invertible.

1.93 Spherical coordinates. For the transformation $\phi : (\rho, \theta, \varphi) \to$ (x, y, z) from $\mathbb{R}^3$ into itself that yields the *spherical coordinates*

$$\begin{cases} x = \rho\cos\theta\sin\varphi, \\ y = \rho\sin\theta\sin\varphi, \\ z = \rho\cos\varphi \end{cases}$$

we have

$$\mathbf{D}\phi(\rho, \theta, \varphi) = \begin{pmatrix} \cos\theta\sin\varphi & -\rho\sin\theta\sin\varphi & \rho\cos\theta\cos\varphi \\ \sin\theta\sin\varphi & \rho\cos\theta\sin\varphi & \rho\sin\theta\cos\varphi \\ \cos\varphi & 0 & -\rho\sin\varphi \end{pmatrix}$$

and $\det \mathbf{D}\phi(\rho, \theta, \varphi) = \rho^2 \sin\varphi$. Therefore $\det \mathbf{D}\phi(\rho, \theta, \varphi) \neq 0$ for all (ρ, θ, φ) in

$$\Omega := \mathbb{R}^3 \setminus \Big\{(\rho, \theta, \varphi) \,\Big|\, \rho \neq 0, \ \varphi \neq (2k+1)\pi, \ k \in \mathbb{Z}\Big\} :$$

we conclude that $\phi_{|\Omega}$ is locally invertible.

The restriction of ϕ to $\Delta :=]0, \infty[\times]0, 2\pi[\times]0, \pi[\subset \Omega$ is instead injective with image $\mathbb{R}^3$ minus the half-plane generated by the z-axis and the positive half-line of abscissa; thus $\phi_{|\Delta}$ is globally invertible with inverse of class C^∞.

When writing $(x, y, z) = \phi(\rho, \theta, \varphi)$, $(\rho, \theta, \varphi) \in \Delta$, the new variables ρ, θ, φ are, respectively, the distance ρ of (x, y, z) from the origin, the longitude θ of (x, y, z) measured in radians from the half-plane generated by the z-axis and the positive half-line of abscissa, and the latitude φ measured in radians from 0 corresponding to the North Pole to π, corresponding to the South Pole, see Figure 1.25.

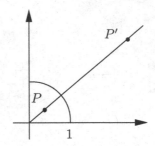

Figure 1.26. The inversion map.

1.94 The inversion in a sphere. This is the transformation that maps each point $P = (x, y) \neq (0, 0)$ in the plane to the point P' that lies in the half-line from 0 through P in such a way that $\overline{OP} \cdot \overline{OP}' = 1$. It maps points inside the circle $x^2 + y^2 = 1$ outside and vice versa. In formula it is given by $\phi : \mathbb{R}^2 \setminus \{(0, 0)\} \to \mathbb{R}^2 \setminus \{(0, 0)\}$,

$$\phi(x, y) = \left(\frac{x}{x^2 + y^2}, \frac{y}{x^2 + y^2} \right),$$

or, in complex coordinates $z = x + iy$,

$$\phi(z) = 1/\overline{z}.$$

It is a globally invertible map from $\mathbb{R}^2 \setminus \{0\}$ onto itself of class C^∞ with inverse of class C^∞.

1.95 ¶ Cofocal elliptic coordinates. Show that the map $\phi : \Omega \subset \mathbb{R}^2 \to \mathbb{R}^2$ given by

$$(x, y) \to (\cosh x \cos y, \sinh x \sin y)$$

with domain $\Omega := \{(x, y) \,|\, x > 0\}$ is locally, but not globally, invertible. Show that its restriction to $\Omega := \{(x, y) \,|\, x > 0, \ 0 < y < 2\pi\}$ is invertible. [*Hint:* Show that vertical segments are taken to ellipses with foci at $\pm(1, 0)$ whereas horizontal half-lines are taken to hyperbolas with the same foci $(\pm 1, 0)$.]

1.96 The exponential map. Let $f : \mathbb{C} \to \mathbb{C}$ be the complex exponential map $f(z) = e^z$, or in Cartesian coordinates $(u, v) = f(x, y) = (e^x \cos y, e^x \sin y)$. Its Jacobian matrix is

$$\mathbf{D}f(x, y) := \begin{pmatrix} e^x \cos y & e^x \sin y \\ -e^x \sin y & e^x \cos y \end{pmatrix} = \begin{pmatrix} u & v \\ -v & u \end{pmatrix}$$

and its Jacobian is $\det \mathbf{D}f(x, y) = u^2 + v^2 = e^{2x}$. Therefore f is locally invertible with C^∞ inverses and

$$\mathbf{D}f^{-1}(u, v) = \frac{1}{u^2 + v^2} \begin{pmatrix} u & -v \\ v & u \end{pmatrix}.$$

Since f is 2π-periodic in y, it is not globally invertible. However, the restriction of f to $\Omega := \mathbb{R} \times]0, 2\pi[$ is injective thus globally invertible, and its inverse is the *principal determination* of the complex logarithm, see [GM2] and Chapter 4.

Vertical segments in Ω are taken by f into circles around the origin, whereas horizontal lines are taken to half-lines from the origin.

1.4.4 A variational proof of the inverse function theorem

We give here an alternative variational proof of Theorem 1.89 in which the fixed point argument is replaced by the Weierstrass theorem.

Another proof of Theorem 1.89. (i) Set $L := \mathbf{D}f(x_0)$ and, as usual, denote by $||A||$ the norm of the matrix A in such a way that $|Ax| \leq ||A|| \, |x|$, $\forall x$. For all $x, y \in \mathbb{R}^n$ we have

$$x - y = L^{-1}L(x - y) \qquad \text{hence} \qquad |x - y| \leq ||L^{-1}|| \, |L(x - y)|. \qquad (1.56)$$

On the other hand, the map $f(x) - L(x - x_0)$ is of class C^1 and its Jacobian matrix $\mathbf{D}f(x) - L$ vanishes at x_0. It follows that for any $\epsilon > 0$ there exists a ball $B(x_0, r)$ such that

$$||\mathbf{D}f - L||_{\infty, B(x_0, r)} = \sup_{x \in B(x_0, r)} ||\mathbf{D}f(x) - L|| < \epsilon$$

and, on account of the mean value theorem, for all $x, y \in B(x_0, r)$ we get

$$|f(x) - Lx - f(y) + Ly| \leq ||\mathbf{D}f - L||_{\infty, B(x_0, r)} |x - y| < \epsilon |x - y|,$$

in particular,

$$|L(x - y)| \leq |f(x) - f(y)| + \epsilon |x - y|, \forall x, y \in B(x_0, r). \qquad (1.57)$$

From (1.56) and (1.57), we easily conclude that for all $x, y \in B(x_0, r)$ we have

$$\left(1 - \epsilon ||L^{-1}|| \right) |x - y| \leq ||L^{-1}|| \, |f(x) - f(y)|,$$

and, choosing ϵ sufficiently small, we can find a constant $C > 0$ and a ball $B(x_0, r)$ such that

$$|x - y| \leq \frac{1}{C} |f(x) - f(y)| \qquad \forall x, y \in B(x_0, r). \qquad (1.58)$$

Therefore, f is injective in $U := B(x_0, r)$.

(ii) Let us show that $V := f(U)$ is an open set, i.e., that every $\overline{y} \in V$ has a neighborhood $V(\overline{y})$ such that $V(\overline{y}) \subset V = f(U)$. By (i) there is a unique $\overline{x} \in U$ with $f(\overline{x}) = \overline{y}$. We now observe that, if $U(\overline{x})$ is a neighborhood of $\overline{x}$ with $U(\overline{x}) \subset\subset U$ and $\Gamma := \partial U(\overline{x})$, since f is injective and $\overline{x} \notin \Gamma$, we have $\overline{y} \notin f(\Gamma)$; moreover, since f is continuous, $f(\Gamma)$ is a compact set, hence the distance of $\overline{y}$ from $f(\Gamma)$ is positive. Let

$$\sigma := \frac{1}{2} \operatorname{dist}(\overline{y}, f(\Gamma)),$$

and $V(\overline{y}) := B(\overline{y}, \sigma)$. We now show that $V(\overline{y}) \subset V$, i.e., that for every $y \in V(\overline{y})$ there is $x \in U(\overline{x})$ such that $y = f(x)$. In order to do that, for any $y \in V_{\overline{y}}$, we show that the function

$$\psi(x) := |y - f(x)|^2, \qquad x \in \overline{U(\overline{x})}$$

has an interior minimum point in $U(\overline{x})$ with value zero. In fact,

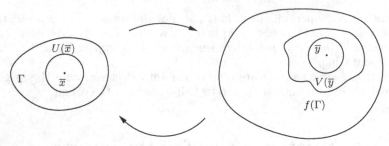

Figure 1.27. Illustration of the proof of Theorem 1.89 in Section 1.4.2.

$$\psi(\overline{x}) = |y - \overline{y}|^2 < \sigma^2$$

while for any $x \in \Gamma$

$$2\sigma = \mathrm{dist}\,(\overline{y}, f(\Gamma)) \le |\overline{y} - f(x)| \le |\overline{y} - y| + |y - f(x)| \le \sigma + |y - f(x)|$$

i.e.,

$$\psi(x) > \sigma^2 \qquad \forall\, x \in \Gamma.$$

Weierstrass's theorem allows us to conclude that ψ has a minimizer $x \in \overline{U(\overline{x})}$ with minimum value less than σ^2. It follows that x necessarily belongs to $U(\overline{x})$ and, by Fermat's theorem $\mathbf{D}\psi(x) = 0$, i.e.,

$$2\,\mathbf{D}f(x)(y - f(x)) = 0.$$

Since $\det \mathbf{D}f(x) \ne 0$ we conclude that $y = f(x)$, that is $\psi(x) = 0$.

Of course, we can repeat the same argument replacing U with any open set $A \subset U$ and V with $f(A)$, concluding that $f(A)$ is open if A is open. In other words, $(f_{|U})^{-1}$ is continuous.

(iii) We shall now show that f^{-1} is differentiable at every point $\overline{y} \in V$ and that (1.53) holds.

We set $g := (f_{|U})^{-1}$ and we assume without loss of generality that $\overline{y} = 0$ and $f(0) = 0$. If $L := \mathbf{D}f(0)$ and for $y \in V$ we set $x := g(y)$, we have

$$g(y) - L^{-1}y = x - L^{-1}f(x) = -L^{-1}(f(x) - Lx).$$

When $y \to 0$ we have $x = g(y) \to 0$ since g is continuous by Step 2. Moreover, from (1.58) we have

$$|f(x)| = |f(x) - f(0)| \ge C\,|x|,$$

for all $x \in U$ hence

$$\frac{|g(y) - g(0) - L^{-1}(y - 0)|}{|y|} \le ||L^{-1}||\,\frac{|f(x) - Lx|}{|f(x)|} \le ||L^{-1}||\,\frac{o(|x|)}{C\,|x|} \to 0 \qquad \text{per } y \to 0,$$

that yields at once the differentiability of g at 0 with $\mathbf{D}g(0) = L^{-1}$.

(iv) Finally, from (1.53) one easily deduces that $(f_{|U})^{-1} \in C^1(V)$ and that $(f_{|U})^{-1}$ is of class $C^k(V)$ if f is of class C^k. $\qquad\square$

1.4.5 Global invertibility

Let $\Omega \subset \mathbb{R}^n$ be an open set and let $f : \Omega \to \mathbb{R}^n$ be of class C^1. We have seen that the nondegeneracy condition $\det \mathbf{D}f \ne 0$ in Ω is equivalent to the local invertibility of f and that local invertible maps are not necessarily globally invertible. Of course injectivity plus local invertibility is equivalent to global invertibility.

In the category of homeomorphisms, the following theorem holds, see [GM3, 8.54].

1.97 Theorem. *Let $\Omega \subset \mathbb{R}^n$ be an open and connected set and let $f : \Omega \to \mathbb{R}^n$ be a local homeomorphism. If f is proper and $f(\Omega)$ is simply connected, then f is injective hence a homeomorphism between Ω and $f(\Omega)$.*

An immediate consequence of this and of Theorem 1.89 is the following.

1.98 Theorem (of global invertibilty). *Let Ω be an open and connected set of $\mathbb{R}^n$ and let $f : \Omega \to \mathbb{R}^n$ be a map of class $C^k(\Omega)$, $k \geq 1$, with $\det \mathbf{D}f(x) \neq 0$ for all $x \in \Omega$. Suppose that f is proper and that $f(\Omega)$ is simply connected. Then f is globally invertible from Ω onto $f(\Omega)$ with inverse of class C^k, and (1.53) holds.*

Another theorem of global invertibility is the following one. We state it without proof since it would need more advanced means.

1.99 Theorem. *Let Ω be an open and connected set of $\mathbb{R}^n$, let $f : \Omega \to \mathbb{R}^n$ be of class $C^0(\overline{\Omega}) \cap C^k(\Omega)$, $k \geq 1$, with positive Jacobian $\det \mathbf{D}f(x) > 0$ $\forall x \in \Omega$, and let $g : \overline{\Omega} \to f(\overline{\Omega})$ be a homeomorphism from $\overline{\Omega}$ onto $f(\overline{\Omega})$. If $f = g$ on $\partial\Omega$, then f is an homeomorphism from $\overline{\Omega}$ onto $f(\overline{\Omega})$, $f_{|\Omega}$ is globally invertible from Ω onto $f(\Omega)$ with inverse of class C^k and (1.53) holds.*

1.5 Differential Calculus in Banach Spaces

The notions of directional derivative and of differential easily extend to mappings between Banach spaces.

1.5.1 Gâteaux and Fréchet differentials

The notions of directional derivatives and of differential extend at once to mappings between normed spaces. But their use is relevant in the setting of complete normed spaces, i.e., Banach spaces, see [GM3].

In this section X and Y will always denote real Banach spaces.

1.100 Definition. *Let $A \subset X$ and $x_0 \in \operatorname{int} A$. We say that $f : X \to Y$ is* Gâteaux-differentiable *at x_0 if there exists a continuous linear map $\partial f(x_0) \in \mathcal{L}(X, Y)$ such that*

$$\lim_{t \to 0} \frac{f(x_0 + tv) - f(x_0)}{t} = \partial f(x_0)(v) \qquad \forall v \in X.$$

$\partial f(x_0)$ is the Gâteaux-derivative *of f at x_0, and $\partial_v f(x_0) := \partial f(x_0)(v) \in Y$ is the* directional derivative *of f in the direction $v \in X$.*

1.101 Definition. *We say that $f : A \subset X \to Y$ is* Fréchet-differentiable *at $x_0 \in \operatorname{int} A$ if there exists a continuous linear map $\ell_{x_0} \in \mathcal{L}(X, Y)$, called the* Fréchet-derivative *or* Fréchet-differential *of f at x_0, such that*

$$\lim_{\substack{h \to 0 \\ h \in X}} \frac{\|f(x_0 + h) - f(x_0) - \ell_{x_0}(h)\|_Y}{\|h\|_X} = 0. \qquad (1.59)$$

The Fréchet-derivative at x_0 is denoted by

$$f'(x_0), \qquad Df(x_0) \qquad or \qquad T_{x_0} f,$$

if we want to emphasize its aspect of linear tangent map to f at x_0.

When A is an open set in X, we say that f is Gâteaux-differentiable (respectively, Fréchet-differentiable) in A if f is Gâteaux-differentiable (respectively, Fréchet-differentiable) at every point of A.

When X is finite dimensional, all norms on X are equivalent and all linear maps $\ell : X \to Y$ are continuous. Thus, when $X = \mathbb{R}^n$, Fréchet-differentiability is just differentiability and Gâteaux-differentiability is the requirement of existence of all directional derivatives and of linearity of the tangent map $h \to \frac{\partial f}{\partial h}$. Therefore the two notions of differentiability in Definitions 1.100 and 1.101 are different even if X has finite dimension, $X = \mathbb{R}^n$, $n \geq 2$, and $Y = \mathbb{R}$, see Example 1.3.

1.102 ¶. Prove the following claims.
  (i) There exist Gâteaux-differentiable maps that are not continuous.
  (ii) There exist continuous and Gâteaux-differentiable functions that are not Fréchet-differentiable,
  (iii) If f is Fréchet-differentiable at x_0, then f is Gâteaux-differentiable at x_0 and the two differentials agree, $T_{x_0} f = \partial f(x_0)$,
  (iv) If f is Fréchet-differentiable at x_0, then f is continuous at x_0.

1.103 ¶. Suppose that f is continuous at x_0 and that (1.59) holds for some linear map $\ell_{x_0} : X \to Y$. Prove that ℓ_{x_0} is continuous and that f is Fréchet-differentiable.

1.104 ¶. Show the following.
  (i) If $f : X \to \mathbb{R}$ has a maximum or a minimum point at x_0 and f is Gâteaux-differentiable at x_0, then $\partial f(x_0) = 0$.
  (ii) Every linear continuous map $\ell : X \to Y$ is Fréchet-differentiable and $T_{x_0} \ell = \ell$.

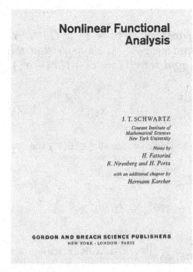
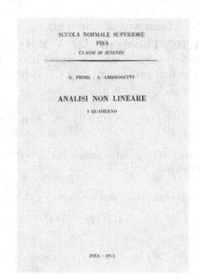

Figure 1.28. Two volumes on calculus in Banach spaces.

(iii) Let X_1, X_2, Y be three Banach spaces and let $b : X_1 \times X_2 \to Y$ be a bi-linear continuous map. Show that b is Fréchet-differentiable at every point $x = (x_1, x_2) \in X_1 \times X_2$ and $T_x b(u, v) = b(v_1, x_2) + b(x_1, v_2)$.

1.105 ¶. State and prove the theorem about Fréchet-differentiability of the composite of functions, see Theorem 1.22.

1.106 ¶ The Dirichlet integral in dimension one. Consider the function

$$\mathcal{D}(u) := \frac{1}{2} \int_0^1 |u'(t)|^2 \, dt$$

defined on the functions u in the Banach space $C^1([0,1], \mathbb{R})$ of functions with continuous derivatives normed by

$$\|u\|_{C^1} := \|u\|_{\infty, [0,1]} + \|u'\|_{\infty, [0,1]}.$$

Show that $\mathcal{D}$ is Gâteaux-differentiable in $C^1([0,1])$ and that for all $u, v \in C^1([0,1])$

$$\partial_v \mathcal{D}(u) = \int_0^1 u'(t) v'(t) \, dt.$$

a. Gradient

Let H be a Hilbert space and let $f : A \subset H \to \mathbb{R}$ be Fréchet-differentiable at $x_0 \in \operatorname{int} A$. Since the linear tangent map $T_{x_0} f$ is a continuous linear map from H to $\mathbb{R}$, by Riesz's theorem, see [GM3], there exists a vector, denoted by $\nabla f(x_0) \in H$ that represents $T_{x_0} f$ via the inner product, i.e., $\nabla f(x_0)$ is defined by

$$T_{x_0} f(v) = v \bullet \nabla f(x_0) \qquad \forall v \in H.$$

$\nabla f(x_0)$ is called the *gradient* of f at x_0.

b. Mean value theorem

1.107 Theorem. *Let X, Y be Banach spaces, $f : A \to Y$ a Gâteaux-differentiable map in an open set $A \subset X$, and $x_1, x_2 \in A$. Suppose that the segment joining x_1 to x_2 is contained in A and set $r(t) := x_1 + t(x_2 - x_1)$, $t \in [0, 1]$. Then*

$$\left\| f(x_2) - f(x_1) \right\|_Y \leq \sup_{t \in [0,1]} \left\| \partial f(r(t)) \right\|_{\mathcal{L}(X,Y)} \left\| x_2 - x_1 \right\|_X. \qquad (1.60)$$

Proof. The proof is simple if Y is a Hilbert space. In this case, for all $\xi \in \mathcal{L}(Y)$ the real function

$$F(t) := <\xi, f(r(t))>, \qquad t \in [0, 1],$$

is differentiable, and for some $\theta \in]0, 1[$ we have

$$<\xi, f(x_2) - f(x_1)> = F(1) - F(0) = <\xi, \partial_{x_2 - x_1} f(r(\theta))>,$$

hence

$$\left| <\xi, f(x_2) - f(x_1)> \right| \leq \sup_{t \in [0,1]} \left\| \partial f(r(t)) \right\|_{\mathcal{L}(X,Y)} \left\| x_2 - x_1 \right\|_X \left\| \xi \right\|_{\mathcal{L}(Y,\mathbb{R})}.$$

Since Y is a Hilbert space, we easily conclude by choosing $<\xi, v> := (f(x_2) - f(x_1)) \bullet v$.

The general case, in which Y is a Banach space, can be treated similarly if we use the Hahn–Banach theorem, see [GM3]. $\qquad \square$

1.108 Integral for mappings $C^0([a,b], Y)$. We define the integral of a map of class C^0 from the interval $[0, 1]$ into a Banach space Y

$$\int_0^1 f(t)\, dt$$

as the limit of the sums $\sum_{i=1}^N f(\tau_i)(t_{i+1} - t_i)$, $\tau_i \in [t_i, t_{i+1}[$, when the lengths of the intervals of the subdivision tend to zero. This way, the integral has the same properties of Riemann's integral, in particular, from

$$\left\| \sum_{i=1}^N f(\tau_i)(t_{i+1} - t_i) \right\| \leq \sum_{i=1}^N \left\| f(\tau_i)(t_{i+1} - t_i) \right\| = \sum_{i=1}^N \left\| f(\tau_i) \right\| |t_{i+1} - t_i|$$

we infer

$$\left\| \int_0^1 f(t)\, dt \right\| \leq \int_0^1 \left\| f(t) \right\|\, dt. \qquad (1.61)$$

1.109 Another proof. If f is Fréchet-differentiable, we have

$$f(x_2) - f(x_1) = \int_0^1 f'(tx_2 + (1-t)x_1)(x_2 - x_1)\, dt.$$

Thus, the use of (1.61) yields another proof of Theorem 1.107, provided we also assume (and, as we have seen, this is not necessary) that $t \to f'(tx_2 + (1-t)x_1)(x_2 - x_1)$ is continuous in $[0, 1]$.

Of course, a consequence of Theorem 1.107 is the analogy of the total derivative theorem.

1.110 Theorem. *Let X and Y be two Banach spaces, $B(x_0, r) \subset X$, $r > 0$, and let $f : B(x_0, r) \to Y$ be Gâteaux-differentiable in $B(x_0, r)$. If $x \to \partial_x f$ as a map from $B(x_0, r)$ into $\mathcal{L}(X, Y)$ is continuous at x_0, then f is Fréchet-differentiable at x_0.*

Proof. In fact, by applying the mean value theorem Theorem 1.107 to $h \to \sigma(h) := f(x_0 + h) - f(x_0) - f'(x_0)h$, we find

$$\|\sigma(h)\| = \|\sigma(h) - \sigma(0)\| \leq \sup_{x \in B(x_0, r)} \left\|\partial f(x)\right\|_{\mathcal{L}(X,Y)} \|v\|.$$

$\square$

c. Higher order derivatives and Taylor's formula

Let $f : \Omega \subset X \to Y$ be Fréchet-differentiable in an open set Ω of a Banach space X and let $f' : \Omega \to \mathcal{L}(X, Y)$ be the map that associates to $x \in \Omega$ the differential of f at x, $f'(x) \in \mathcal{L}(X, Y)$.

1.111 Definition. *Let $f : \Omega \subset X \to Y$ be Fréchet-differentiable in an open set Ω of a Banach space X . If f' is in turn Fréchet-differentiable, we say that f has second derivatives f'', $f'' := (f')'$; in this case f'' defines a map from Ω into $\mathcal{L}(X, \mathcal{L}(X, Y))$.*

If we observe that $\mathcal{L}(X, \mathcal{L}(X, Y))$ may be identified with the space of bilinear continuous forms from $X \times X$ into Y denoted by $\mathcal{L}_2(X, Y)$, via the identification map

$$L \in (X, \mathcal{L}(X, Y)) \to B \in \mathcal{L}_2(X, Y), \qquad B(u, v) := (Lu)(v),$$

we conclude that the second derivative of f is a bilinear form in $\mathcal{L}_2(X, Y)$.

Similarly to the finite-dimensional case, we define the derivatives of order k. The k-derivative is then a k-linear map from $X \times X \cdots \times X$ to Y.

1.112 Definition. *Let X, Y be two Banach spaces and let $\Omega \subset X$ be an open set of X. We say that $f : \Omega \subset X \to Y$ is of class C^k (respectively, of class C^∞) if f has Fréchet-derivatives up to order k included, (respectively, of any order), and those derivatives as maps from X into $\mathcal{L}(X, \mathcal{L}(X, \ldots))\ldots)$ are continuous.*

By induction and as in the finite-dimensional case, one proves the following.

1.113 Theorem (Schwarz). *The k-derivative of $f \in C^k(X, Y)$ is a k-linear symmetric form.*

1.114 Theorem (Taylor's formula). *Let X, Y be two Banach spaces, let $\Omega \subset X$ be an open set of X, and let x_0, x_1 be two points in Ω such that the segment joining them is contained in Ω. If $f : \Omega \to Y$ is of class C^k, then we have*

$$f(x) = P_{k-1}(x - x_0; x_0)$$
$$+ \left(\int_0^1 \frac{(1-t)^{k-1}}{(k-1)!} f^{(k)}(tx_0 + (1-t)(x_1 - x_0)) \, dt \right)(x - x_0)^k$$

and

$$\|f(x) - P_{k-1}(x - x_0; x_0)\| = o(\|x - x_0\|^k) \qquad as \; x \to x_0$$

where

$$P_{k-1}(t; x_0) := \sum_{h=1}^{k-1} \frac{1}{h!} f^{(h)}(x_0) t^{(h)}$$

and $t^{(h)}$ is the h-tuple $t^{(h)} := (t, t, \ldots, t)$.

1.5.2 Local invertibility in Banach spaces

Going through the proof of Theorem 1.89, it is easily seen that it can be repeated word by word, replacing $\mathbf{D}f$ with the Fréchet-differential to get the following.

1.115 Theorem (Local invertibility). *Let X, Y be two Banach spaces, $\Omega \subset X$ an open subset of X, $x_0 \in \Omega$, and $f : \Omega \to Y$ a map of class C^k, $k \geq 1$. Suppose that the linear tangent map $f'(x_0) : X \to Y$ is a continuous isomorphism, i.e., it is continuous and invertible with continuous inverse.[2] Then there exists an open neighborhood U of x_0 such that, setting $V := f(U)$, we have*

   (i) *$f_{|U}$ is a continuous bijection from U onto V,*
   (ii) *$(f_{|U})^{-1} : V \to U$ is continuous,*
   (iii) *$(f_{|U})^{-1}$ is of class $C^k(V)$ and $[(f_{|U})^{-1}]'(y) = [f'(x)]^{-1}$ for all $y \in V$ and $x = (f_{|U})^{-1}(y)$.*

1.6 Exercises

1.116 ¶. When possible compute the partial and directional derivatives and the differential of the following functions

---

[2] Actually, the continuity of the inverse of $f'(x_0)$ follows from a theorem of Banach stating that the continuity of the inverse L^{-1} of a linear map L between Banach spaces follows from the continuity and invertibility of L, see [GM3].

$$(x \mid x_0),$$

$$|x|^2,$$

$$\frac{xy}{1+x^2},$$

$$e^{x^2+y} + \int_0^x \sin(yt^2)\,dt,$$

$$|x|,$$

$$xe^{x^2+y^2},$$

$$x^{3/2}y,$$

$$\begin{cases} \frac{xy}{x^2+y^2} & \text{if } (x,y) \neq (0,0), \\ \lambda & \text{if } (x,y) = (0,0), \end{cases}$$

$$\begin{cases} \frac{x^2 y}{x^2+y^2} & \text{if } (x,y) \neq (0,0), \\ 0 & \text{if } (x,y) = (0,0), \end{cases}$$

$$\begin{cases} \frac{x^3 y}{x^2+y^2} & \text{if } (x,y) \neq (0,0), \\ 0 & \text{if } (x,y) = (0,0), \end{cases}$$

$$\begin{cases} \frac{xy^2}{x^2+y^4} & \text{if } (x,y) \neq (0,0), \\ 0 & \text{if } (x,y) = (0,0), \end{cases}$$

$$\begin{cases} \frac{y \sin(x^2)}{x^2+y^2} & \text{if } (x,y) \neq (0,0), \\ 0 & \text{if } (x,y) = (0,0), \end{cases}$$

$$\begin{cases} \left(\frac{xy^2}{x^2+y^4}\right)^2 & \text{if } (x,y) \neq (0,0), \\ 0 & \text{if } (x,y) = (0,0). \end{cases}$$

1.117 ¶. Write the equation of the tangent plane at $(0,0)$ to the graphs of the following functions

$$xy, \qquad e^{xy}, \qquad (x^2+y^2)\log(x^2+y^2).$$

1.118 ¶. Find the points on the surface $z = x^4 - 4xy^3 + 6y^2 - 2$ in $\mathbb{R}^3$ with horizontal tangent plane.

1.119 ¶. Write $\dfrac{\partial}{\partial v}\dfrac{\partial f}{\partial v}$ in terms of the components of v and of the second derivatives of f.

1.120 ¶. Show that, if $\Lambda \subset M_{m,n}$, then $|\mathbf{A}x| = O(|x|)$ as $x \to 0$.

1.121 ¶. Show that $\mathbf{H}(|x|)(x) = \dfrac{1}{|x|}\left(\mathrm{Id} - \dfrac{xx^T}{|x|^2}\right) \forall x \in \mathbb{R}^n, x \neq 0$.

1.122 ¶. Let $\Omega \subset \mathbb{R}^n$ be open, $f \in C^1(\Omega)$ and $K \subset \Omega$ be compact. Show that f is Lipschitz-continuous in K.

1.123 ¶ Peano example. Let $\varphi(t,x) = (t^2 - x^2)/(t^2 + x^2)$ and $f(t,x) := xt\varphi(x,t)$. Show that

$$\frac{\partial^2 f}{\partial x \partial t}(0,0) \neq \frac{\partial^2 f}{\partial t \partial x}(0,0).$$

1.124 ¶. Study the critical points and the graphs of the following functions

$$x^3 - 3xy^2, \qquad x^2 y^2,$$

$$e^{-x^2-y^2}\sin^2 x, \qquad \frac{1}{\sqrt{x^2+y^2}},$$

$$2y^2 - x(x-1)^2.$$

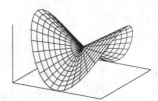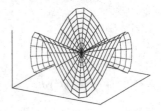

Figure 1.29. The graph of two homogeneous functions of degree 1. From the left: (i) $z = \rho\cos(2\theta)$ and (ii) $z = \rho\cos(3\theta)$.

1.125 ¶. Let $a, b, c \in \mathbb{R}^n$. Find the critical points of the following functions (if any)

$$f(x) = \frac{1}{|x - a|} + \frac{1}{|x - b|}, \qquad x \in \mathbb{R}^n \setminus \{a, b\},$$

$$f(x) = |x - a| + |x - b| + |x - c|, \qquad x \in \mathbb{R}^n,$$

$$f(x) = |x - a|^2 + |x - b|^2 + |x - c|^2, \qquad x \in \mathbb{R}^n,$$

$$f(x) = \max(|x - a|, |x - b|, |x - c|), \qquad x \in \mathbb{R}^n.$$

1.126 ¶. Find the variation of the intensity of a field E in $\mathbb{R}^3$ in time measured by an observer that moves along the trajectory described by the law $x = \cos t$, $y = \sin t$, $z = t$.

1.127 ¶. Let $y(x)$ be a differentiable function such that $F(x, y(x)) = 0 \ \forall x$ where $F(x, y)$ is a differentiable function, too. Show that

$$\frac{dy}{dx}(x) = -\frac{\dfrac{\partial F}{\partial x}(x, y(x))}{\dfrac{\partial F}{\partial y}(x, y(x))}$$

if $\dfrac{\partial F}{\partial y}(x, y(x)) \neq 0$. Find similar formulas in case $y_1(x)$, $y_2(x)$, $F_1(x, y_1, y_2)$ and $F_2(x, y_1, y_2)$ are differentiable and

$$\begin{cases} F_1(x, y_1(x), y_2(x)) = 0, \\ F_2(x, y_1(x), y_2(x)) = 0. \end{cases}$$

1.128 ¶ Homogeneous functions. Let $\alpha \in \mathbb{R}$. We say that f is *homogeneous* of degree α, or α-homogeneous, if

$$f(tx) = t^\alpha f(x) \qquad \forall \, x \neq 0 \text{ and } t > 0.$$

Of course, the domain of a homogeneous function is a cone with the origin as vertex; the origin may or may not belong to the domain of f. Show that

(i) the function $f : \mathbb{R}^2 \setminus \{(0,0)\} \to \mathbb{R}$, $f(x, y) := \frac{2xy}{x^2+y^2}$ is 0-homogeneous,

(ii) the function $f : \mathbb{R}^3 \setminus \{yz - y^2 = 0\} \to \mathbb{R}$, $f(x, y, z) := \frac{x+y-z}{yz-y^2}$ is homogeneous of degree -1,

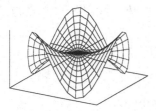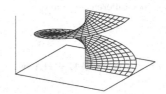

Figure 1.30. From the left: (i) the graph of the 2-homogeneous function $z = \rho^2 \cos(2\theta)$ and (ii) the graph of the 0-homogeneous function $z = \theta = \frac{xy}{x^2+y^2}$.

(iii) every quadratic form $Q(x) = \mathbf{A}x \bullet x = \sum_{i,j=1}^{n} A_{ij} x^i x^j$ is homogeneous of degree 2; hence, if $\mathbf{A}$ is positive definite, the function $Q(x)^{\alpha/2}$ is homogeneous of degree α.

1.129 ¶ Euler's formula. Show the following.

Theorem (Euler). *Let f be a function of class C^1 in $\mathbb{R}^n \setminus \{0\}$. Then f is α-homogeneous if and only if*

$$\nabla f(x) \bullet x = \alpha f(x) \qquad \forall x \neq 0.$$

1.130 ¶ Some useful inequalities. Let A be an open set in $\mathbb{R}^n$, $x_0 \in A$, and let $f : A \to \mathbb{R}^m$ be differentiable at x_0. Prove the following claims, see Theorem 1.89.

(i) For any $\epsilon > 0$ there exists a neighborhood U_{x_0} of x_0 such that

$$|f(x) - f(x_0)| \leq \Big(||\mathbf{D}f(x_0)|| + \epsilon\Big)|x - x_0| \qquad \forall\, x \in U_{x_0}.$$

(ii) For any $\epsilon > 0$ there exists a neighborhood U_{x_0} of x_0 such that

$$|f(x) - f(y)| \leq \Big(||\mathbf{D}f(x_0)|| + \epsilon\Big)|x - y| \qquad \forall\, x, y \in U_{x_0}.$$

(iii) If $\mathbf{D}f(x_0)$ is nonsingular for any $\epsilon > 0$ there exists a neighborhood U_{x_0} of x_0 such that

$$|f(x) - f(y)| \geq \left(\frac{1}{||\mathbf{D}f(x_0)^{-1}||} - \epsilon\right)|x - y| \qquad \forall\, x, y \in U_{x_0}.$$

[*Hint:* Notice that, if $L = \mathbf{D}f(x_0)$, then $|s - t| = |L^{-1}(Ls - Lt)| \leq ||L^{-1}||\,|Ls - Lt|$.]

1.131 ¶ Differentiability of functions of matrices. It is also convenient to consider the differentiation of matrix-valued functions via the calculus of functions between Banach spaces.

(i) Show that the product map, $m : M_{r,m} \times M_{m,n} \to M_{r,n}$, $m(\mathbf{A}, \mathbf{B}) := \mathbf{A}\mathbf{B}$ is Fréchet-differentiable in $M_{r,m} \times M_{m,n}$ and

$$T_{(\mathbf{A},\mathbf{B})}m(\mathbf{H}, \mathbf{K}) = \mathbf{A}\mathbf{K} + \mathbf{H}\mathbf{B} \qquad \forall \mathbf{A}, \mathbf{H} \in M_{r,m}, \; \forall \mathbf{B}, \mathbf{K} \in M_{m,n},$$

(ii) Using (i) show that $m^k : M_{n,n} \to M_{n,n}$, $m^k(\mathbf{A}) := \mathbf{A}^k$ is Fréchet-differentiable in $M_{n,n}$ and

$$T_{\mathbf{A}}m^k(\mathbf{H}) = \sum_{i=0}^{k-1} \mathbf{A}^i \mathbf{H} \mathbf{A}^{k-i-1};$$

in particular $T_{\mathrm{Id}}m^k(\mathbf{H}) = k\mathbf{H}$.

(iii) Show that the exponential map $\exp : M_{n,n} \to M_{n,n}$, $\exp(\mathbf{A}) := \sum_{k=0}^{\infty} \frac{1}{k!}\mathbf{A}^k$ is Fréchet-differentiable and

$$T_{\mathbf{A}}\exp(\mathbf{H}) = \sum_{k,\ell \geq 0} \frac{\mathbf{A}^k \mathbf{H} \mathbf{A}^\ell}{(k+\ell+1)!};$$

in particular, if $\mathbf{AH} = \mathbf{HA}$, we have

$$T_{\mathbf{A}}\exp(\mathbf{H}) = \mathbf{H}\exp(\mathbf{A}).$$

(iv) Noticing that $\det(\mathrm{Id} + \epsilon\mathbf{H}) = 1 + \epsilon\,\mathrm{tr}(\mathbf{H}) + o(\epsilon)$ as $\epsilon \to 0$, show that $\mathbf{A} \to \det\mathbf{A}$ is Fréchet-differentiable in the open set of nonsingular matrices and that

$$T_A \det(\mathbf{H}) = (\det\mathbf{A})\,\mathrm{tr}(\mathbf{A}^{-1}\mathbf{H}),$$

in particular we have $T_{\mathrm{Id}}\det(\mathbf{H}) = \mathrm{tr}(\mathbf{H})$.

1.132 ¶. Let X be a Banach space. The family of isomorphisms of X denoted $\mathrm{Isom}(X)$, is an open set of $\mathcal{L}(X,X)$, see [GM3, Exercise 1.50 Chapter 9]. Show that the map $\mathrm{inv} : \mathrm{Isom}(X) \to \mathrm{Isom}(X)$ that to $\ell \in \mathrm{Isom}(X)$ associates its inverse ℓ^{-1}, is Fréchet-differentiable and
$$T_\ell \,\mathrm{inv}\,(\mu) := -\ell^{-1} \circ \mu \circ \ell^{-1}.$$
This formula generalizes the formula $D(1/x) = -1/x^2$. [*Hint:* Compute first $(T_{\mathrm{Id}}\mathrm{inv})$ using the expansion of the inverse.]

1.133 ¶. Let X and Y be two Banach spaces, Ω an open set in X, and $f : \Omega \to Y$ an invertible map. Show that, if f and f^{-1} are Fréchet-differentiable respectively, at $x_0 \in \Omega$ and $f(x_0)$, then $(T_{f(x_0)}f^{-1}) = (T_{x_0}f)^{-1}$.

1.134 ¶. Let $f : C^0([0,1]) \to C^0([0,1])$ be a map of class C^1. Of course, $f : C^1([0,1]) \subset C^0([0,1]) \to C^0([0,1])$ is also Fréchet-differentiable. Show that the map f, regarded as a map between the Banach spaces $C^1([0,1])$ and $C^0([0,1])$, is Fréchet-differentiable.

2. Integral Calculus

The problems of characterizing the class of functions that are Riemann integrable and of discussing discontinuous functions, in particular of understanding for which functions the fundamental theorem of calculus is valid, as well as the need of integrating new functions, led to a new definition of integral due to Henri Lebesgue (1875–1941). Though the main ideas of Lebesgue's integration theory go back to Henri Lebesgue (1875–1941) and Giuseppe Vitali (1875–1932) at the beginning of the 1900's, applications as well as generalizations and extensions followed each other during the past century giving *measure* and *integration theory* a fundamental role in mathematical analysis.

Here we follow the approach of first introducing *Lebesgue's measure* and accordingly *Lebesgue's integral*. In Section 2.1.1 we collect the main results of the theory without proofs,[1] and in the following sections we develop its basic features.

2.1 Lebesgue's Integral

2.1.1 Definitions and properties: a short summary

The area of the subgraph of a nonnegative function can be computed in at least two different ways, see Figure 2.1. We compute the area of trapezoidal approximations determined by subdivisions of the x axis and then we pass to the limit when the lengths of the intervals of the subdivision tend to zero: this leads to Riemann's integral, compare [GM1]. Alternatively, we may subdivide the y axis and proceed similarly. In this second case, by taking equidistributed subdivisions, we may define

---

[1] The reader may find these proofs in, e.g., M. Giaquinta, G. Modica, *Mathematical Analysis: Foundations and Advanced Techniques for Functions of Several Variables*, Birkhäuser, to which in the sequel we shall refer to as [GM5].

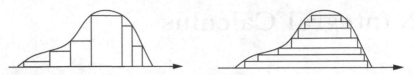

Figure 2.1. The integral: on the left Riemann's approach and on the right Lebesgue's approach.

$$\int_a^b f(x)\, dx := \lim_{N\to\infty} \frac{1}{2^N} \sum_{k=1}^{\infty} |E_{f, k\, 2^{-N}}|, \qquad (2.1)$$

where

$$E_{f,t} := \left\{ x \,\Big|\, f(x) > t \right\}, \qquad t \in \mathbb{R},$$

and $|E_{f,t}|$ denotes the "measure" of $E_{f,t}$. Since $t \to |E_{f,t}|$ is nondecreasing, (2.1) defines Lebesgue's integral of f via *Cavalieri's formula* as

$$\text{Lebesgue} \int_a^b f(x)\, dx := \text{Riemann} \int_0^{\infty} |E_{f,t}|\, dt. \qquad (2.2)$$

However, in order to proceed this way, we need a "good" notion of "measure" in $\mathbb{R}^n$ that allows us to measure rather wild sets as the sets $E_{f,t}$ may be. This is the role of *Lebesgue's measure*.

a. Lebesgue's measure

An *interval* I in $\mathbb{R}^n$, $n \geq 1$, is the product of n intervals, which for convenience we take left-open and right-closed, $I = \prod_{i=1}^{n}]a_i, b_i]$. The elementary n-dimensional *volume* of the interval I is by definition $|I| := \prod_{i=1}^{n}(b_i - a_i)$. The *outer* or *external measure* of an arbitrary subset E of $\mathbb{R}^n$ is defined by

$$\mathcal{L}^{n*}(E) := \inf\left\{ \sum_{k=1}^{\infty} |I_k| \,\Big|\, I_k \text{ intervals}, E \subset \bigcup_{k=1}^{\infty} I_k \right\}. \qquad (2.3)$$

Of course, $\mathcal{L}^{n*}$ defines a map $\mathcal{L}^{n*} : \mathcal{P}(\mathbb{R}^n) \to \overline{\mathbb{R}}_+$. It is easy to see that $\mathcal{L}^{n*}$ extends the elementary volume, in the sense that for every interval I we have $\mathcal{L}^{n*}(I) = |I|$. Intuitively $\mathcal{L}^{n*}(E)$ is computed by covering E in an "optimal" way with intervals $\{I_k\}$ and computing the sum of the series of the volumes of these intervals.

At this point, we would be done were if not for the fact that the outer measure $\mathcal{L}^{n*}$ is not additive: there exist disjoint subsets E, F of $\mathbb{R}^n$ such that $\mathcal{L}^{n*}(E \cup F) < \mathcal{L}^{n*}(E) + \mathcal{L}^{n*}(F)$.[2] We avoid this by selecting a class of special subsets, the class of *Lebesgue measurable sets*, and we define *Lebesgue's measure* as the restriction of $\mathcal{L}^{n*}$ to measurable sets.

---

[2] *Banach's paradox*: We can divide a ball in two parts each of the measure of the entire initial ball.

2.1 Definition. *A subset $E \subset \mathbb{R}^n$ is said to be* Lebesgue's measurable *or simply* measurable *if, given $\epsilon > 0$, there exists a set P_ϵ that is the union of at most a denumerable set of intervals such that*

$$P_\epsilon \supset E \qquad and \qquad \mathcal{L}^{n*}(P_\epsilon \setminus E) < \epsilon.$$

The class of all Lebesgue measurable subsets of $\mathbb{R}^n$ will be denoted by $\mathcal{M}$. The exterior measure of a measurable set E is its (Lebesgue) measure and denoted by $\mathcal{L}^n(E)$ or simply by $|E|$.

Intervals and countable union of intervals, as well as sets for which $\mathcal{L}^{n*}(E) = 0$, are clearly in $\mathcal{M}$. One can also easily see that the interior and the closure of an interval, as well as the countable union of open or closed intervals, are measurable sets. Since *every open set is the denumerable union of disjoint intervals*, we then infer that open sets are measurable. Moreover, though we can show that there exist nonmeasurable sets, see, e.g., [GM5], one shows that $\mathcal{M}$ has the following *closure properties* and that Lebesgue's measure is well behaved on measurable sets.

2.2 Theorem. *We have*

(i) *$\mathcal{M}$ is a σ-algebra, i.e., if $E, F \in \mathcal{M}$, then $E \cup F$, $E \setminus F$ and $E \cap F$ are in $\mathcal{M}$ and, if $\{E_k\}$ is a sequence of measurable subsets, then $\cup_k E_k$ e $\cap_k E_k$ are measurable.*

(ii) *$\mathcal{L}^n$ is σ-additive, i.e., if $E, F \subset \mathbb{R}^n$ are measurable, then $|E \cup F| + |E \cap F| = |E| + |F|$ and, if $\{E_k\}$ is a sequence of measurable pairwise disjoint subsets of $\mathbb{R}^n$, then*

$$\left| \bigcup_{k=1}^{\infty} E_k \right| = \sum_{k=1}^{\infty} |E_k|.$$

(iii) *$\mathcal{L}^n$ is continuous on nondecreasing sequences of measurable sets, i.e., if $\{E_k\}$ is a sequence of measurable sets in $\mathbb{R}^n$ such that $E_k \subset E_{k+1}$ $\forall k$, then $|E_k| \to |\cup_h E_h|$ as $k \to \infty$.*

(iv) *$\mathcal{L}^n$ is continuous on nonincreasing sequences of measurable subsets with finite measure, i.e., if $\{E_k\}$ is a sequence of measurable subsets such that $E_k \supset E_{k+1}$ $\forall k$ and if $|E_1| < +\infty$, then $|E_k| \to |\cap_h E_h|$ as $k \to \infty$.*

For arbitrary sequences of subsets $\{E_k\}$, one shows:

(i) $\mathcal{L}^{n*}(\cup_k E_k) \le \sum_{k=1}^{\infty} \mathcal{L}^{n*}(E_k)$,

(ii) if $E_k \subset E_{k+1}$ $\forall k$, then $\mathcal{L}^{n*}(E_k) \to \mathcal{L}^{n*}(\cup_k E_k)$.

Since open sets are measurable, Theorem 2.2 (i) yields that closed sets are measurable, too. Finally, one shows that a measurable set is the countable intersection of open sets except for a set of zero measure. One also shows that it is a countable union of closed sets union a set of zero measure, compare [GM5].

Figure 2.2. Henri Lebesgue (1875–1941) and Giuseppe Vitali (1875–1932).

b. Measurable functions

Starting from Lebesgue's measure in $\mathbb{R}^n$ and from the class of Lebesgue's measurable sets $\mathcal{M}$ we are now able to build a theory of integration for functions $f : E \subset \mathbb{R}^n \to \overline{\mathbb{R}}$, where E is a measurable set and $\overline{\mathbb{R}} = \mathbb{R} \cup \{+\infty, -\infty\}$.

We begin by selecting the class of *measurable functions*, with respect to the $\mathcal{L}^n$ Lebesgue measure.

A function $f : \mathbb{R}^n \to \overline{\mathbb{R}}$ is said to be $\mathcal{L}^n$-measurable, in short *measurable* when the measure is understood, if for every $t \in \mathbb{R}$ the set

$$E_{f,t} := f^{-1}(]t, +\infty]) = \left\{ x \in \mathbb{R}^n \mid f(x) > t \right\}$$

is $\mathcal{L}^n$-measurable. We then say that $f : E \subset \mathbb{R}^n \to \overline{\mathbb{R}}$ is *measurable on* E if E is measurable and the extension of f to $\mathbb{R}$ as $-\infty$ outsides E produces a function $\widetilde{f} : \mathbb{R}^n \to \overline{\mathbb{R}}$ that is measurable. Of course, a continuous function in $\mathbb{R}^n$ is measurable, and actually, if $E \subset \mathbb{R}^n$ is measurable, a function $f : E \to \mathbb{R}$ continuous in E is measurable.

As there exist nonmeasurable sets, there also exist nonmeasurable functions. However, on the ground of the fact that measurable sets form a σ-algebra, one shows that *all algebraic operations on measurable functions as well as taking pointwise limits of measurable functions produce measurable functions.*

Finally, the possibility of approximating in measure a Lebesgue measurable set from inside with closed sets and from outside with open sets yields the following characterization of Lebesgue measurable functions, see, e.g., [GM5].

2.3 Theorem (Lusin). *Let $f : E \to \mathbb{R}$ be a function defined on a measurable set $E \subset \mathbb{R}^n$. Then f is $\mathcal{L}^n$-measurable if and only if for any $\epsilon > 0$ there exists a closed set $F_\epsilon \subset \mathbb{R}^n$ such that $|E \setminus F_\epsilon| < \epsilon$ and the restriction of f to F_ϵ is continuous.*

c. Lebesgue's integral

We are now ready to define Lebesgue's integral of a measurable nonnegative function via Cavalieri's formula (2.2), using Riemann's integral and Lebesgue's measure:

$$\int_E f(x)\,dx := \int_0^{+\infty} \mathcal{L}^n(\{x \mid f(x) > t\})\,dt.$$

However, we prefer to follow a more direct approach and recover Cavalieri's formula later.

Recall that the *characteristic* (or *indicator*) *function* of a subset A of a set X is defined by

$$\chi_A(x) := \begin{cases} 1 & \text{if } x \in A, \\ 0 & \text{if } x \in X \setminus A. \end{cases}$$

A *simple function* is a measurable function that assumes only a finite number of values all of which are finite. We denote the class of simple functions by $\mathcal{S}$. If $a_1, a_2, \ldots, a_k$ are the distinct values of a simple function φ, we can write

$$\varphi(x) = \sum_{j=1}^{k} a_j \chi_{E_j}(x), \qquad E_j := \left\{ x \,\middle|\, \varphi(x) = a_j \right\}$$

where the E_j are measurable and pairwise disjoint sets. If φ is a simple nonnegative function, as suggested by intuition, the *integral of* φ is

$$I(\varphi) := \sum_{j=1}^{k} a_j |E_j|$$

with the agreement that $a_j|E_j| = 0$ if $a_j = 0$ and $|E_j| = +\infty$. The *Lebesgue integral* of a generic measurable and nonnegative function $f : E \to \overline{\mathbb{R}}$ is then defined by

$$\int_E f(x)\,d\mathcal{L}^n(x) \tag{2.4}$$

$$:= \sup\left\{ I(\varphi) \,\middle|\, \varphi \in \mathcal{S},\ \varphi(x) \leq f(x) \forall x \in E,\ \varphi(x) = 0\ \forall x \in E^c \right\}.$$

We also write when necessary $\int_E f(x)\,d\mathcal{L}^n(x)$ instead of $\int_E f(x)\,dx$.

Finally, if $f : E \to \overline{\mathbb{R}}$ is measurable (but not of a constant sign) we decompose f as difference of its positive and negative parts, $f(x) = f_+(x) - f_-(x)$ where

$$f_+(x) := \max(f(x), 0), \qquad f_-(x) := \max(-f(x), 0),$$

and we set the following.

Figure 2.3. Beppo Levi (1875–1962) and Guido Fubini (1879–1943).

2.4 Definition. *Let* $f : E \to \overline{\mathbb{R}}$ *be measurable on the measurable set* $E \subset \mathbb{R}^n$. *We say that* f *is (Lebesgue-)integrable if at least one of the two integrals* $\int_E f_+(x)\,dx$ *and* $\int_E f_-(x)\,dx$ *is finite. If* f *is integrable, its Lebesgue integral is defined by*

$$\int_E f(x)\,d\mathcal{L}^n(x) := \int_E f_+(x)\,d\mathcal{L}^n(x) - \int_E f_-(x)\,d\mathcal{L}^n(x).$$

When no confusion may arise, we write

$$\int_E f(x)\,dx \qquad instead\ of \qquad \int_E f(x)\,d\mathcal{L}^n(x).$$

Finally, we say that f *is (Lebesgue-)summable if both the integrals of* f_+ *and of* f_- *are finite. We denote the class of summable functions in* E *by* $\mathcal{L}^1(E)$.

Of course, $\int_{\mathbb{R}^n} \varphi(x)\,dx = I(\varphi)$ if $\varphi \in \mathcal{S}$. Also, notice that the difference in the definitions of the Riemann and Lebesgue integrals consists merely in the choice of the class of simple functions: finite combinations of characteristic functions of *intervals* in Riemann's theory, finite combinations of characteristic functions of *Lebesgue's measurable sets* in Lebesgue's theory.

2.5 ¶. Let $f : E \subset \mathbb{R}^n \to \overline{\mathbb{R}}$ be integrable, and let $\widetilde{f} : \mathbb{R}^n \to \overline{\mathbb{R}}$ denote its extension with zero values outside E. Show that

$$\int_E f(x)\,dx = \int_{\mathbb{R}^n} \widetilde{f}(x)\,dx.$$

d. Basic properties of Lebesgue's integral

The basic properties of Lebesgue's integral are easily inferred from the analogous properties of the integral of simple functions using the denumerable additivity of the Lebesgue measure and the following approximation lemma.

2.6 Lemma. *Let $f : \mathbb{R}^n \to \overline{\mathbb{R}}$ be a nonnegative and measurable function. Then, there exists an increasing sequence $\{\psi_k\}$ of simple functions such that*

$$\psi_k \to f \quad pointwise, \qquad and \qquad \int_{\mathbb{R}^n} \psi_k(x)\, dx \to \int_{\mathbb{R}^n} f(x)\, dx.$$

In this way we can prove the following.

2.7 Theorem. *We have*

(i) (MONOTONICITY) *If f and g are integrable on E, and $f \le g$, then*

$$\int_E f(x)\, dx \le \int_E g(x)\, dx.$$

(ii) (LINEARITY) *$\mathcal{L}^1(E)$ is a real vector space, and the integral as a map from $\mathcal{L}^1(E)$ into $\mathbb{R}$ is a linear operator,*

$$\int_E (\alpha f(x) + \beta g(x))\, dx = \alpha \int_E f(x)\, dx + \beta \int_E g(x)\, dx$$

for all $\alpha, \beta \in \mathbb{R}$ and all $f, g \in \mathcal{L}^1(E)$.

(iii) (CONTINUITY) *If f is integrable on E, then*

$$\left| \int_E f(x)\, dx \right| \le \int_E |f(x)|\, dx.$$

(iv) (BEPPO LEVI THEOREM) *Let E be a measurable set, let $f_k : E \to \mathbb{R}_+$ be an increasing sequence of nonnegative and measurable functions in E, and let $f(x) := \lim_{k \to \infty} f_k(x)$ be the pointwise limit of $\{f_k\}$. Then we have*

$$\int_E f(x)\, dx = \lim_{k \to +\infty} \int_E f_k(x)\, dx.$$

Beppo Levi's theorem is also referred to as the monotone convergence theorem for nonnegative functions.

The following claims are easy consequences of the above.

(i) If f is integrable on E, $|f(x)| \le M$ for all $x \in E$ and $|E| < +\infty$, then f is summable on E and $\int_E |f(x)|\, dx \le M\, |E|$.
(ii) $f \in \mathcal{L}^1(E)$ if and only if f is measurable and $\int_E |f(x)|\, dx < +\infty$.
(iii) If E and F are measurable sets, and f is integrable in $E \cup F$, then

$$\int_E f(x)\, dx + \int_F f(x)\, dx = \int_{E \cup F} f(x)\, dx + \int_{E \cap F} f(x)\, dx.$$

e. The integral as area of the subgraph

The Lebesgue integral can be equivalently defined as the area of the subgraph or via Cavalieri's formula. In fact the following holds.

2.8 Theorem. *Let $f : E \subset \mathbb{R}^n \to \overline{\mathbb{R}}$ be a nonnegative function. Then, f is measurable on E if and only if its subgraph*

$$SG_{f,E} := \Big\{(x,t) \,\Big|\, x \in E, \ 0 < t < f(x)\Big\} \subset \mathbb{R}^{n+1}$$

is a measurable subset of $\mathbb{R}^{n+1}$. Moreover,

$$\int_E f(x)\,dx = \mathcal{L}^{n+1}(SG_{f,E}). \tag{2.5}$$

2.9 Theorem (Cavalieri's formula). *Let $f : E \subset \mathbb{R}^n \to \overline{\mathbb{R}}$ be a nonnegative measurable function. Then*

$$\int_E f(x)\,dx = \int_0^{+\infty} \mathcal{L}^n(\{x \in E \mid f(x) > t\})\,dt.$$

f. Chebyshev's inequality

Let $f : E \to \overline{\mathbb{R}}$ be measurable in $E \subset \mathbb{R}^n$ and nonnegative. Set $E_{f,t} := \{x \in E \mid f(x) > t\}$. From the monotonicity of the integral we infer

$$|E_{f,t}| \leq \frac{1}{t}\int_{E_{f,t}} f(x)\,dx \qquad \forall t > 0 \tag{2.6}$$

which for its wide use in several contexts has got various names: *weak estimate, Markov's inequality, Chebyshev's inequality*. It estimates the "size" of f in terms of the integral of f. The nondecreasing function $t \to |E_{f,t}|$ is called the *repartition function* of f.

g. Negligible sets and the integral

We say that the predicate $p(x)$, $x \in E \subset \mathbb{R}^n$ is true *for almost every $x \in E$*, or *almost everywhere in E* (in short a.e.), if the Lebesgue measure of the set

$$\Big\{x \in E \,\Big|\, p(x) \text{ is not true}\Big\}$$

is zero. For instance, if $f : E \subset \mathbb{R}^n \to \overline{\mathbb{R}}$ is a function, we say "$f = 0$ a.e. in E" or "$f(x) = 0$ for a.e. $x \in E$" if $\mathcal{L}^n(\{x \mid f(x) \neq 0\}) = 0$. Similarly, we say that "$|f(x)| < \infty$ a.e. in E" or "$|f(x)| < +\infty$ for a.e. $x \in E$" if $\mathcal{L}^n(\{x \mid |f(x)| \notin \mathbb{R}\}) = 0$. From the denumerable additivity of the Lebesgue measure we can easily deduce the following.

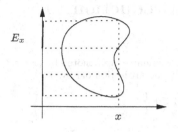 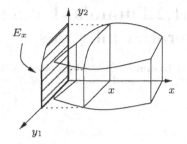

Figure 2.4. The slice E_x of E over x.

2.10 Proposition. *We have*

(i) *If* $f : E \to \mathbb{R}$ *is summable,* $f \in \mathcal{L}^1(E)$, *then* $|f(x)| < +\infty$ *for a.e.* $x \in E$.

(ii) *If* $f : E \to \mathbb{R}$ *is nonnegative, then* $\int_E f(x)\, dx = 0$ *if and only if* $f(x) = 0$ *for a.e.* $x \in E$.

Let $f : E \subset \mathbb{R}^n \to \overline{\mathbb{R}}$ be measurable. The *essential supremum of* f is the number (possibly $+\infty$) defined by

$$\|f\|_{\infty,E} = \operatorname*{ess\,sup}_E f := \inf\Big\{ t \in \mathbb{R} \,\Big|\, f(x) < t \text{ for a.e. } x \in E \Big\}. \qquad (2.7)$$

Of course, $\|f\|_{\infty,E} = \sup_E |f(x)|$ if f is continuous on E, and

$$\int_E |f(x)|\, dx \le \|f\|_{\infty,E}\, |E| \qquad \forall f \in \mathcal{L}^1(E).$$

h. Riemann integrable functions

The Lebesgue integral extends the Riemann integral. In fact, (generalized) Riemann integrable functions are Lebesgue integrable and the Lebesgue integral and Rieamnn integral of one of these functions agree, see [GM5]. This remark gives us a way to compute the Lebesgue integral of a large class of functions. For instance,

$$\int_0^1 \frac{1}{x}\, dx = +\infty, \qquad \int_{-\infty}^{+\infty} \frac{1}{1+x^2}\, dx = \pi, \qquad \text{ctc.}$$

On the other hand, the long-standing problem of characterizing Riemann integrable functions was solved by Giuseppe Vitali (1875–1932) in terms of Lebesgue integral: *A bounded function* $f : [a, b] \to \mathbb{R}$ *is Riemann integrable if and only if it is* $\mathcal{L}^1$*-almost-everywhere continuous.*

2.1.2 Fubini's theorem and reduction to iterated integrals

2.11 Example. Let E be a subset of $\mathbb{R}^2$ whose coordinates are denoted by (x, y). For every $x \in \mathbb{R}$, we define the *slice of E at x* (actually the projection of) by

$$E_x := \left\{ y \in \mathbb{R} \mid (x, y) \in E \right\}.$$

If $E =]a, b] \times]c, d]$, then

$$E_x := \begin{cases}]c, d] & \text{if } x \in]a, b], \\ \emptyset & \text{otherwise.} \end{cases}$$

In particular $|E_x| = 0$ if $x \notin]a, b]$ and $|E_x| = d - c$ if $x \in]a, b]$; consequently

$$\mathcal{L}^2(]a, b] \times]c, d]) = (b - a)(d - c) = \int_a^b |E_x| \, dx.$$

Fubini's theorem extends the remark of the previous example to arbitrary measurable subsets of Euclidean spaces. Split the coordinate variables in $\mathbb{R}^{n+k}$ in two groups, for instance the first n coordinates and the remaining ones, which we denote by $x \in \mathbb{R}^n$ and $y \in \mathbb{R}^k$ respectively, so that (x, y) denotes the coordinate variables in $\mathbb{R}^{n+k}$. Let E be a subset of $\mathbb{R}^{n+k}$, and for $x \in \mathbb{R}^n$, let

$$E_x := \left\{ y \in \mathbb{R}^k \mid (x, y) \in E \right\}$$

denote the *slice* of E over x (projected into the coordinate space $\mathbb{R}^k$), see Figure 2.4.

2.12 Theorem (Fubini). *Let $E \subset \mathbb{R}^{n+k}$ be $\mathcal{L}^{n+k}$-measurable in $\mathbb{R}^{n+k}$. Then the following hold:*

(i) *For a.e. $x \in \mathbb{R}^n$ the set $E_x \subset \mathbb{R}^k$ is $\mathcal{L}^k$-measurable.*
(ii) *The function $x \to \mathcal{L}^k(E_x)$ is $\mathcal{L}^n$-measurable.*
(iii) *We have*

$$\mathcal{L}^{n+k}(E) = \int_{\mathbb{R}^n} \mathcal{L}^k(E_x) \, d\mathcal{L}^n(x).$$

A very useful variant of Fubini's theorem is the following theorem that provides a formula that allows us to compute a multiple integral as the iteration of simple integrals.

2.13 Theorem (Reduction to iterated integrals). *Let $f : E \to \overline{\mathbb{R}}$, $E \subset \mathbb{R}^{n+k}$, be an $\mathcal{L}^{n+k}$-integrable function. Then*

(i) *for a.e. $x \in E$ the function $y \to f_x(y) := f(x, y)$ is $\mathcal{L}^k$-integrable in E,*
(ii) *the function $x \to \int_{E_x} f(x, y) \, dy$ is $\mathcal{L}^n$-measurable,*

(iii) *we have*

$$\int_E f(x,y)\, d\mathcal{L}^{n+k}(x,y) := \int_{\mathbb{R}^n} \left(\int_{E_x} f(x,y)\, d\mathcal{L}^k(y) \right) d\mathcal{L}^n(x).$$

We emphasize the fact that the only assumptions in the previous theorems are the $\mathcal{L}^{n+k}$ integrability of f in E in Theorem 2.12 and the $\mathcal{L}^{n+k}$-measurability of E in Theorem 2.13. We recall once again that f is integrable in E in each of the following cases:

(i) f is measurable and has constant sign;
(ii) f is summable in E; this happens in particular if f is measurable in E, $|f|$ is bounded, and $|E| < +\infty$.

We observe that Theorem 2.13 reduces in particular the calculus of a *double* integral to successively computing two simple integrals, the order being irrelevant.

$$\iint_E f(x,y)\, dx\, dy = \int_{-\infty}^{+\infty} \left(\int_{E_x} f(x,y)\, dy \right) dx$$
$$\iint_E f(x,y)\, dx\, dy = \int_{-\infty}^{+\infty} \left(\int_{E_y} f(x,y)\, dx \right) dy$$

where

$$E_x := \left\{ y \in \mathbb{R}^k \,\middle|\, (x,y) \in E \right\}, \qquad E_y := \left\{ x \in \mathbb{R}^n \,\middle|\, (x,y) \in E \right\}.$$

Of course, Theorem 2.13 can be used iteratively, thus reducing the calculus of the integral of an integrable function of n-variables to successively computing n integrals in one variable, the order of them being irrelevant. In other words we can also state the following.

2.14 Theorem (Tonelli). *Let $f : E \subset \mathbb{R}^2 \to \overline{\mathbb{R}}$ be integrable in E. Then, the three integrals*

$$\int_E f(x,y)\, d\mathcal{L}^{n+k}(x,y),$$

$$\int_{\mathbb{R}^n} \left(\int_{E_x} f(x,y)\, d\mathcal{L}^k(y) \right) d\mathcal{L}^n(x), \qquad \int_{\mathbb{R}^k} \left(\int_{E_y} f(x,y)\, d\mathcal{L}^n(x) \right) d\mathcal{L}^n(y)$$

exist and are equal.

2.1.3 Change of variables

The exterior Lebesgue measure $\mathcal{L}^{n*}$ is invariant under isometries; even more, if $T : \mathbb{R}^n \to \mathbb{R}^n$ is linear, then T *maps $\mathcal{L}^n$-measurable sets into $\mathcal{L}^n$-measurable sets, and*

$$\mathcal{L}^{n*}(T(E)) = |\det T| \, \mathcal{L}^{n*}(E), \qquad \forall E \subset \mathbb{R}^n, \tag{2.8}$$

in particular, *linear maps map set of measure zero into sets of measure zero.* Lipschitz-continuous functions, and consequently C^1 functions do the same, however continuous functions do not; in fact, continuous maps may map null sets into sets of positive Lebesgue measure.

The formula (2.8) extends to diffeomorphisms, i.e., one-to-one transformations of class C^1 with inverse of class C^1, as follows.

2.15 Theorem (Change of variables). *Let A be an open set in $\mathbb{R}^n$, and let $\varphi : A \to \mathbb{R}^n$ be a map of class C^1. Then φ maps measurable sets into measurable sets and negligible sets into negligible sets. Moreover, if $E \subset A$ is measurable, and φ is injective in E, then:*

(i) *We have*

$$\mathcal{L}^n(\varphi(E)) = \int_E |\det \mathbf{D}\varphi(x)| \, dx.$$

(ii) *If $f : \varphi(E) \to \overline{\mathbb{R}}$ is any function, then f is integrable on $\varphi(E)$ if and only if $x \to f(\varphi(x)) \, |\det \mathbf{D}\varphi(x)|$ is integrable on E and*

$$\int_{\varphi(E)} f(y) \, dy = \int_E f(\varphi(x)) \, |\det \mathbf{D}\varphi(x)| \, dx.$$

Notice that there is no need to assume $\det \mathbf{D}\varphi(x) \neq 0$, yet another relevant consequence of the Lebesgue integrability.

2.1.4 Differentiation and primitives

Let $f, g : \mathbb{R}^n \to \overline{\mathbb{R}}$ be two nonnegative and measurable functions. Of course, $\int_A f(x) \, dx = \int_A g(x) \, dx \; \forall A \subset \mathbb{R}^n$ if and only if $f(x) = g(x)$ a.e. $x \in \mathbb{R}^n$. Is there a way to characterize $f(x)$ in terms of integrals, or more precisely in terms of the map $A \to \int_A f(x) \, dx$? The theory of *differentiation of integrals* answers this important question in measure theory.

Recall that, if $f : \mathbb{R} \to \mathbb{R}$ is continuous, then the integral mean value theorem yields

$$f(x_0) = \lim_{r \to 0} \frac{1}{2r} \int_{x_0 - r}^{x_0 + r} f(t) \, dt \qquad \forall x_0 \in \mathbb{R}.$$

We also have the following.

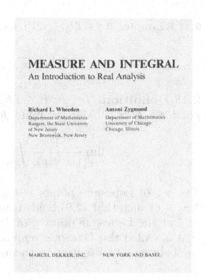

Figure 2.5. Two classic books on Lebesgue's integration.

2.16 Theorem (Lebesgue's differentiation). *Let $f : \mathbb{R}^n \to \mathbb{R}$ be such that $\int |f|^p \, dx < +\infty$ for some $1 \le p < +\infty$. Then, for a.e. $x \in \mathbb{R}^n$ we have*

$$\frac{1}{|B(x_0, r)|} \int_{B(x_0, r)} |f(y) - f(x)|^p \, dy \to 0 \qquad as \ r \to 0^+;$$

in particular, for a.e. $x \in \mathbb{R}^n$,

$$\frac{1}{|B(x, r)|} \int_{B(x, r)} f(y) \, dy \to f(x) \qquad for \ a.e. \ x \in \mathbb{R}^n.$$

Notice that if $f \in \mathcal{L}^1(E)$, E being measurable in $\mathbb{R}^n$, by applying the previous theorem to the function

$$\widetilde{f}(x) = \begin{cases} f(x) & \text{if } x \in E, \\ 0 & \text{if } x \in E^c \end{cases}$$

we get that for a.e. $x \in E$

$$\frac{1}{|B(x_0, r)|} \int_{E \cap B(x_0, r)} |f(y) - f(x)|^p \, dy \to 0 \qquad as \ r \to 0^+;$$

in particular, for a.e. $x \in \mathbb{R}^n$,

$$\frac{1}{|B(x, r)|} \int_{E \cap B(x, r)} f(y) \, dy \to \begin{cases} f(x) & \text{for a.e. } x \in E, \\ 0 & \text{for a.e. } x \in E^c. \end{cases}$$

2.17 Example. If $f \in \mathcal{L}^1(]-1,1[)$, then for $\mathcal{L}^1$-a.e. $x \in]-1,1[$ we have

$$\lim_{r \to 0} \frac{1}{2r} \int_{-r}^{+r} f(y)\, dy = f(x).$$

2.18 Definition. *Let $f : E \subset \mathbb{R}^n \to \overline{\mathbb{R}}$ be summable in E. We say that a point $x \in E$ is a* Lebesgue point *for f if there exists $\lambda \in \mathbb{R}$ such that*

$$\lim_{r \to 0} \frac{1}{|B(x,r)|} \int_{E \cap B(x,r)} |f(y) - \lambda|\, dy \to 0. \qquad (2.9)$$

The set of Lebesgue points is then denoted by $\mathcal{L}_f$. Moreover, the value $\lambda = \lambda(x)$ such that (2.9) holds is unique and it is called the *Lebesgue value* of f at the Lebesgue point x of f. Therefore, we have a map $\lambda : \mathcal{L}_f \to \mathbb{R}$ that is called the *Lebesgue representative* of f and, with these notations, the Lebesgue differentiation theorem, Theorem 2.16, reads as follows.

2.19 Theorem (Lebesgue's differentiation). *Let $f \in \mathcal{L}^1(E)$ and let $\mathcal{L}_f$ be the set of Lebesgue points of f. Then $E \setminus \mathcal{L}_f$ has zero Lebesgue measure, $\mathcal{L}^n(E \setminus \mathcal{L}_f) = 0$.*

2.20 Asymmetric differentiation. In the differentiation theorem, Theorem 2.16, we can replace balls with cubes, and actually differentiate with respect to bounded sets A such that for instance

$$A \subset B(0,100), \qquad |A| = c|B_1|.$$

For $x \in \mathbb{R}^n$ and $r > 0$, we set $A_{x,r} := x + rA$. Trivially $A_{x,r} \subset B(x,100r)$ and $|A_{x,r}| = r^n|A| = cr^n|B_1| = c|B(x,r)|$.

Theorem. *Let $f : E \subset \mathbb{R}^n \to \overline{\mathbb{R}}$ be measurable with $\int_E |f|^p\, dx < \infty$ for some $0 \leq p < +\infty$. Then for a.e. $x \in E$ we have*

$$\frac{1}{|A_{x,r}|} \int_{E \cap A_{x,r}} |f(y) - f(x)|^p\, dy \to 0 \qquad \text{as } r \to 0^+.$$

Example. If $f \in \mathcal{L}^1(\mathbb{R})$, then for a.e. $x \in \mathbb{R}$ we have

$$\lim_{r \to 0^+} \frac{1}{r} \int_0^r f(y)\, dy = \lim_{r \to 0^+} \frac{1}{r} \int_{-r}^0 f(y)\, dy = f(x)$$

and also

$$\lim_{r \to 0^+} \frac{1}{8r} \int_{2r}^{10r} f(y)\, dy = f(x).$$

We conclude by collecting a few relevant consequences of the differentiation theorem.

2.21 Theorem (Vitali). *Monotone real-valued functions* $f : \mathbb{R} \to \mathbb{R}$ *are a.e. differentiable in the classic sense. Moreover,* $h' \in \mathcal{L}^1((a,b)) \ \forall a, b \in \mathbb{R}$, h' *is nonnegative if* h *is nondecreasing and*

$$0 \le \int_x^y h'(t)\, dt \le h(y) - h(x) \qquad \forall x < y.$$

A function $f : \mathbb{R} \to \mathbb{R}$ is said to be *absolutely continuous* if for any $\epsilon > 0$ there exists $\delta > 0$ such that $\sum_{k=1}^\infty |f(x_k) - f(y_k)| < \epsilon$ whenever $\{x_k\}$ and $\{y_k\}$ are such that $\sum_{k=1}^\infty |x_k - y_k| < \delta$. Trivially Lipschitz-continuous functions are absolutely continuous, absolutely continuous functions are continuous, and there exist functions that are continuous but not absolutely continuous. A celebrated example is the so-called Cantor–Vitali function, see [GM5]. We have the following.

2.22 Theorem (Vitali). *A function* $f : [a, b] \to \mathbb{R}$ *is absolutely continuous in* $[a, b]$ *if and only if* f *is a.e. differentiable in* $[a, b]$, $f' \in \mathcal{L}^1([a, b])$ *and*

$$\int_x^y h'(t)\, dt = h(y) - h(x) \qquad \forall x, y \in [a, b], \ x < y. \tag{2.10}$$

The above implies that Lipschitz-continuous functions from $\mathbb{R}$ into $\mathbb{R}$ are a.e. differentiable and that the equality (2.10) holds for them. For Lipschitz-continuous functions of several variables we state the following.

2.23 Theorem (Rademacher). *Every Lipschitz-continuous function* $f : \mathbb{R}^n \to \mathbb{R}$ *is differentiable in the classic sense for a.e.* $x \in \mathbb{R}^n$. *Moreover, the components of the map* $x \to \mathbf{D}f(x)$ *are measurable and*

$$\|\mathbf{D}f(x)\|_{\infty, \mathbb{R}^n} = \mathrm{Lip}\,(f).$$

2.2 Convergence Theorems

In many respects and especially for the applications, the main results of Lebesgue's integration theory are contained in Beppo Levi's monotone convergence theorem, Theorem 2.7 (iv), and in Proposition 2.10. In this section we discuss some important, useful consequences.

a. Monotone convergence

First we state in a more general form Beppo Levi's theorem, weakening the positivity assumption and taking advantage of the fact that a.e. equal functions have the same integral.

Come già in altra mia nota (*), noi diremo che un gruppo di intervalli presi sopra una medesima retta è un gruppo di intervalli distinti, quando due qualsiansi di questi intervalli non hanno punti interni comuni, ed ampiezza di un gruppo di intervalli la somma delle lunghezze dei singoli intervalli del gruppo.

Sia $F(x)$ una funzione finita della variabile reale x in un intervallo (a, b), ed $a < b$.

Se (a, β) è un intervallo parziale di (a, b), ed $a \leq a < \beta \leq b$, noi chiameremo incremento di $F(x)$ in (a, β) la differenza $F(\beta) - F(a)$. Diremo poi incremento di $F(x)$ in un gruppo di intervalli parziali di (a, b) distinti la somma, se è determinata e finita, degli incrementi di $F(x)$ nei singoli intervalli.

Se per ogni numero $\sigma > 0$ esiste un numero $u > 0$ tale che sia minore di σ il modulo dell'incremento di $F(x)$ in ogni gruppo di ampiezza minore di u di intervalli parziali di (a, b) distinti, si dirà che $F(x)$ è assolutamente continua.

Infine diremo che $F(x)$ è in (a, b) una funzione integrale se e soltanto se esiste in (a, b) una funzione $f(x)$ finita e sommabile (**), per cui $F(x) - F(a) = \int_a^x f(x)dx$, per ogni x tale che $a \leq x \leq b$.

Io dimostrerò che:

CONDIZIONE NECESSARIA E SUFFICIENTE PERCHÈ UNA FUNZIONE $F(x)$ SIA IN (a, b) UNA FUNZIONE INTEGRALE È CHE ESSA SIA ASSOLUTAMENTE CONTINUA IN (a, b).

(*) G. VITALI, Sui gruppi di punti, § 2. * Rend. del Circolo matem. di Palermo ,, tomo XVIII. 1904.
(**) La parola sommabile ed il simbolo ∫ (integrale) sono usati nel senso di LEBESGUE. Vedi Leçons sur l'intégration etc. par Henri LEBESGUE. Paris, Gauthier-Villars, 1904.

Sul problema della misura
dei gruppi di punti di una retta

—

NOTA

G. VITALI

BOLOGNA
TIP. GAMBERINI E PARMEGGIANI
1905

Figure 2.6. The first page of the paper *Sulle funzioni integrali*, Acad. Sci. Torino 1905, by Giuseppe Vitali (1875–1932) and the frontispiece of the paper, again by Giuseppe Vitali, where for the first time the example of a set that is not Lebesgue measurable is presented.

2.24 Theorem (Beppo Levi). *Let* $\{f_k\}$ *be a nondecreasing sequence of integrable functions on* $E \subset \mathbb{R}^n$ *such that* $f_k(x) \to f(x)$ *for a.e.* $x \in E$. *If there exists a function* $\phi \in \mathcal{L}^1(E)$ *such that* $f_k(x) \geq \phi(x)$ *for all* k *and a.e.* $x \in E$, *then*

$$\int_E f(x)\,dx = \lim_{k\to\infty} \int_E f_k(x)\,dx.$$

Proof. We apply Beppo Levi's theorem to the nondecreasing sequence of nonnegative functions $\{f_k - \phi\}$ to get

$$\int_E (f_k(x) - \phi(x))\,dx \to \int_E (f(x) - \phi(x))\,dx.$$

The result then follows on account of the fact that ϕ has finite integral. $\square$

2.25 ¶. Notice that the assumption $f_k \geq \phi$, $\phi \in \mathcal{L}^1(E)$, that is, the assumption that the lower envelope of the $f_k's$ is summable, cannot be omitted, as shown by the sequence

$$f_k(x) := \begin{cases} -1 & \text{if } x > k, \\ 0 & \text{otherwise.} \end{cases}$$

As a trivial consequence of Beppo Levi's theorem we can state the following.

2.26 Corollary (Total convergence of series). *Let* $f_k : E \to \overline{\mathbb{R}}$, $k = 1, 2 \ldots$, *be nonnegative measurable functions on* E. *Then*

Figure 2.7. Frontispiece of the first edition of the treatise on integration by Henri Lebesgue (1875–1941) and a page from the second edition of 1928.

$$\int_E \sum_{k=1}^{\infty} f_k(x)\, dx = \sum_{k=1}^{\infty} \int_E f_k(x)\, dx.$$

2.27 Corollary. *Let* $f_k : E \to \overline{\mathbb{R}}$, $k = 1, 2, \ldots$ *be measurable functions on* E. *If* $\{f_k\}$ *is nonincreasing and there exists* $\phi \in \mathcal{L}^1(E)$ *such that* $f_k(x) \leq \phi(x)$ *for all* k *and a.e.* $x \in E$, *then*

$$\int_E \lim_{k \to \infty} f_k(x)\, dx = \lim_{k \to \infty} \int_E f_k(x)\, dx.$$

2.28 ¶. Notice that the assumption $f_k \leq \phi$, $\phi \in \mathcal{L}^1(E)$ cannot be omitted as shown by the sequence

$$f_k(x) := \begin{cases} 1 & \text{if } x < -k, \\ 0 & \text{otherwise.} \end{cases}$$

2.29 ¶. Let $f : E \to \overline{\mathbb{R}}$ be integrable on E and let $\{E_k\}$, $k = 1, 2, \ldots$, be a sequence of denumerable pairwise disjoint measurable subsets such that $E = \cup_k E_k$. Show that

$$\int_E f(x)\, dx = \sum_{k=1}^{\infty} \int_{E_k} f(x)\, dx.$$

b. Dominated convergence

2.30 Lemma (Fatou). *Let* $\{f_k\}$ *be a sequence of nonnegative and measurable functions on* E. *Then*

$$\int_E \liminf_{k\to\infty} f_k(x)\,dx \le \liminf_{k\to\infty} \int_E f_k(x)\,dx.$$

Proof. The functions $g_n(x) := \inf_{k\ge n} f_k(x)$, are nonnegative, measurable on E, and form a nondecreasing sequence; moreover

$$0 \le g_n(x) \le f_k(x),\ k \ge n, \qquad \liminf_{\kappa\to\infty} f_k(x) := \lim_{n\to\infty} g_n(x).$$

Thus $\int_E g_n(x)\,dx \le \inf_{k\ge n} \int_E f_k(x)\,dx$, and we infer, using Beppo Levi's theorem,

$$\int_E \liminf_{k\to\infty} f_k(x)\,dx = \lim_{n\to\infty} \int_E g_n(x)\,dx \le \lim_{n\to\infty} \inf_{k\ge n} \int_E f_k(x)\,dx =: \liminf_{k\to\infty} \int_E f_k(x)\,dx.$$

$\square$

As previously, we can weaken the positivity condition to get the following result.

2.31 Corollary (Fatou lemma). *Let $\{f_k\}$ be a sequence of integrable functions on E and let $\phi \in \mathcal{L}^1(E)$.*

(i) *If $f_k(x) \ge \phi(x)$ for all k and a.e. $x \in E$, then*

$$\int_E \liminf_{k\to\infty} f_k(x)\,dx \le \liminf_{k\to\infty} \int_E f_k(x)\,dx.$$

(ii) *If $f_k(x) \le \phi(x)$ for all k and a.e. $x \in E$, then*

$$\limsup_{k\to\infty} \int_E f_k(x)\,dx \le \int_E \limsup_{k\to\infty} f_k(x)\,dx.$$

2.32 Theorem (Lebesgue dominated convergence theorem). *Let $\{f_k\}$ be a sequence of measurable functions on $E \subset \mathbb{R}^n$. If*

(i) *$f_k(x) \to f(x)$ for a.e. $x \in E$,*
(ii) *there exists $\phi \in \mathcal{L}^1(E)$ such that $|f_k(x)| \le \phi(x)$ for all k and a.e. $x \in E$,*

then

$$\int_E |f_k(x) - f(x)|\,dx \to 0,$$

in particular

$$\int_E f_k(x)\,dx \to \int_E f(x)\,dx.$$

Proof. By the assumptions $|f_k(x) - f(x)| \to 0$ for a.e. $x \in E$ and $|f_k(x) - f(x)| \le 2\phi(x)$ for all k and a.e. $x \in E$. Fatou's lemma, Corollary 2.31 (ii), then yields

$$\limsup_{k\to\infty} \int_E |f_k(x) - f(x)|\,dx \le \int_E \limsup_{k\to\infty} |f_k(x) - f(x)|\,dx = \int_E 0\,dx = 0.$$

The second part of the claim follows since

$$\left| \int_E f_k(x)\,dx - \int_E f(x)\,dx \right| = \left| \int_E (f_k(x) - f(x))\,dx \right| \le \int_E |f_k(x) - f(x)|\,dx.$$

$\square$

2.33 ¶. Notice that the assumption (ii) amounts to requiring that the *envelope* of the functions $|f_k|$, defined as $\phi(x) := \sup_k |f_k(x)|$ is a summable function on E. Notice that (ii) cannot be omitted as it is shown by the sequence

$$
f_k(x) = \begin{cases} k & \text{if } 0 < x < 1/k, \\ 0 & \text{otherwise.} \end{cases}
$$

Finally, we state the following important convergence theorem for series of functions.

2.34 Theorem (Lebesgue). *Let $\{f_n\}$ be a sequence of measurable functions on E such that*

$$
\sum_{k=0}^{\infty} \int_E |f_n(x)| \, dx < +\infty.
$$

Then the series of functions $\sum_{k=0}^{\infty} f_n(x)$ converges absolutely for a.e. $x \in E$ to a function $f \in \mathcal{L}^1(E)$ and

$$
\int_E \left| f(x) - \sum_{k=0}^{p} f_k(x) \right| dx \to 0 \qquad p \to \infty. \tag{2.11}
$$

In particular,

$$
\int_E f(x) \, dx = \sum_{k=0}^{\infty} \int_E f_k(x) \, dx.
$$

Proof. For all $x \in E$, we let $g(x) \in \overline{\mathbb{R}}_+$ be the sum of the series $\sum_{k=0}^{\infty} |f_k(x)|$ with positive terms. From Beppo Levi's theorem and the assumptions we have

$$
\int_E g(x) \, dx = \sum_{k=0}^{\infty} \int_E |f_k(x)| \, dx < +\infty.
$$

Hence g is summable on E. Proposition 2.10 yields $g(x) < +\infty$ for a.e. $x \in E$. Therefore, for these x the series $\sum_{k=0}^{\infty} f_k(x)$ converges absolutely to a real-valued function $f(x) := \sum_{k=0}^{\infty} f_k(x)$ and for all integers $p \geq 1$ we have

$$
\left| \sum_{k=p}^{\infty} f_k(x) \right| \leq \sum_{k=p}^{\infty} |f_k(x)|, \tag{2.12}
$$

hence

$$
|f(x)| \leq \sum_{k=0}^{\infty} |f_k(x)| = g(x)
$$

for a.e. $x \in E$. This yields $f \in \mathcal{L}^1(E)$. Integrating (2.12) we also infer

$$
\int_E \left| f(x) - \sum_{k=0}^{p-1} f_k(x) \right| dx = \int_E \left| \sum_{k=p}^{\infty} f_k(x) \right| dx \leq \int_E \sum_{k=p}^{\infty} |f_k(x)| \, dx = \sum_{k=p}^{\infty} \int_E |f_k(x)| \, dx,
$$

hence the first part of the claim, when p tends to infinity. The second part easily follows as

$$
\left| \int_E f(x) \, dx - \sum_{k=0}^{p-1} \int_E f_k(x) \, dx \right| \leq \int_E \left| f(x) - \sum_{k=0}^{p-1} f_k(x) \right| dx \to 0 \qquad \text{as } p \to \infty.
$$

$\square$

c. Absolute continuity of the integral

2.35 Theorem (Absolute continuity of the integral). *Suppose $f \in \mathcal{L}^1(E)$. Then for every $\epsilon > 0$ there exists $\delta > 0$ such that for every measurable subset $F \subset E$ with $|F| < \delta$ we have $\int_F |f| \, dx < \epsilon$. Equivalently*

$$\int_E f(x) \, dx \to 0 \qquad as \ |E| \to 0.$$

Proof. Let $\epsilon > 0$. We set

$$f_k(x) = \begin{cases} k & \text{if } f(x) > k, \\ f(x) & \text{if } -k \le f(x) \le k, \\ -k & \text{if } f(x < -k). \end{cases}$$

Trivially $|f_k(x) - f(x)| \to 0$ for every $x \in E$ as $k \to \infty$, and $|f_k(x) - f(x)| \le 2|f(x)| \in \mathcal{L}^1(E)$; the theorem of dominated convergence, Theorem 2.32, then yields that there exists $N = N_\epsilon$ such that

$$\int_E |f(x) - f_N(x)| \, dx < \epsilon/2.$$

We now choose $\delta := \epsilon/(2N)$; clearly for any $F \subset E$ with $|F| \le \delta$ we find

$$\int_F |f_N(x)| \, dx \le N |F| \le N \frac{\epsilon}{N} = \epsilon/2$$

hence

$$\int_F |f(x)| \, dx \le \int_F |f_N(x)| \, dx + \int_E |f - f_N| \, dx \le \frac{\epsilon}{2} + \frac{\epsilon}{2} = \epsilon.$$

$\square$

2.36 ¶. Let f be summable in $\mathbb{R}^n$. Show that the function

$$F(x, r) := \int_{B(x,r)} f(t) \, dt, \qquad x \in \mathbb{R}^n, \ r \ge 0,$$

is continuous on $\mathbb{R}^n \times [0, +\infty[$.

d. Differentiation under the integral sign

Let E be a measurable set in $\mathbb{R}^n$ and let A be an open set in $\mathbb{R}^k$. If $f(t, x)$ is a function defined in $A \times E$ and integrable on E for each fixed $t \in A$, we may consider the function

$$F(t) := \int_E f(t, x) \, dx, \qquad t \in A.$$

2.37 Proposition. *Let $A \subset \mathbb{R}^k$ be open and $E \subset \mathbb{R}^n$ measurable. If $f : A \times E \to \mathbb{R}$ is such that*

(i) *for a.e. $x \in E$ the function $t \to f(t, x)$ is continuous on A,*
(ii) *$\forall t \in A$ the function $x \to f(t, x)$ is summable on E,*
(iii) *there exists $\phi \in \mathcal{L}^1(E)$ such that*

$$|f(t, x)| \le \phi(x) \qquad for \ all \ t \in A \ and \ a.e. \ x \in E, \qquad (2.13)$$

then the function

$$F(t) := \int_E f(t, x)\, dx, \qquad t \in A,$$

is continuous on A.

Proof. Let $t_0 \in A$ and let $\{t_k\}$ be a sequence in A converging to t_0. If $g_k(x) := f(t_k, x)$, then $g_k(x) \to f(t_0, x)$ as $k \to \infty$ for a.e. $x \in E$ and $|g_k(x)| \le \phi(x) \in \mathcal{L}^1(E)$ for all k and a.e. $x \in E$. The dominated convergence theorem, Theorem 2.32, then yields

$$F(t_k) = \int_E f(t_k, x)\, dx = \int_E g_k(x)\, dx \to \int_E f(t_0, x)\, dx = F(t_0),$$

i.e., the conclusion since the point t_0 and the sequence $\{t_k\}$ were arbitrary. $\square$

2.38 ¶. Notice the following:

(i) The hypotheses of Proposition 2.37 hold if A and E are bounded domains and $f \in C^0(\overline{A \times E})$.

(ii) Consider the family of functions $x \to f_t(x) := f(t, x)$ when t varies in A. The estimate (2.13) amounts to the summability of the envelope $h(x) := \sup_{t \in A} |f_t(x)|$ of the family $\{|f_t(x)|\}_{t \in A}$.

(iii) The assumption (2.13) cannot be omitted. Indeed, if

$$f(t, x) = \begin{cases} \frac{|t|-|x|}{t^2} & \text{if } |x| < t, \\ 0 & \text{if } |x| \ge t, \end{cases}$$

we have $F(t) = 1$ for $t \ne 0$ and $F(0) = 0$.

The following claim is a simple extension of Proposition 2.37.

2.39 Proposition. *Let* $A \subset \mathbb{R}^k$ *be open and let* $f : A \times]c, d[\to \mathbb{R}$ *be a function such that*

(i) $x \to f(t, x)$ *is summable for all* $t \in A$,

(ii) $t \to f(t, x)$ *is continuous on* A *for a.e.* x,

(iii) *there exists* $\phi \in L^1(]c, d[)$ *such that* $|f(t, x)| \le \phi(x)$ *for all* $t \in A$ *and a.e.* $x \in]c, d[$,

Then the function $F : A \times]c, d[\times]c, d[\to \mathbb{R}$ *defined by*

$$F(t, r, s) := \int_r^s f(t, x)\, dx$$

is continuous on $A \times]c, d[\times]c, d[$.

Proof. Let $t, t_0 \in A$ and $r, s, r_0, s_0 \in]c, d[$. According to Proposition 2.37 we have

$$F(t, r_0, s_0) - F(t_0, r_0, s_0) = o(1) \qquad \text{as } t \to t_0$$

while

$$|F(t, r, s) - F(t, r_0, s_0)| \le \left| \int_{r_0}^r |f(t, x)|\, dx \right| + \left| \int_{s_0}^s |f(t, x)|\, dx \right|$$

$$\le \left| \int_{r_0}^r \phi(x)\, dx \right| = \left| \int_{s_0}^s \phi(x)\, dx \right| = o(1),$$

uniformly in t as $r \to r_0$ and $s \to s_0$ by the absolute continuity of the integral, Theorem 2.35. Therefore we conclude

$$|F(t,r,s) - F(t_0, r_0, s_0)|$$
$$\leq |F(t,r,s) - F(t,r_0,s_0)| + |F(t,r_0,s_0) - F(t_0,r_0,s_0)| \to 0$$

as $(t,r,s) \to (t_0, r_0, s_0)$.   □

Now let us state the theorem of derivation under the integral sign.

2.40 Theorem. *Let $A \subset \mathbb{R}^k$ be open and $E \subset \mathbb{R}^n$ be measurable. Denote by $t = (t_1, t_2, \ldots, t_k)$ and $x = (x_1, x_2, \ldots, x_n)$ the coordinates in A and E respectively. Suppose that $f : A \times E \to \mathbb{R}$, $f = f(t,x)$, satisfies the following:*

(i) $x \to f(t,x)$ *is $\mathcal{L}^n$-summable on E for all $t \in A$,*
(ii) f *has a partial derivative in the variable t_j at (t,x) for all t and for a.e. $x \in E$,*
(iii) *there exists $\phi \in \mathcal{L}^1(E)$ such that*

$$\left| \frac{\partial f}{\partial t_j}(t,x) \right| \leq \phi(x) \qquad \text{for all } t \in A \text{ and a.e. } x \in E. \tag{2.14}$$

Then the function

$$F(t) := \int_E f(t,x)\,dx, \qquad t \in A,$$

has a partial derivative with respect to t_j at t for all $t \in A$ and

$$\frac{\partial F}{\partial t_j}(t) = \int_E \frac{\partial f}{\partial t_j}(t,x)\,dx \qquad \forall t \in A.$$

Proof. Let $t_0 \in A$ and let $t_k \to t_0$. Since A is open, we assume without loss of generality that $t_k \in B(t_0, \delta)$ for some $\delta > 0$. We have

$$\frac{F(t_k) - F(t_0)}{t_k - t_0} = \int_E \frac{f(t_k, x) - f(t_0, x)}{t_k - t_0}\,dx.$$

Also

$$\frac{f(t_k, x) - f(t_0, x)}{t_k - t_0} \to \frac{\partial f}{\partial t_j}(t_0, x) \qquad \text{as } k \to \infty$$

for a.e. $x \in E$, thus $\frac{\partial f}{\partial t_j}(t_0, x)$ is measurable on E. Applying Lagrange's theorem and the assumption (iii), we find $\xi_k \in A$ such that

$$\left| \frac{f(t_k, x) - f(t_0, x)}{t_k - t_0} \right| = \left| \frac{\partial f}{\partial t}(\xi_k, x) \right| \leq \phi(x)$$

for all k and for a.e. $x \in E$. Therefore, the dominated convergence theorem yields that $x \to \frac{\partial f}{\partial t}(t_0, x)$ is summable on E and that

$$\frac{F(t_k) - F(t_0)}{t_k - t_0} = \int_E \frac{f(t_k, x) - f(t_0, x)}{t_k - t_0}\,dx \to \int_E \frac{\partial f}{\partial t}(t_0, x)\,dx,$$

i.e., the conclusion, since the point t_0 and the sequence $\{t_k\}$ were arbitrary.   □

2.41 Corollary. *Let $A \subset \mathbb{R}^k$ be open and let $f : A \times]c, d[\to \mathbb{R}$ be a continuous function such that*

(i) *for all $t \in A$, $x \to f(t, x)$ is summable on $]c, d[$,*
(ii) *$t \to f(t, x)$ is of class $C^1(A)$ for a.e. $x \in]c, d[$,*
(iii) *there exists $\phi \in L^1(]c, d[)$ such that*

$$\sum_{j=1}^{k} \left| \frac{\partial f}{\partial t_j}(t, x) \right| \le g(x) \qquad \text{for all } t \text{ and a.e. } x \in]c, d[,$$

Then the function

$$F(t, r, s) := \int_r^s f(t, x) \, dx$$

is of class $C^1(A \times]c, d[\times]c, d[)$. In particular, if $\alpha, \beta \in C^1(A)$ take value in $]c, d[$, then the map

$$G(t) := \int_{\alpha(t)}^{\beta(t)} f(t, x) \, dx$$

is of class $C^1(A)$ and, for $j = 1, \ldots, k$ we have

$$\frac{\partial G}{\partial t_j}(t) = \int_{\alpha(t)}^{\beta(t)} \frac{\partial f}{\partial t_j}(t, x) \, dx + f(t, \beta(t)) \frac{\partial \beta}{\partial t_j}(t) - f(t, \alpha(t)) \frac{\partial \alpha}{\partial t_j}(t).$$

Proof. Theorem 2.40 yields the existence of the partial derivatives of $F(t, r, s)$ with respect to the t's variables, which are continuous by Proposition 2.39. On the other hand, by the fundamental theorem of calculus,

$$\frac{\partial F}{\partial s}(t, r, s) = f(t, s), \qquad \frac{\partial F}{\partial r} = -f(t, r)$$

that are continuous by assumptions. Thus $F(t, r, s)$ is of class $C^1(A \times]c, d[\times]c, d[)$. The chain rule yields the second part of the claim, since

$$G(t) = F(t, \alpha(t), \beta(t)), \qquad \forall t \in A.$$

$\square$

2.3 Mollifiers and Approximations

a. C^0-approximations and Lusin's theorem

From Theorem 2.3 and Tietze's extension theorem, see [GM3], we readily infer the following.

2.42 Theorem (Lusin). *Let $f : E \subset \mathbb{R}^n \to \mathbb{R}$ be measurable on E. For every $\epsilon > 0$ there exists a continuous function $g_\epsilon : \mathbb{R}^n \to \mathbb{R}$ such that*

$$\mathcal{L}^n\left(\left\{x \in E \,\middle|\, f(x) \neq g_\epsilon(x)\right\}\right) < \epsilon \qquad and \qquad ||g||_{\infty,E} \leq ||f||_{\infty,E}.$$

Moreover, if $f = 0$ outside an open set Ω of finite measure, then for every $\epsilon > 0$ there exists a function $g \in C_c^0(\Omega)$ such that

$$|\{x \in \Omega \,|\, f(x) \neq g(x)\}| < \epsilon, \qquad and \qquad ||g||_{\infty,\Omega} \leq ||f||_{\infty,\Omega}.$$

Proof. Theorem 2.3 yields a closed set $C \subset E$ such that $f_{|C}$ is continuous and $\mathcal{L}^n(E \setminus C) < \epsilon$. By Tietze's theorem $f_{|C}$ admits a continuous extension $g : \mathbb{R}^n \to \mathbb{R}$ with

$$||g||_{\infty,\mathbb{R}^n} \leq \sup_{x \in C} |f(x)| = ||f||_{\infty,C} \leq ||f||_{\infty,E}$$

and, since $\{x \in E \,|\, f(x) \neq g(x)\} \subset E \setminus C$, we have

$$\left|\left\{x \in E \,\middle|\, f(x) \neq g(x)\right\}\right| < \epsilon.$$

The second part of the claim can be proved similarly. Since Ω has finite measure, Lusin's theorem, Theorem 2.3, yields a compact set $K \subset\subset \Omega$ such that $|\Omega \setminus K| < \epsilon$ and $f_{|K}$ is continuous. If $\epsilon_0 > 0$ is such that $K \subset\subset \Omega_{\epsilon_0}$, where

$$\Omega_{\epsilon_0} := \left\{x \in \Omega \,\middle|\, \text{dist}\,(x, \partial\Omega) > \epsilon_0\right\},$$

the function $\overline{f} : K \cup \Omega_{\epsilon_0}^c \to \mathbb{R}$ defined by

$$\overline{f}(x) = \begin{cases} f(x) & x \in K, \\ 0 & se\ x \in \Omega_{\epsilon_0}^c \end{cases}$$

is continuous on the closed set $K \cup \Omega_{\epsilon_0}^c$, hence by Tietze's extension theorem admits a continuous extension to the whole of $\mathbb{R}^n$ with

$$||g||_{\infty,\mathbb{R}^n} \leq ||\overline{f}||_{\infty,\Omega_{\epsilon_0}^c \cup K} \leq ||f||_{\infty,\Omega}.$$

Clearly $g \in C_c^0(\Omega)$ and, since $\{x \in \Omega \,|\, \overline{f}(x) \neq g(x)\} \subset \Omega \setminus K$, we conclude that

$$\left|\left\{x \in A \,\middle|\, f(x) \neq g(x)\right\}\right| < \epsilon.$$

$\square$

As a consequence we find the following.

2.43 Theorem. *Let Ω be an open set in $\mathbb{R}^n$ and let $f : \Omega \to \mathbb{R}$ be summable on Ω. There exists a sequence $\{\varphi_n\}$ of functions of class $C_c^0(\Omega)$ such that*

$$\int_\Omega |f(x) - \varphi_n(x)|\, dx \to 0 \qquad as\ n \to \infty.$$

Proof. It suffices to show that for any $\epsilon > 0$, there exists $g \in C_c^0(\Omega)$ such that $\int_\Omega |f-g| < 2\epsilon$. Given $\epsilon > 0$ we choose N large so that, by setting

$$
f_N(x) := \begin{cases} N & \text{if } f(x) > N \text{ and } |x| \le N, \\ f(x) & \text{if } |f(x)| \le N \text{ and } |x| \le N, \\ -N & \text{if } f(x) < -N \text{ and } |x| \le N, \\ 0 & \text{if } |x| > N, \end{cases}
$$

we have $\int_\Omega |f - f_N|\, dx < \epsilon$. This can be done since $\int_\Omega |f - f_N|\, dx \to 0$ as $N \to \infty$ by the dominated convergence theorem.

According to Theorem 2.42 there exists a function $g \in C_c^0(\Omega)$ such that

$$
||g||_\infty \le ||f_N||_\infty \le N \qquad \text{and} \qquad |\{x \in \Omega \,|\, g(x) \ne f_N(x)\}| < \frac{\epsilon}{2N}.
$$

Consequently we find

$$
\int_\Omega |f - g|\, dx \le \int_\Omega |f - f_N|\, dx + \int_\Omega |f_N - g|\, dx \le \epsilon + 2N \frac{\epsilon}{2N} = 2\epsilon.
$$

$\square$

2.44 Proposition (Mean continuity). *Let $f \in \mathcal{L}^1(\mathbb{R}^n)$. Then*

$$
\int_{\mathbb{R}^n} |f(x + h) - f(x)|\, dx \to 0 \qquad \text{as } h \to 0.
$$

Proof. (i) If $f \in C_c^0(\mathbb{R}^n)$, spt $f \subset B(0, R)$, and $|h| < 1$, we have

$$
|f(x + h) - f(x)| \to 0 \qquad \text{for all } x \in E
$$
$$
|f(x + h) - f(x)| \le 2||f||_\infty \chi_{B(0, R+1)}(x).
$$

Therefore, in this case, the claim follows from the dominated convergence theorem.

(ii) In the general case we proceed by approximation. Given $\epsilon > 0$, there exists $g \in C_c^0(\mathbb{R}^n)$ such that $\int_{\mathbb{R}^n} |f - g|\, dx < \epsilon$. Since

$$
\int_{\mathbb{R}^n} |f(x + h) - f(x)|\, dx \le \int_{\mathbb{R}^n} |g(x + h) - g(x)|\, dx + 2\int_{\mathbb{R}^n} |f - g|\, dx
$$

when $h \to 0$, by (i) we conclude

$$
\limsup_{h \to 0} \int_{\mathbb{R}^n} |f(x + h) - f(x)|\, dx \le 2\epsilon.
$$

$\square$

b. Mollifying in $\mathbb{R}^n$

A function $k(x) \in C^\infty(\mathbb{R}^n)$ such that

$$
k(x) = k(-x), \quad k(x) \ge 0, \quad k(x) = 0 \text{ if } |x| \ge 1 \quad \text{and} \quad \int_{\mathbb{R}^n} k(x)\, dx = 1,
$$

is called a *mollifying* (or *regularizing*) *kernel*. The family

$$
k_\epsilon(x) := \epsilon^{-n} k\left(\frac{x}{\epsilon}\right), \qquad \epsilon > 0,
$$

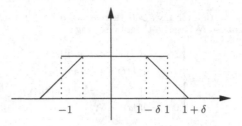

Figure 2.8. The function $\int_{\mathbb{R}} \chi_{[-1,1]}(y)\chi_{[-\delta,\delta]}(x-y)\,dx$, $\delta \leq 1$.

is the family of *mollifiers* generated by k. Clearly $k_\epsilon(x) = k_\epsilon(-x)$, $k_\epsilon(x) \geq 0$ and $k_\epsilon(x) = 0$ outside $B(0,\epsilon)$. Moreover, by the change of variables $y = x/\epsilon$,

$$\int_{\mathbb{R}^n} k_\epsilon(y)\,dy = 1 \qquad \forall \epsilon > 0.$$

2.45 Example. The function $\varphi(x) := g(|x|)$, $x \in \mathbb{R}^n$, where

$$\varphi(x) := \begin{cases} \exp\left(-\dfrac{1}{1-|x|^2}\right) & \text{if } |x| < 1, \\ 0 & \text{otherwise} \end{cases}$$

is symmetric, nonnegative, of class $C^\infty(B(0,1))$, and nonzero exactly on $B(0,1)$. Consequently, if

$$C := \int_{\mathbb{R}^n} \varphi(x)\,dx,$$

the function $k(x) := \frac{1}{C}\varphi(x)$ is a mollifying kernel in $\mathbb{R}^n$.

A function $f : \mathbb{R}^n \to \mathbb{R}$ is *locally summable* and we write $f \in \mathcal{L}^1_{loc}(\mathbb{R}^n)$, if $f \in \mathcal{L}^1(A)$ for any bounded set $A \subset \mathbb{R}^n$. If f is locally summable in $\mathbb{R}^n$, the function

$$f_\epsilon(x) = f * k_\epsilon(x) := \int_{\mathbb{R}^n} k_\epsilon(x-y)f(y)\,dy$$

is called the *ϵ-regularized*, or *ϵ-mollified*, of f, and the operators $S_\epsilon(f) := f_\epsilon$ are called the *regularizing operators* associated to k. Notice that

$$f * k_\epsilon(x) = \int_{\mathbb{R}^n} k_\epsilon(x-y)f(y)\,dy = \int_{B(x,\epsilon)} f(y)k_\epsilon(x-y)\,dy$$

$$= \int_{B(0,\epsilon)} f(x-z)k_\epsilon(z)\,dz$$

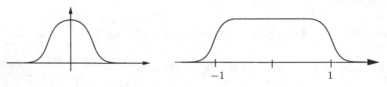

Figure 2.9. A convolution kernel and a regularized of $f(x) = \chi_{[-1,1]}(x)$.

since k_ϵ vanishes outside $B(0, \epsilon)$; the last inequality follows by changing y into $z := x - y$.

Finally, given $\Omega \subset \mathbb{R}^n$ and $\epsilon > 0$, we define

$$\Omega_\epsilon := \Big\{ x \in \Omega \,\Big|\, \mathrm{dist}\,(x, \partial\Omega) > \epsilon \Big\}, \qquad \Omega_{-\epsilon} := \Big\{ x \in \mathbb{R}^n \,\Big|\, \mathrm{dist}\,(x, \Omega) < \epsilon \Big\}.$$

Of course Ω_ϵ is nonempty for ϵ small if and only if Ω has nonempty interior.

2.46 Proposition. *Let $f : \mathbb{R}^n \to \mathbb{R}$ be a summable function. Then*

(i) *For any $\epsilon > 0$, the function $f_\epsilon(x) := f * k_\epsilon(x)$, $x \in \mathbb{R}^n$, is of class $C^\infty(\mathbb{R}^n)$. If f is constant in Ω, $f(x) = c$, then $f_\epsilon(x) = c$ in Ω_ϵ. In particular, if f vanishes outside Ω, then f_ϵ vanishes outside $\Omega_{-\epsilon}$.*

(ii) *We have*

$$\int_{\mathbb{R}^n} |f_\epsilon(x)|\, dx \leq \int_{\mathbb{R}^n} |f(x)|\, dx \qquad and \qquad \int_{\mathbb{R}^n} |f_\epsilon - f|\, dx \to 0.$$

(iii) *For every compact $K \subset \mathbb{R}^n$ we have*

$$\sup_{x \in K} |f_\epsilon(x)| \leq ||f||_{\infty, K_{-\epsilon}}.$$

Proof. (i) The theorem of differentiation under the integral sign yields that $f * k_\epsilon(x)$ has continuous partial derivatives and for $i = 1, \ldots, n$,

$$D_i(f * k_\epsilon)(x) = \int_{\mathbb{R}^n} f(y) D_i k_\epsilon(x - y)\, dy.$$

By induction we conclude that $f_\epsilon = f * k_\epsilon \in C^\infty(\mathbb{R}^n)$. The second part of the claim follows since k_ϵ has support in $\overline{B(0, \epsilon)}$.

(ii) Changing the order of integration, Fubini's theorem, we infer

$$\int_{\mathbb{R}^n} |g_\epsilon(x)|\, dx \leq \int_{\mathbb{R}^n} dx \int_{\mathbb{R}^n} |g(y)|\, k_\epsilon(x - y)\, dy = \int_{\mathbb{R}^n} |g(y)| \left(\int_{\mathbb{R}^n} k_\epsilon(x - y)\, dx \right) dy.$$

Since $\int_{\mathbb{R}^n} k_\epsilon(x - y)\, dx = 1 \; \forall y \in \mathbb{R}^n$, we conclude that

$$\int_{\mathbb{R}^n} |g_\epsilon(x)|\, dx \leq \int_{\mathbb{R}^n} |g(y)|\, dy.$$

To prove the second part of the claim, first, we notice that we have

$$|f_\epsilon(x) - f(x)| = \left| \int_{\mathbb{R}^n} f(y) k_\epsilon(x - y)\, dy - f(x) \right| = \left| \int_{\mathbb{R}^n} (f(y) - f(x)) k_\epsilon(x - y)\, dy \right|$$

$$\leq \int_{\mathbb{R}^n} k_\epsilon(z) |f(x - z) - f(x)|, \tag{2.15}$$

and, integrating we conclude

$$\int_{\mathbb{R}^n} |f_\epsilon(x) - f(x)|\, dx \leq \int_{\mathbb{R}^n} k_\epsilon(z) \left(\int_{\mathbb{R}^n} |f(x - z) - f(x)|\, dx \right) dz.$$

Given $\sigma > 0$, by Proposition 2.44 there exists $\epsilon_0 > 0$ such that

$$\int_{\mathbb{R}^n} |f(x - z) - f(x)|\, dx < \sigma$$

for all z with $|z| < \epsilon_0$. Therefore, for every $\epsilon < \epsilon_0$ we have

$$\int_{\mathbb{R}^n} k_\epsilon(z) \left(\int_{\mathbb{R}^n} |f(x-z) - f(x)| \, dx \right) dz \leq \sigma \int_{\mathbb{R}^n} k_\epsilon(z) \, dz = \sigma.$$

(iii) For all $x \in \mathbb{R}^n$ we have

$$|f_\epsilon(x)| \leq \int_{B(x,\epsilon)} |f(y)| k_\epsilon(x-y) \, dy \leq ||f||_{\infty, K_{-\epsilon}} \int_{B(x,\epsilon)} k_\epsilon(x-y) \, dy = ||f||_{\infty, K_{-\epsilon}}.$$

□

c. Mollifying in Ω

Let $\Omega \subset \mathbb{R}^n$ be an open set and let $f : \Omega \to \mathbb{R}$ be summable. We can extend f as a function $\overline{f}$ defined on all of $\mathbb{R}^n$ and summable on $\mathbb{R}^n$ in several ways, for example as

$$\overline{f}(x) = \begin{cases} f(x) & \text{if } x \in \Omega, \\ 0 & \text{if } x \in \Omega^c. \end{cases} \tag{2.16}$$

The mollified of $\overline{f}$, *a priori* depend on the value of $\overline{f}$ on Ω^c; however, for every $\epsilon > 0$, if $\Omega_\epsilon \neq \emptyset$, the value of the ϵ-mollified of $\overline{f}$ at point $x \in \Omega_\epsilon$ depends merely on f, since $f = \overline{f}$ on $B(x, \epsilon)$ and

$$\overline{f}_\epsilon(x) = \int_{B(x,\epsilon)} \overline{f}(y) k_\epsilon(x-y) \, dy = \int_\Omega \overline{f}(y) k_\epsilon(x-y) \, dy. \tag{2.17}$$

We therefore define the *ϵ-mollified*, or *ϵ-regularized*, of f in Ω by setting for $x \in \Omega_\epsilon$

$$f_\epsilon(x) := \int_\Omega f(y) k_\epsilon(x-y) \, dy = \int_{B(x,\epsilon)} f(y) k_\epsilon(x-y) \, dy$$

so that (2.17) writes also as $f_\epsilon(x) = \overline{f}_\epsilon(x) \ \forall x \in \Omega_\epsilon$.

2.47 Proposition. *Let Ω be an open set in $\mathbb{R}^n$ and let $f : \Omega \to \mathbb{R}$ be a summable function. For all $\epsilon > 0$, the ϵ-mollified $f_\epsilon := f * k_\epsilon$ is well defined in Ω_ϵ and we have the following.*

(i) *If $\widetilde{\Omega} \subset\subset \Omega$, then*

$$\int_{\widetilde{\Omega}} |f_\epsilon| \, dx \leq \int_\Omega |f| \, dx, \qquad ||f_\epsilon||_{\infty, \widetilde{\Omega}} \leq ||f||_{\infty, \Omega} \qquad \forall \epsilon < \text{dist}\,(\widetilde{\Omega}, \partial\Omega)$$

and

$$\int_{\widetilde{\Omega}} |f_\epsilon(x) - f(x)| \, dx \to 0 \qquad \text{as } \epsilon \to 0.$$

(ii) *If $f \in C^0(\Omega)$, then $f_\epsilon \to f$ uniformly on compact sets of Ω.*

(iii) *If $f \in C^k(\Omega)$, then for every α, $|\alpha| \le k$,*

$$D^\alpha(f * k_\epsilon)(x) = (D^\alpha f) * k_\epsilon(x) \qquad \forall x \in \Omega_\epsilon,$$

and $D^\alpha f_\epsilon \to D^\alpha f$ uniformly on compact sets of Ω.
(iv) *If $f \in Lip(\Omega)$, then f_ϵ is Lipschitz-continuous in Ω_ϵ and*

$$\sup_{x,y \in \Omega_\epsilon} \frac{|f_\epsilon(x) - f_\epsilon(y)|}{|x - y|} \le \mathrm{Lip}\,(f, \Omega).$$

(v) *If $\varphi : E \to \mathbb{R}$, $||\varphi||_{\infty,E} < +\infty$ and spt $\varphi \subset \Omega_{2\epsilon}$, then $f\varphi_\epsilon$ and $f_\epsilon \varphi$ are summable on Ω and*

$$\int_\Omega f(x)\varphi_\epsilon(x)\,dx = \int_\Omega f_\epsilon(x)\varphi(x)\,dx.$$

Proof. Let $\overline{f}$ be as in (2.16).

(i) Trivially, it follows from Proposition 2.46.

(ii) Let K be compact and let $\epsilon_0 := \mathrm{dist}\,(K, \partial\Omega)$. The set $\overline{K_{\epsilon_0/2}}$ is again a compact in Ω, and using (2.15) we infer

$$|f_\epsilon(x) - f(x)| = |\overline{f}_\epsilon(x) - \overline{f}(x)| \le \sup_{z \in B(0,\epsilon)} |\overline{f}(x - z) - \overline{f}(x)|$$
$$= \sup_{z \in B(0,\epsilon)} |f(x - z) - f(x)| \le \sup_{\substack{x \in K, y \in K_{\epsilon_0/2} \\ |x-y| < \epsilon}} |f(y) - f(x)| \qquad (2.18)$$

for all $\epsilon < \epsilon_0/2$. The uniform continuity of f on $\overline{K_{\epsilon_0/2}}$ yields, for $\sigma > 0$, a $\delta > 0$ such that $|f(x) - f(y)| < \sigma$ if $x, y \in \overline{K_{\epsilon_0/2}}$ and $|x - y| < \delta$. Therefore we find

$$|f_\epsilon(x) - f(x)| \le \sigma \qquad \forall \epsilon \le \min(\delta, \epsilon_0/2) \text{ and } \forall x \in K,$$

i.e., $f_\epsilon \to f$ uniformly on K.

(iii) Changing variables, $z = x - y$, we find

$$f_\epsilon(x) = \int_{B(0,\epsilon)} f(x - z)k_\epsilon(z)\,dz,$$

and differentiating under the integral sign,

$$D_i(f * k_\epsilon)(x) = \int_{B(0,\epsilon)} D_i f(x-z)k_\epsilon(z)\,dz = \int_{B(x,\epsilon)} D_i f(y)k_\epsilon(x-y)\,dy = (D_i f) * k_\epsilon(x).$$

From (ii) we then infer that $D_i f_\epsilon \to D_i f$ uniformly on the compact sets of Ω.

(iv) In fact, if $x, y \in \Omega_\epsilon$, then

$$|f_\epsilon(x) - f_\epsilon(y)| = \left| \int_{B(0,\epsilon)} (f(x - z) - f(y - z))k_\epsilon(z)\,dz \right|$$
$$\le \int_{B(0,\epsilon)} |f(x - z) - f(y - z)|k_\epsilon(z)\,dz \le \mathrm{Lip}\,(f, \Omega)|x - y|.$$

(v) Using (i) we find

$$\int_\Omega |f(x)|\,|\varphi_\epsilon(x)|\,dx = \int_{\Omega_\epsilon} |f(x)|\,|\varphi_\epsilon(x)|\,dx \le ||\varphi_\epsilon||_{\infty,\Omega_\epsilon} \int_\Omega |f|\,dx$$

$$\le ||\varphi||_{\infty,\Omega} \int_\Omega |f|\,dx < +\infty$$

and

$$\int_\Omega |f_\epsilon(x)|\,|\varphi(x)|\,dx = \int_{\Omega_{2\epsilon}} |f_\epsilon(x)|\,|\varphi(x)|\,dx \le ||\varphi||_{\infty,\Omega} \int_\Omega |f|\,dx < +\infty.$$

The function $f(x)\varphi(y)k_\epsilon(x-y)$, $(x,y) \in \Omega \times \Omega$, is summable; therefore, by changing the order of integration, we find

$$\int_\Omega f(x)\varphi_\epsilon(x)\,dx = \int_\Omega f(x)\left(\int_\Omega \varphi(y)k_\epsilon(x-y)dy\right)dx$$

$$= \int_\Omega dx \int_\Omega f(x)\varphi(y)k_\epsilon(x-y)\,dy$$

$$= \int_\Omega \varphi(y)\left(\int_\Omega f(x)k_\epsilon(x-y)\,dx\right)dy$$

$$= \int_\Omega \varphi(y)\left(\int_\Omega f(x)k_\epsilon(y-x)\,dx\right)dy$$

$$= \int_\Omega \varphi(y)f_\epsilon(y)\,dy.$$

□

2.4 Calculus of Integrals

The aim of this section is to familiarize the reader with the calculus of multiple integrals and with the theorem of derivation under the integral sign.

2.4.1 Calculus of multiple integrals

As we have seen, the calculus of a *double integral*, i.e., of the integral of a function of two independent variables, can be reduced to the successive calculus of two simple integrals, i.e., of a function of one variable, and this can be done in two different ways that are equivalent. Moreover, if it is useful, we may at each stage change variables. For the calculus of a *triple integral*, i.e., the integral of a function of three variables, there are 12 different ways of using the formula of reduction of integrals, *a priori* all praticable, and at each step we can change variables. In short, any strategy that uses all possible combinations of Fubini's theorem in one of its forms and of the theorem of change variables, even the most unlikely, is possible as long as it leads to the end.

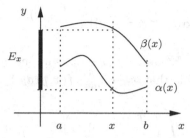

Figure 2.10. A normal subset in $\mathbb{R}^2$.

The aim of exercises is that of learning how to choose an optimal strategy for the calculus of integrals on the basis, for instance, of symmetries of the domain of integration and/or of the function to be integrated.

We recall that, for the formula of reduction of integrals to be valid, see Tonelli's theorem, Theorem 2.14, the summability of the involved functions is required.

2.48 ¶. Show that the assumption of integrability in Tonelli's theorem is essential. For instance, show that the following iterated integrals

$$\int_0^1 dy \int_1^\infty (e^{-xy} - 2e^{-2xy})\, dx, \qquad \int_1^\infty dx \int_0^1 (e^{-xy} - 2e^{-2xy})\, dy$$

both exist and are different.

We repeat that f is integrable on E in each of the following two cases.

(i) f is measurable on E and has constant sign, for instance if E is measurable and f is a.e. continuous on E and nonnegative. This applies in particular for $|f|$.

(ii) f is summable in E, $f \in \mathcal{L}^1(E)$, in particular if $|E| < +\infty$ and f is bounded on E; for instance if E is compact and f is continuous on E.

In other cases the measurability of f and the application of the reduction formula to $|f|$ or to f_+ and f_- (that are nonnegative) suffice to decide on the integrability of f.

a. Normal sets

2.49 Normal sets in $\mathbb{R}^2$. We say that a set $E \subset \mathbb{R}^2$ is *normal with respect to the y axis* if E can be written as

$$E := \{(x, y)\,|\, a < x < b, \alpha(x) < y < \beta(x)\}$$

where $\alpha, \beta :\,]a, b[\to \mathbb{R}$ are functions with $\alpha(x) < \beta(x)\ \forall x \in\,]a, b[$, see Figure 2.10. What makes normal sets useful is the fact that the slice of E over x is a possibly empty interval

$$E_x := \{y \in \mathbb{R} \mid \alpha(x) < y < \beta(x)\} = \begin{cases}]\alpha(x), \beta(x)[& \text{if } x \in A, \\ \emptyset & \text{otherwise.} \end{cases}$$

If E is measurable in $\mathbb{R}^2$ and $f : E \to \overline{\mathbb{R}}$ is an integrable function on E, Fubini's theorem yields that $x \to \int_{\alpha(x)}^{\beta(x)} f(x, y) \, dy$ is measurable on $]a, b[$ and

$$\iint_E f(x, y) \, dx \, dy = \int_{\mathbb{R}} \left(\int_{E_x} f(x, y) \, dy \right) dx = \int_A dx \int_{\alpha(x)}^{\beta(x)} f(x, y) \, dy.$$

Notice that one also proves that E is measurable if $\alpha, \beta : A \to \overline{\mathbb{R}}$ are measurable functions, for instance if α and β are continuous, see, e.g., [GM5].

2.50 Normal sets in $\mathbb{R}^n$. Similarly, we say that a set $E \subset \mathbb{R}^n$ is *normal* with respect to a coordinate axis, say x_n, if E can be written as

$$E := \left\{ x = (x', x_n) \in \mathbb{R}^{n-1} \times \mathbb{R} \,\middle|\, x' \in A, \ \alpha(x') < x_n < \beta(x') \right\}.$$

where $A \subset \mathbb{R}^{n-1}$ and $\alpha, \beta : A \to \overline{\mathbb{R}}$ are functions with $\alpha(x) < \beta(x) \ \forall x \in A$. The slice of E over $x' \in \mathbb{R}^{n-1}$ is a possibly empty interval

$$E_{x'} := \{t \in \mathbb{R} \mid \alpha(x') < t < \beta(x')\} = \begin{cases}]\alpha(x'), \beta(x')[& \text{if } x' \in A \\ \emptyset & \text{otherwise.} \end{cases}$$

If E is $\mathcal{L}^n$-measurable and $f : E \to \overline{\mathbb{R}}$ is an integrable function on E, Fubini's theorem then yields that $x' \to \int_{\alpha(x')}^{\beta(x')} f(x', t) \, dt$ is measurable on A and

$$\int_E f(x) \, dx = \int_{\mathbb{R}^{n-1}} dx' \left(\int_{E'_x} f(x', t) \, dt \right) = \int_A dx' \int_{\alpha(x')}^{\beta(x')} f(x', t) \, dt.$$

Notice that one also proves that E is $\mathcal{L}^n$-measurable if A is $\mathcal{L}^{n-1}$-measurable and α and β are measurable functions on A, see, e.g., [GM5]. A typical case would be the one in which A is an open or closed set in $\mathbb{R}^{n-1}$ and α, β are continuous functions on A.

2.51 Example. Compute $\int_T x^2 \, dxdy$ where T is the triangle in $\mathbb{R}^2$ of vertices $(0, 0)$, $(0, 2)$, and $(1, 0)$.

The function x^2 is continuous and nonnegative, the domain T is compact, thus x^2 is summable on T so that we can use the reduction formulas. The domain T is normal both with respect to the x-axis and the y-axis, as

$$T = \left\{ 0 \le x \le 1, 0 \le y \le -2x + 2 \right\} = \left\{ 0 \le y \le 1, 0 \le x \le -y/2 + 1 \right\}.$$

Since the function to be integrated, x^2, depends only on the variable x, it is convenient to leave integration in x as the last, and look at T as a normal domain with respect to the x-axis to obtain

$$\iint_T x^2 \, dxdy = \int_0^1 dx \int_0^{-2x+2} x^2 \, dy = \int_0^1 x^2(-2x + 2) dx = \frac{1}{6}.$$

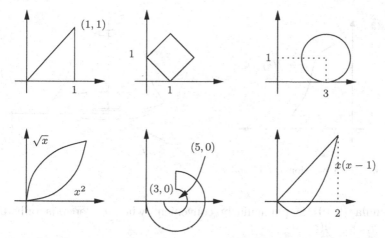

Figure 2.11. Some normal sets or union of normal sets in $\mathbb{R}^2$.

2.52 ¶. Integrate $f(x, y) = x^2$ on each of the domains $E \subset \mathbb{R}^2$ in Figure 2.11.

b. Rotational figures

2.53 Rotational solids. Let $f :]a, b[\subset \mathbb{R} \to \mathbb{R}_+$ be a nonnegative and measurable (for example, continuous) function. By rotating in $\mathbb{R}^3$ the graph of $x = f(z)$ around the z-axis we find the solid

$$E := \left\{ (x, y, z) \,\middle|\, x^2 + y^2 < f^2(z) \right\}.$$

The slice of E by the plane through $(0, 0, z)$ and orthogonal to the z-axis is

$$E_z := \begin{cases} \{(x, y) \in \mathbb{R}^2 \,|\, x^2 + y^2 < f^2(z)\} & \text{if } a < z < b, \\ \emptyset & \text{otherwise}, \end{cases}$$

i.e., E_z is the disk on the plane (x, y) of radius $f(z)$ around the origin if $z \in]a, b[$ and the empty set otherwise. If E is measurable and g is integrable on E, Fubini's theorem yields that $z \to \iint_{E_z} g(x, y, z)\, dz$ is measurable and

$$\int_E g(x, y, z)\, dxdydz = \int_a^b dz \iint_{E_z} g(x, y, z)\, dx\, dy. \qquad (2.19)$$

Notice that, since $x^2 + y^2 - f^2(z)$ is measurable on $\mathbb{R}^3$ if f is measurable on $]a, b[$, E is $\mathcal{L}^3$-measurable if $f :]a, b[\to \overline{\mathbb{R}}_+$ is measurable.

If $g = 1$ we get in particular that $z \to \mathcal{L}^2(E_z)$ is measurable and

$$\mathcal{L}^3(E) = \int_E 1\, dxdydz = \int_{-\infty}^{+\infty} \mathcal{L}^2(E_z)\, dz = \int_a^b \pi f^2(z)\, dz = \pi \int_a^b f^2(z)\, dz.$$

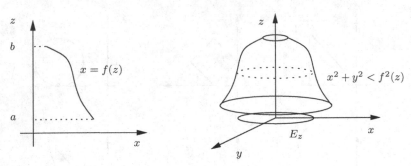

Figure 2.12. A rotational solid and E_z.

Formula (2.19) is particularly convenient when g depends only on z, $g(x, y, z) := g(z)$, as

$$\int_E g(z)\, dx dy dz = \int_a^b g(z) \iint_{E_z} 1\, dx dy$$

$$= \int_a^b g(z)\mathcal{L}^2(E_z)\, dz = \pi \int_a^b g(z)f^2(z)\, dz.$$

c. Changes of coordinates

2.54 Polar coordinates in $\mathbb{R}^2$. The map $\varphi(\rho, \theta) := (\rho \cos \theta, \rho \sin \theta)$ is of class $C^1(\mathbb{R}^2)$ with $|\det \mathbf{D}\varphi(\rho, \theta)| = \rho$ and is injective on the set $A :=$ $]0, +\infty[\times]0, 2\pi[$. Moreover $\mathcal{L}^2(\partial A) = 0$ and $\mathcal{L}^2(\varphi(\partial A)) = 0$. Therefore, for every measurable $E \subset \overline{A} = [0 \times +\infty[\times[0, 2\pi]$ and every integrable function f on $\varphi(E)$ we have

$$\int_{\varphi(E)} f(x, y)\, dx\, dy = \int_E f(\rho \cos \theta, \rho \sin \theta) \rho\, d\rho\, d\theta.$$

Since φ is injective on each interval $]0, +\infty[\times]a, a + 2\pi[$, $a \in \mathbb{R}$, the same conclusion holds for E measurable, $E \subset]0, \infty[\times[a, a + 2\pi[$.

2.55 Polar coordinates in $\mathbb{R}^3$. The map $\phi : \mathbb{R}^3 \to \mathbb{R}^3$ given by

$$\phi(\rho, \theta, \varphi) := \begin{cases} x = \rho \sin \varphi \cos \theta, \\ y = \rho \sin \varphi \sin \theta, \\ z = \rho \cos \varphi \end{cases}$$

is of class $C^1(\mathbb{R}^3)$ with $|\det \mathbf{D}\phi(\rho, \theta, \varphi)| = \rho^2 \sin \varphi$ and injective on $A :=]0, +\infty[\times]0, 2\pi[\times]0, \pi[$. Moreover $\mathcal{L}^3(\partial A) = 0$ and $\mathcal{L}^3(\phi(\partial A)) = 0$. Therefore, for every measurable set $E \subset \overline{A} = [0 \times +\infty[\times[0, 2\pi] \times [0, \pi]$ and every integrable function f on $\phi(E)$ we have

$$\int_{\phi(E)} f(x,y) \, dx \, dy \, dz$$

$$= \int_E f(\rho \sin\theta \cos\varphi, \rho \sin\theta \sin\varphi, \rho \cos\theta)\rho^2 \sin\varphi \, d\rho \, d\theta \, d\varphi.$$

2.56 Cylindrical coordinates in $\mathbb{R}^3$. The map $\phi : \mathbb{R}^3 \to \mathbb{R}^3$ given by

$$\begin{cases} x = \rho \cos\theta, \\ y = \rho \sin\theta, \\ z = z \end{cases}$$

is of class $C^1(\mathbb{R}^3)$ with $|\det \mathbf{D}\phi(\rho,\theta,z)| = \rho$ and injective on $A :=$ $]0,+\infty[\times]a, a+2\pi[\times\mathbb{R}$. Moreover, $\mathcal{L}^3(\partial A) = 0$ and $\mathcal{L}^3(\varphi(\partial A)) = 0$. Therefore, for every measurable set $E \subset \overline{A} = [0 \times +\infty[\times [a, a+2\pi] \times \mathbb{R}$ and every integrable function f on $\phi(E)$ we have

$$\int_{\phi(E)} f(x,y,z) \, dx \, dy \, dz = \int_E f(\rho \cos\theta, \rho \sin\theta, z)\rho \, d\rho \, d\theta \, dz.$$

2.57 Example. Compute $\iint_E \sqrt{x^2 + y^2} \, dx \, dy$ where $E \subset \mathbb{R}^2$ is the disk of radius 1 around $(1,0)$.

We notice that E is compact and $\sqrt{x^2 + y^2}$ is summable on E, therefore we can use both Fubini's and the change of variables theorems. The disk E has equation $(x - 1)^2 + y^2 \leq 1$ that is, $x^2 + y^2 - 2x \leq 0$, and in polar coordinates, we get

$$\rho \geq 0, \quad \theta \in [-\pi, \pi], \quad \rho^2 - 2\rho\cos\theta \leq 0,$$

i.e.,

$$\rho \geq 0, \quad \theta \in [-\pi/2, \pi/2], \quad \rho \leq 2\cos\theta.$$

If φ denotes the polar coordinates map and

$$F := \Big\{ (\rho,\theta) \, \Big| -\pi/2 \leq \theta \leq \pi/2, 0 \leq \rho \leq 2\cos\theta \Big\},$$

then the set F is contained in a strip of periodicity of φ and $E = \varphi(F)$. Therefore by a change of variables and taking into account that F is normal with respect to ρ, we find

$$\iint_E \sqrt{x^2 + y^2} \, dx \, dy = \iint_F \rho^2 \, d\rho \, d\theta = \int_{-\pi/2}^{\pi/2} d\theta \int_0^{2\cos\theta} \rho^2 \, d\rho$$

$$= \frac{8}{3} \int_{-\pi/2}^{\pi/2} \cos^3\theta \, d\theta = \frac{32}{9}.$$

2.58 Example (Rotational solids). Let $f : [a,b] \to \mathbb{R}_+$ be a measurable function and let E be the set obtained by rotating the subgraph of f in the plane (y,z) around the z-axis,

$$E := \Big\{ (x,y,z) \, \Big| \, a \leq z \leq b, \, x^2 + y^2 \leq f^2(z) \Big\}.$$

By parameterizing E with the cylindrical coordinates,

$$\phi(r,\theta,z) := \begin{cases} x = \rho \cos\theta, \\ y = \rho \sin\theta, \\ z = z \end{cases}$$

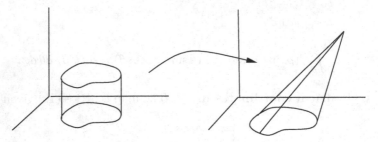

Figure 2.13. Conical coordinates.

so that $E \setminus \{(0,0,z)\}$ is the one-to-one image of the set

$$F := \Big\{ (\rho, \theta, z) \,\Big|\, 0 \le \theta < 2\pi, \ a \le z \le b, \ 0 < \rho \le f(z) \Big\}$$

and changing variables, we find that E is measurable and

$$\mathcal{L}^3(E) = \mathcal{L}^3(E \setminus \{(0,0,z)\}) = \int_F |\det \mathbf{D}\phi| \, d\rho d\theta dz$$

$$= \int_a^b dz \int_0^{2\pi} d\theta \int_0^{f(z)} \rho \, d\rho = \pi \int_a^b f^2(z) \, dz.$$

2.59 Example (Guldin's formula). Let $f, g : [a, b] \to \mathbb{R}_+$ be measurable functions with $g \le f$. The set

$$E := \Big\{ (x, y, z) \,\Big|\, z \in [a, b], \ g(z) \le \sqrt{x^2 + y^2} \le f(z) \Big\},$$

obtained by rotating the set

$$A := \Big\{ (x, y, z) \,\Big|\, x = 0, \ z \in [a, b], \ f(z) \le y \le g(z) \Big\}$$

around the z-axis, has as volume

$$\mathcal{L}^3(E) = \pi \int_a^b (g^2(z) - f^2(z)) \, dz.$$

The *center of mass*, or *barycenter*, of A (the density is assumed to be one) is the point $(\overline{x}, \overline{y}) \in \mathbb{R}^2$ given by

$$\overline{y} := \frac{1}{\mathcal{L}^2(A)} \int_A y \, dydz, \qquad \overline{z} := \frac{1}{\mathcal{L}^2(A)} \int_A z \, dydz.$$

Guldin's formula writes as: *The volume of the rotational solid E is the product of section A times the length of the circle of revolution of the barycenter of section A*, i.e.,

$$\mathcal{L}^3(E) = \mathcal{L}^2(A) \, 2\pi \, \overline{y}.$$

In fact, we have

$$2\pi \overline{y} \mathcal{L}^2(A) = 2\pi \int_A y \, dydz = \int_a^b dz \int_{f(z)}^{g(z)} y dy = \pi \int_a^b (g^2(z) - f^2(z)) \, dz = \mathcal{L}^3(E).$$

2.60 Conical coordinates in $\mathbb{R}^3$. Consider $\mathbb{R}^2$ as the coordinate plane $z = 0$ of $\mathbb{R}^3$, $\mathbb{R}^2 = \{(x, y, z) \,|\, z = 0\}$, let A be an open set in $\mathbb{R}^2$ and let $P_0 := (x_0, y_0, z_0)$. By definition, a point P is in the cone $C(P_0, A)$ of vertex P_0 and basis A if there are $(\alpha, \beta, 0) \in A$ and $t \in [0, 1]$ such that

$$\begin{pmatrix} x \\ y \\ z \end{pmatrix} = (1 - t) \begin{pmatrix} \alpha \\ \beta \\ 0 \end{pmatrix} + t \begin{pmatrix} x_0 \\ y_0 \\ z_0 \end{pmatrix}.$$

The function $\phi(\alpha, \beta, t) : \mathbb{R}^3 \to \mathbb{R}^3$ defined by

$$\begin{cases} x = (1 - t)\alpha + t\,x_0, \\ y = (1 - t)\beta + t\,y_0, \\ z = t\,z_0 \end{cases}$$

is a map of class $C^1(\mathbb{R}^3)$ with $\det \mathbf{D}\varphi(\alpha, \beta, t)| = (1 - t)^2$ and one-to-one from $A \times [0, 1[$ onto $C(P_0, A) \setminus \{P_0\}$. Consequently,

$$\int_{C(P_0, A)} f(x, y, z)\, dx\, dy\, dz = \int_{A \times [0, 1]} f(\phi(\alpha, \beta, t))(1 - t)^2\, d\alpha\, d\beta\, dt$$

$$= \int_0^1 (1 - t^2) \left(\int_A f(\phi(\alpha, \beta, t))\, d\alpha\, d\beta \right) dt.$$

In particular, if $f = 1$, we get the 3-dimensional measure of the cone $C(P_0, A)$,

$$\mathcal{L}^3(C(P_0, A)) = \frac{z_0}{3} \mathcal{L}^2(A).$$

2.61 Example. Suppose we want to compute the measure of

$$E := \left\{ (x, y) \in \mathbb{R}^2 \,\middle|\, 0 < x < y < 2x,\ 1 < xy < 2 \right\}.$$

We set $u = xy$ and $v = y/x$; then we have

$$E = \varphi(F)$$

where $\varphi(u, v) := (\sqrt{u/v}, \sqrt{uv})$ and

$$F := \left\{ (u, v) \in \mathbb{R}^2 \,\middle|\, 1 < u < 2,\ 1 < v < 2 \right\}.$$

Since $\det \mathbf{D}\varphi = \frac{1}{2v} > 0$ on F, we find

$$\mathcal{L}^2(E) = \mathcal{L}^2(\varphi(F)) = \int_F \frac{1}{2v}\, du\, dv = \frac{1}{2} \int_1^2 du \int_1^2 \frac{dv}{v} = \frac{1}{2} \log 2.$$

d. Measure of the n-dimensional ball

Let ω_n be the n-dimensional measure of the n-dimensional ball

$$\omega_n := \mathcal{L}^n(B^n(0,1)), \qquad B^n(0,1) := \{x \in \mathbb{R}^n \mid |x| \le 1\}.$$

2.62 Proposition. *We have*

$$\omega_{2k} = \frac{\pi^k}{k!}, \qquad \omega_{2k+1} = \frac{2^{k+1}\pi^k}{(2k+1)!!}. \tag{2.20}$$

Proof. We split the coordinates $x = (x_1, x_2, \ldots, x_n)$ of $\mathbb{R}^n$ as $x = (y,t)$ where $y := (x_1, x_2, \ldots, x_{n-1}) \in \mathbb{R}^{n-1}$ and $t = x_n \in \mathbb{R}$. The unit ball is then described as

$$B^n(0,1) := \{(y,t) \in \mathbb{R}^{n-1} \times \mathbb{R} \mid |y|^2 + t^2 < 1\}.$$

Now we slice $\mathbb{R}^{n-1} \times \mathbb{R}$ with $(n-1)$-planes perpendicular to the t-axis. The slice of $B^n(0,1)$ at the level t is then

$$E_t := \left\{y \in \mathbb{R}^{n-1} \mid |y|^2 < 1 - t^2\right\} = \begin{cases} B^{n-1}(0, \sqrt{1-t^2})) & \text{if } t \in [-1,1], \\ \emptyset & \text{if } |t| > 1. \end{cases}$$

By homogeneity $B^{n-1}(0, \sqrt{1-t^2}) = \omega_{n-1}(1-t^2)^{(n-1)/2}$, and, since $B^n(0,1)$ is open, Fubini's theorem yields

$$\omega_n = \int_{-\infty}^{+\infty} \mathcal{L}^{n-1}(E_t)\, dt = \omega_{n-1}\int_{-1}^{1}(1-t^2)^{\frac{n-1}{2}}\, dt = 2\omega_{n-1}\int_0^1 (1-t^2)^{\frac{n-1}{2}}\, dt.$$

Since $\int_0^1 (1-t^2)^{\frac{n-1}{2}}\, dt = \int_0^{\pi/2}\cos^n(t)\, dt$, and

$$\int_0^{\pi/2}\cos^n(t)\, dt = \frac{n-1}{n}\int_0^{\pi/2}\cos^{n-2}(t)\, dt,$$

we find, see [GM2],

$$\int_0^{\pi/2}\cos^{2k}(t)\, dt = \frac{(2k-1)!!}{(2k)!!}\frac{\pi}{2}, \qquad \int_0^{\pi/2}\cos^{2k+1}(t)\, dt = \frac{(2k)!!}{(2k+1)!!}.$$

As $\omega_1 = 2$ and $\omega_2 = \pi$, we get the result. $\qquad\square$

As a curiosity, notice that $\omega_n \to 0$ as $n \to \infty$. On the other hand the measure of the n-dimensional cube of side 2 that circumscribes the unit ball is 2^n and tends to infinity as $n \to \infty$.

The measure of the n-ball is tied to Euler's Γ function, see Example 2.67.

2.63 ¶. Let $a > 0$. Compute the measure of the n-dimensional set

$$E := \left\{x = (x_1, x_2, \ldots, x_n) \mid \sum_{i=1}^{n} x_i \le a,\ x_i \ge 0\ \forall i\right\}.$$

e. Isodiametric inequality

2.64 Proposition (Isodiametric inequality). *Let E be a measurable bounded set in $\mathbb{R}^n$. Then*

$$\mathcal{L}^n(E) \le \omega_n \left(\frac{\operatorname{diam} E}{2}\right)^n.$$

Notice that, whereas in $\mathbb{R}$ every set E is contained in an interval of radius half the diameter of E, this is not true anymore if $n \ge 2$: think of the equilateral triangle in $\mathbb{R}^2$. Of course every set E is contained in a ball of radius the diameter of E so that

$$\mathcal{L}^n(E) \le \omega_n (\operatorname{diam} E)^n. \tag{2.21}$$

But proving the isodiametric inequality requires some effort. However, it is trivial for special sets. For instance, if E is symmetric with respect to the origin, that is $x \in E$ iff $-x \in E$, then we have $2|x| = |x-(-x)| \le \operatorname{diam} E$, hence $E \subset B(0, \operatorname{diam} E/2)$ which yields the isodiametric inequality.

For generic sets, we shall use *Steiner's symmetrization method*. Given a direction $a \in S^{n-1}$, we denote by $P(a)$ the $(n-1)$-dimensional subspace of $\mathbb{R}^n$ orthogonal to a so that every $x \in \mathbb{R}^n$ writes uniquely as $x = y + ta$ with $y \in P(a)$ and $t \in \mathbb{R}$. For every $y \in P(a)$ we then set

$$E_{a,y} = \left\{t \in \mathbb{R} \,\middle|\, ta + y \in E\right\} \qquad e \qquad \ell_a(y) := \mathcal{L}^1(E_{a,y})$$

and define the *Steiner symmetrization* of E in the direction a by

$$S_a(E) := \left\{(y,t) \in \mathbb{R}^{n-1} \times \mathbb{R} \,\middle|\, |t| \le \frac{\ell_a(y)}{2}\right\}.$$

We have

2.65 Lemma. *If E is bounded and measurable, then*

(i) *$S_a(E)$ is measurable,*
(ii) *if E is symmetric with respect to a k-plane orthogonal to a, $1 \le k \le n-1$, then $S_a(E)$ has the same symmetry,*
(iii) *$|S_a(E)| = |E|$,*
(iv) *$\operatorname{diam}(S_a(E)) \le \operatorname{diam}(E)$.*

Proof. After a rotation that does not change the measurability, the measure, and the diameter of E, see (2.8), we can assume $a = (0,0,\ldots,1)$. Consequently $P(a) = \{x = (y,0), \ y \in \mathbb{R}^{n-1}\}$, every point $x \in \mathbb{R}^n$ writes as $x = (y,t)$, and $E_{a,y}$ is the slice of E over y. Fubini's theorem then yields that $E_{a,y}$ is measurable for a.e. $y \in \mathbb{R}^{n-1}$ and $y \to \ell_a(y) := \mathcal{L}^1(E_{a,y})$ is a measurable function, hence $S_a(E)$ is measurable, see Theorem 2.8, and

$$|E| = \int_{\mathbb{R}^{n-1}} \mathcal{L}^1(E_{a,y})\,dy = \int_{\mathbb{R}^{n-1}} \left(\int_{-\ell_a(y)/2}^{\ell_a(y)/2} 1\,dt\right) dy = |S_a(E)|.$$

A symmetry of E with respect to a k-plane orthogonal to $(0,0,\ldots,1)$ yields a similar symmetry for the function $\ell_a(y)$ hence of $S_a(E)$. Finally, from the elementary inequality

$$\mathcal{L}^1(I_1) + \mathcal{L}^1(I_2) \le \operatorname{diam}(I_1 \cup I_2)$$

for subsets of $\mathbb{R}$, we readily infer that $\operatorname{diam}(S_a(E)) \le \operatorname{diam}(E)$.   □

Proof of Proposition 2.64. Let $(e_1, e_2, \ldots, e_n)$ be the standard basis of $\mathbb{R}^n$ and let $E_1 := S_{e_1}(E)$, $E_2 := S_{e_2}(E_1), \ldots, E_n := S_{e_n}(E_{n-1})$. Applying iteratively Lemma 2.65, we deduce that

$$|E| = |E_1| = \ldots |E_n|, \qquad \mathrm{diam}\,(E_n) \leq \mathrm{diam}\,(E_{n-1}) \leq \cdots \leq \mathrm{diam}\,E,$$

and E_1 is symmetric with respect to the plane perpendicular to e_1, E_2 is symmetric with respect to the plane perpendicular to e_1 and $e_2, \ldots$, E_n is symmetric with respect to the coordinate axes, hence with respect to the origin. Therefore, E_n is contained in a ball of radius $\mathrm{diam}\,E_n/2$, thus concluding

$$|E| = |E_n| \leq \omega_n \left(\frac{\mathrm{diam}\,E_n}{2}\right)^n \leq \omega_n \left(\frac{\mathrm{diam}\,E}{2}\right)^n.$$

$\square$

f. Euler's Γ function

2.66 Example. We have

$$\int_{-\infty}^{+\infty} e^{-x^2}\,dx = \sqrt{\pi}. \tag{2.22}$$

In fact, since $e^{-x^2-y^2}$ is integrable on $\mathbb{R}^2$, using Fubini's theorem and passing to polar coordinates, we find

$$\left(\int_{-\infty}^{+\infty} e^{-x^2}\,dx\right)^2 = \int_{-\infty}^{\infty} e^{-x^2}\,dx \int_{-\infty}^{\infty} e^{-y^2}\,dy = \iint_{\mathbb{R}^2} e^{-x^2-y^2}\,dx\,dy$$

$$= \int_0^{2\pi} d\theta \int_0^{\infty} e^{-\rho^2}\rho\,d\rho = 2\pi\frac{1}{2}\int_0^{\infty} e^{-\sigma}\,d\sigma = \pi.$$

If we change variable in (2.22), we also get

$$\int_{-\infty}^{+\infty} e^{-\lambda x^2}\,dx = \sqrt{\frac{\pi}{\lambda}}, \qquad \lambda > 0. \tag{2.23}$$

2.67 Example (Euler's Γ function and the measure of $B^n(0,1)$). The function Γ was defined by Euler in 1729,

$$\Gamma(\alpha) := \int_0^{\infty} t^{\alpha-1}e^{-t}\,dt, \qquad \alpha > 0, \tag{2.24}$$

It is an important *special function* that surprisingly appears in many contexts.

Trivially $\Gamma(1) = 1$ and, on account of Example 2.66,

$$\Gamma(1/2) = 2\int_0^{\infty} e^{-s^2}\,ds = \sqrt{\pi}.$$

Integrating by parts we see that

$$\Gamma(\alpha+1) = \alpha\,\Gamma(\alpha) \qquad \forall \alpha > 0.$$

It follows by induction

$$\Gamma(n+1) = n!, \qquad \Gamma(n+1/2) = \frac{(2n-1)!!}{2^n}\Gamma(1/2) = \sqrt{\pi}\frac{(2n-1)!!}{2^n} = \sqrt{\pi}\frac{(2n)!}{4^n n!},$$

that, by comparison with (2.20) yields

$$\omega_n = \mathcal{L}^n(B^n(0,1)) = \frac{\pi^{n/2}}{\Gamma\left(\frac{n}{2}+1\right)} \qquad \forall n \geq 1. \tag{2.25}$$

We presented some of the properties of the Γ-function in [GM2]. Further properties of the Γ-function will be discussed in the following Example 2.68 and Section 2.4.3.

2.68 Example (Euler's Beta function). *Euler's Beta* function is defined by

$$B(p, q) := \int_0^1 x^{p-1}(1 - x)^{q-1} \, dx, \qquad p, q > 0.$$

Changing variables $y = 1 - x$, we see that

$$B(p, q) = B(q, p) \qquad \forall p, q > 0, \tag{2.26}$$

while, writing

$$x^{p-1} = \frac{x^{p+q-2}}{x^{q-1}}, \qquad x^{p+q-2} = D\frac{x^{p+q-1}}{p+q-1}$$

and integrating by parts, we find

$$B(p, q) = \frac{q-1}{p+q-1} B(p, q-1) \qquad \forall p > 0, \ q > 1, \tag{2.27}$$

and, because of the symmetry,

$$B(p, q) = \frac{p-1}{p+q-1} B(p-1, q) \qquad \forall p > 1, \ q > 0. \tag{2.28}$$

Changing variables, $x = z/(1 + z)$, we also find

$$B(p, q) = \int_0^\infty \frac{z^{p-1}}{(1+z)^{p+q}} \, dz. \tag{2.29}$$

We can compute the B-function in terms of the Γ-function as

$$B(p, q) = \frac{\Gamma(p)\Gamma(q)}{\Gamma(\mu + q)}, \qquad p, q > 0. \tag{2.30}$$

To prove this, we begin by noticing that if we change variables $x = \lambda z$, $\lambda > 0$, then

$$\frac{\Gamma(\alpha)}{\lambda^\alpha} = \lambda^{-\alpha} \int_0^\infty x^{\alpha-1} e^{-x} \, dx = \int_0^\infty z^{\alpha-1} e^{-\lambda z} \, dz, \qquad \alpha > 0 \tag{2.31}$$

Now, applying Fubini's theorem, changing variables $(\lambda, y) \to x = \lambda y, y = y$, and taking into account (2.31) and (2.29), we find

$$\Gamma(p)\Gamma(q) = \int_0^\infty \int_0^\infty x^{p-1} y^{q-1} e^{-(x+y)} \, dx \, dy$$

$$= \int_0^\infty \lambda^{p-1} \left(\int_0^\infty y^{p+q-1} e^{-(1+\lambda)y} \, dy \right) d\lambda$$

$$= \int_0^\infty \lambda^{p-1} \frac{\Gamma(p+q)}{(1+\lambda)^{p+q}} \, d\lambda = \Gamma(p+q)B(p, q).$$

The beta function is useful when computing several interesting integrals. For instance, if

$$I_\alpha := \int_{-1}^1 (1 - x^2)^\alpha \, dx, \qquad \alpha > -1,$$

and we change variables, we find

$$I_\alpha = \int_0^1 (1 - t)^\alpha t^{-1/2} \, dt = B(1/2, \alpha + 1) = \sqrt{\pi} \frac{\Gamma(\alpha + 1)}{\Gamma(\alpha + 3/2)}. \tag{2.32}$$

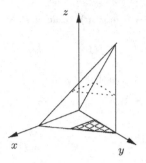

Figure 2.14. A tetrahedron.

2.69 Example. Let $p \geq 1$ and $\|x\|_p := \left(\sum_{i=1}^n |x_i|^p \right)^{1/p}$, $x \in \mathbb{R}^n$. We want to compute $\gamma_{n,p} := \mathcal{L}^n(\{x \mid \|x\|_p \leq 1\})$.

By slicing with planes orthogonal to a chosen coordinate axis, we find the following recursive relation for $\gamma_{n,p}$,

$$\gamma_{n,p} = \gamma_{n-1,p} \cdot 2 \int_0^1 (1 - t^p)^{(n-1)/p} \, dt.$$

By (2.32) we get

$$\frac{\gamma_{n,p}}{\gamma_{n-1,p}} = \frac{2}{p} B\left(\frac{n+p-1}{p}, \frac{1}{p} \right) = \frac{2}{p} \Gamma\left(\frac{1}{p} \right) \frac{\Gamma\left(\frac{n-1+p}{p} \right)}{\Gamma\left(\frac{n+p}{p} \right)},$$

hence, since $\gamma_1 = 2$,

$$\gamma_{n,p} = \gamma_{1,p} \frac{\gamma_{2,p}}{\gamma_{1,p}} \cdots \frac{\gamma_{n-1,p}}{\gamma_{n-2,p}} \frac{\gamma_{n,p}}{\gamma_{n-1,p}} = \gamma_{1,p} \prod_{i=1}^n \frac{\gamma_{i,p}}{\gamma_{i-1,p}}$$

$$= 2 \left(\frac{2}{p} \right)^{n-1} \Gamma\left(\frac{1}{p} \right)^{n-1} \prod_{i=2}^n \frac{\Gamma\left(\frac{i-1+p}{p} \right)}{\Gamma\left(\frac{i+p}{p} \right)} = 2 \left(\frac{2}{p} \right)^{n-1} \Gamma\left(\frac{1}{p} \right)^{n-1} \frac{\Gamma\left(\frac{p+1}{p} \right)}{\Gamma\left(\frac{n+p}{p} \right)}$$

$$= 2 \left(\frac{2}{p} \right)^{n-1} \Gamma\left(\frac{1}{p} \right)^{n-1} \frac{\frac{1}{p} \Gamma\left(\frac{1}{p} \right)}{\frac{n}{p} \Gamma\left(\frac{n}{p} \right)} = \frac{p}{n} \left(\frac{2}{p} \right)^n \frac{\Gamma(1/p)^n}{\Gamma(n/p)}.$$

g. Tetrahedrons

2.70 Example (Tetrahedrons, I). Consider the tetrahedron $T \subset \mathbb{R}^3$ of vertices $(0,0,0)$, $(1,0,0)$, $(0,1,0)$, and $(0,2,2)$, see Figure 2.14. Let us compute

$$\int_T \frac{z}{1+z} \, dz.$$

A face of the tetrahedron is on the plane $z = 0$ and, if we slice the tetrahedron with planes parallel to the basis, we get slices that are congruent to the basis; moreover, the function to be integrated depends only on the variable z. Therefore, we decide to slice with planes orthogonal to the z-axis. If T_z is the slice of T at the level z, we see that $T_z \neq \emptyset$ if and only if $0 \leq z \leq 2$. Since T is measurable and $z/(1 + z)$ is continuous on $\overline{T}$, Fubini's theorem yields

$$\int_T \frac{z}{1+z}\, dz = \int_0^2 \frac{z}{1+z}\mathcal{L}^2(T_z)\, dz.$$

Since, by Thales theorem, $\mathcal{L}^2(T_z) = \mathcal{L}^2(T_0)(\frac{2-z}{2})^2$, we conclude

$$\int_T \frac{z}{1+z}\, dz = \frac{1}{8}\int_0^2 \frac{z(2-z)^2}{1+z}\, dz = \dots$$

2.71 Example (Tetrahedrons, II). Consider the tethrahedron $T \subset \mathbb{R}^3$ of Example 2.70 and let us compute

$$\int_T \frac{xz}{1+z}\, dz$$

that is well defined since T is compact and the integrand is continuous on $\overline{T}$. We slice as in Example 2.70 and, with the same notation, we find

$$\int_T \frac{xz}{1+z}\, dz = \int_0^2 \frac{z}{1+z}\iint_{T_z} x\, dxdy.$$

Now we compute $\iint_{T_z} x\, dxdy$. The domain T_z is a triangle in $\mathbb{R}^2$ congruent to the basis of T. Its vertices $P(z)$, $Q(z)$, $R(z)$ are the projections on the (x,y)-plane of the intersections of the plane perpendicular to the z-axis through $(0,0,z)$ and the straight line respectively through $(0,2,2)$ and $(0,0,0)$, $(0,2,2)$ and $(1,0,0)$, and $(0,2,2)$ and $(0,1,0)$. Again by Thales theorem, the coordinates $x(z)$ and $y(z)$ of $P(z)$ depend linearly on z, i.e.,

$$\begin{cases} x(z) = mz + q, \\ x(0) = 0, x(2) = 0, \end{cases} \qquad \begin{cases} y(z) = mz + q, \\ y(0) = 0, y(2) = 2, \end{cases}$$

hence $P(z) = (0,z)$. Similarly, one computes $Q(z) = (1-z/2, z)$ and $R(z) = (0, 1+z/2)$. Points $P(z)$ and $R(z)$ have the same abscissa, hence the triangle T_z is normal with respect to the x-axis. Writing the equation for the straight line through $P(z)$ and $Q(z)$, and $R(z)$ and $Q(z)$ respectively,

$$\alpha_z(x) = z, \qquad \beta_z(x) = -(x + z/2 - 1) + z = 1 + z/2 - x,$$

we find

$$T_z := \Big\{ (x,y) \in \mathbb{R}^2 \,\Big|\, 0 \le x \le 1 - z/2,\ \alpha_z(x) \le y \le \beta_z(x) \Big\}$$

$$\int_{T_z} xdxdy = \int_o^{1-z/2} x\, dx \int_{\alpha_z(x)}^{\beta_z(x)} dy = \int_0^{1-z/2} x(1 - z/2 - x)\, dx = \frac{1}{6}\Big(1 - \frac{z}{2}\Big)^3.$$

In conclusion

$$\int_T \frac{z}{1+z}\, dx\, dy\, dz = \frac{1}{6}\int_0^2 \frac{z(1-z/2)^3}{1+z}\, dz = \dots$$

We may proceed differently. We regard the tetrahedron as a cone over a face and let the formula of change of variables operate the details. The map $\varphi(t,a,b): \mathbb{R}^3 \to \mathbb{R}^3$ given by

$$\begin{cases} x = ta + (1-t)\,0, \\ y = tb + (1-t)\,2, \\ z = t\cdot 0 + (1-t)\,2 \end{cases}$$

maps the prism $T_0 \times [0,1]$ onto the cone-tetrahedron T with basis T_0 defined by the vertices $(0,0,0)$, $(1,0,0)$, $(0,1,0)$ and vertex $(0,2,2)$. It is easily seen that φ is one-to-one from $T_0 \times]0,1]$ onto $T \setminus \{(0,2,2)\}$ and that $\det \mathbf{D}\varphi(t,a,b) = -2t^2$. Thus,

$$\int_T \frac{xz}{1+z}\, dx\, dy\, dz = \int_{T_0 \times [0,1]} \frac{2ta(1-t)}{1 + 2(1-t)} 2t^2\, da\, db\, dt = \int_0^1 \frac{4t^3(1-t)}{3-2t}\iint_{T_0} a\, da\, db$$

$$= \int_0^1 \frac{4t^3(1-t)}{3-2t}\int_0^1 a\Big(\int_0^{1-a} db\Big)\, da = \frac{2}{3}\int_0^1 \frac{t^3(1-t)}{3-2t}\, dt = \dots$$

Figure 2.15. Pierre Fatou (1878–1929) and Felix Hausdorff (1869–1942).

2.4.2 Monte Carlo method

Suppose we want to evaluate

$$f_Q := \int_Q f(x)\,dx, \qquad Q = [0,1]^n$$

for a function $f \in C^0(Q)$. We may use the analog of the one-dimensional *Simpson's rule*, see [GM2]. We subdivide the cube Q into k^n subcubes of side $1/k$, on each of those cubes we choose a point x_i and then compute $\frac{1}{k^n} \sum_{i=1}^{k^n} f(x_i)$; in particular, we need to compute f in k^n points: an enormous value already if $k = 100$ and $n = 4$.

During the Second World War, Enrico Fermi (1901–1954), John von Neumann (1903–1957), and Stanislaw Ulam (1909–1984) invented a probabilistic method, nowadays known as the *Monte Carlo method*. This method with probability close to 1 allows us to compute the value of the integral except for a small error by means of relatively few cubes.

Notice that Lebesgue's measure $\mathcal{L}^n$ on Q is a probability measure on Q and actually the equidistributed probability measure. Let $\{X_k\}$ be a sequence of points that are equidistributed and independently chosen on Q, i.e., a sequence on independent random variables on Q. If $f : Q \to \mathbb{R}$, then the *expectation* and the *variance* are defined respectively by

$$\mathbf{E}\left(f(X_j)\right) = \int_Q f(x)\,d\mathcal{L}^n(x) = f_Q, \qquad \mathrm{Var}\left(f(X_j)\right) = \int_Q |f(x) - f_Q|^2\,dx$$

for all integers j. Since the variables $\{X_j\}$ are independent

$$\mathrm{Var}\left(\sum_{j=1}^k f(X_k)\right) = \sum_{j=1}^k \mathrm{Var}\left(f(X_j)\right) = k \int_Q |f(x) - f_Q|^2\,dx \le 4kM^2$$

where $M := ||f||_\infty$. Hence

$$\int_{Q^k} \left(\frac{1}{k}\sum_{j=1}^k f(x_j) - f_Q\right)^2 dx_1 \ldots dx_k = \mathrm{Var}\left(\frac{1}{k}\sum_{j=1}^k f(X_j)\right) \le \frac{4}{k}||f||_\infty^2.$$

If $A \subset Q \times \cdots \times Q = Q^k$ is the event

$$A := \left\{ (X_1, \ldots, X_k) \,\Big|\, \Big| \frac{1}{k} \sum_{i=1}^{k} f(X_i) - f_Q \Big| > \epsilon \right\}$$

then Chebyshev's inequality yields

$$P(A_k) := \mathcal{L}^{nk}(A_k) \leq \frac{1}{\epsilon^2} \int_{Q^k} \left(\frac{1}{k} \sum_{j=1}^{k} f(x_j) - f_Q \right)^2 dx_1 \ldots dx_k \leq \frac{4M^2}{\epsilon^2 \, k},$$

i.e., the probability that, choosing randomly k equidistributed points $\{X_i\}$, the event that $\frac{1}{k} \sum_{i=1}^{k} f(X_i)$ has distance from f_Q more than ϵ has a probability to happen less than $4M^2/k\epsilon^2$. For instance, if $M \leq 1$ and we choose $k = 10^6$, in 99% of the cases we find an error less than 2%.

2.4.3 Differentiation under the integral sign

2.72 Example. Let us compute

$$F(t) := \int_0^{+\infty} \exp\left(-x^2 - t^2/x^2\right) dx, \qquad t \in \mathbb{R}.$$

It is easily seen that F is even, $F(0) = \int_0^\infty e^{-x^2/2}\, dx = \sqrt{\pi}/2$, see Example 2.66, and we have

$$|f(t, x)| \leq e^{-x^2} \qquad \forall t \in \mathbb{R}, \ \forall x \geq 0.$$

Therefore $F(t)$ is continuous in $\mathbb{R}$, see Proposition 2.37. Moreover, for $t > 0$ we have

$$\left| \frac{\partial f}{\partial t}(t, x) \right| = \frac{2e^{-x^2}}{t} \frac{t^2}{x^2} e^{-t^2/x^2} \leq \frac{2}{t} e^{-x^2} \sup_{\mathbb{R}_+} (se^{-s}) = \frac{2}{e\,t} e^{-x^2},$$

thus

$$\left| \frac{\partial f}{\partial t}(t, x) \right| \leq \frac{2}{e\,\epsilon} e^{-x^2} \in \mathcal{L}^1(\mathbb{R}_+)$$

for all $t > \epsilon > 0$ and $x \geq 0$. Theorem 2.40 then yields that $F(t)$ is differentiable for all $t > \epsilon$, and therefore for all $t > 0$, since ϵ is arbitrary, and

$$F'(t) = \int_0^\infty \frac{\partial f}{\partial t}(t, x)\, dx = -2t \int_0^\infty \frac{1}{x^2} \exp\left(-x^2 - t^2/x^2\right) dx = -2t\, F(t) \qquad \forall t > 0,$$

where the last equality follows by changing variables $y = t/x$. It follows

$$F(t) = F(0)e^{-2t}, \qquad t > 0,$$

i.e.,

$$F(t) = \frac{\sqrt{\pi}}{2} e^{-2|t|}.$$

2.73 Example. We shall write in terms of elementary functions the following oscillatory integral

$$g(\omega) = \int_{-\infty}^{+\infty} e^{-x^2/2} \cos(\omega x)\, dx, \qquad \omega \in \mathbb{R}.$$

As usual, it is convenient to use the complex notation. Since

$$\int_{-\infty}^{+\infty} e^{-x^2/2} \sin(\omega x)\, dx = 0,$$

we have

$$g(\omega) = \int_{-\infty}^{+\infty} e^{-x^2/2} e^{-i\omega x}\, dx.$$

Since

$$\left| \frac{\partial}{\partial \omega} \left(e^{-x^2/2} e^{-i\omega x} \right) \right| = \left| -ix e^{-x^2/2} e^{-i\omega x} \right| \le |x| e^{-x^2/2} \in \mathcal{L}^1(\mathbb{R}),$$

the function $g(\omega)$ is differentiable and

$$g'(\omega) = -i \int_{-\infty}^{+\infty} x e^{-x^2/2} e^{-i\omega x}\, dx.$$

Writing $-x e^{-x^2/2} = D(e^{-x^2/2})$ and integrating by parts, we find

$$g'(\omega) = -\omega\, g(\omega),$$

hence, by integration,

$$g(\omega) = g(0)\, e^{-\omega^2/2} = \sqrt{2\pi}\, e^{-\omega^2/2}.$$

Alternatively, we may also proceed as follows. Since

$$\cos(\omega x) = \sum_{n=0}^{\infty} (-1)^n \frac{(\omega x)^{2n}}{(2n)!}, \quad x \in \mathbb{R},$$

we consider the functions

$$f_n(x) := (-1)^n \frac{\omega^{2n} x^{2n}}{(2n)!} e^{-x^2/2}$$

and compute

$$(-1)^n \int_{-\infty}^{+\infty} f_n(x)\, dx = \frac{\omega^{2n}}{(2n)!}\, 2 \int_0^{\infty} e^{-x^2} x^{2n}\, dx = \text{(by changing variables } y = x^2)$$

$$= \sqrt{2}\, \omega^{2n} \frac{\Gamma(n+1/2)}{(2n)!} = \omega^{2n} \sqrt{2\pi}\, \frac{(2n-1)!!}{(2n)!}$$

$$= \sqrt{2\pi}\, \frac{(\omega^2/2)^n}{n!}.$$

We infer

$$\sum_{n=0}^{\infty} \int_{-\infty}^{+\infty} |f_n(x)|\, dx < +\infty$$

and, on account of Lebesgue's theorem,

$$\int_{-\infty}^{+\infty} e^{-x^2/2} \cos(\omega x)\, dx = \int_{-\infty}^{+\infty} \sum_{n=0}^{\infty} f_n(x)\, dx = \sum_{n=0}^{\infty} \int_{-\infty}^{+\infty} f_n(x)\, dx$$

$$= \sum_{n=0}^{\infty} (-1)^n \sqrt{2\pi}\, \frac{(\omega^2/2)^n}{n!} = \sqrt{2\pi} \sum_{n=0}^{\infty} \frac{(-\omega^2/2)^n}{n!}$$

$$= \sqrt{2\pi}\, e^{-\omega^2/2}.$$

2.74 Example (Derivatives of Γ). We already observed, see Examples 2.66 and 2.67, that for Euler's Γ-function

$$\Gamma(\alpha) = \int_0^\infty t^{\alpha-1} e^{-t}\, dt, \qquad \alpha > 0,$$

we have

$$\begin{cases} \Gamma(\alpha+1) = \alpha\Gamma(\alpha) & \forall \alpha > 0, \\ \Gamma(n+1) = n!, \\ \Gamma(1/2) = \sqrt{\pi}. \end{cases}$$

Moreover, we discussed some characteristic properties of Euler functions in [GM2]. Here we want to compute the derivatives of Γ.

We prove that Γ is of class C^∞ in its domain $E := \{\alpha \mid \alpha > 0\}$. Choose $\alpha_0 > 0$ and set $h(t) := \max(1, t^{\alpha_0/2-1}, t^{2\alpha_0-1})$, $t > 0$. For $k = 0, 1, \ldots$, the functions $h(t)|\log t|^k e^{-t}$ are summable on E and, for all $\alpha \in]\alpha_0/2, 2\alpha_0[$, we have

$$t^{\alpha-1} \leq h(t) \qquad \forall t > 0, \forall \alpha,\ \alpha_0/2 < \alpha < 2\alpha_0.$$

If follows for $f(\alpha, t) := t^{\alpha-1}e^{-t}$, $t > 0$, that

$$\left| \frac{\partial f}{\partial \alpha}(\alpha, t) \right| \leq h(t)\, |\log t|\, e^{-t}$$

and by induction

$$\left| \frac{\partial^k f}{\partial \alpha^k}(\alpha, t) \right| \leq h(t)|\log t|^k\, e^{-t}$$

for all $t > 0$ and for all $\alpha \in]\alpha_0/2, 2\alpha_0[$. Applying the theorem of differentiation under the integral sign, we conclude that Γ has derivatives of any order at α_0, and

$$\Gamma^{(k)}(\alpha_0) = \int_0^\infty t^{\alpha_0-1}(\log t)^k e^{-t}\, dt, \qquad (2.33)$$

consequently,

$$\Gamma'(\alpha) = \int_0^\infty t^{\alpha-1} \log t\, e^{-t}\, dt \qquad \forall \alpha > 0.$$

Since $\Gamma'(\alpha) > 0$ for $\alpha \geq 2$, Γ is increasing for $\alpha \geq 2$. Since $\Gamma(n) \to +\infty$ as $n \to \infty$, also $\Gamma(\alpha) \to +\infty$ as $\alpha \to +\infty$. On the other hand, from $\Gamma(\alpha+1) = \alpha\Gamma(\alpha)$ we infer that $\Gamma(u) \sim 1/u$ as $u \to 0^+$. Moreover,

$$\Gamma''(\alpha) = \int_0^\infty t^{\alpha-1}(\log t)^2 e^{-t}\, dt > 0 \qquad \forall \alpha > 0,$$

thus Γ is strictly convex on $[0, \infty[$. Since $\Gamma(1) = \Gamma(2) = 1$, we conclude that Γ has a unique minimum point and it is contained in the interval $]1, 2[$. Moreover, as $|\log t| \leq 1 + \log^2 t\ \forall t > 0$, we also get $(\Gamma')^2(x) \leq \Gamma(x)\Gamma''(x)$, that is, $\log \Gamma(x)$ is convex.

2.75 Example. We have

$$\Gamma(\alpha)\Gamma(1-\alpha) = \frac{\pi}{\sin \pi\alpha}, \qquad 0 < a < 1. \qquad (2.34)$$

In fact, in terms of the Beta function, see (2.29)(2.30), for $0 < \alpha < 1$ we have

$$\Gamma(\alpha)\Gamma(1-\alpha) = \Gamma(1)\, B(\alpha, 1-\alpha) = \int_0^\infty \int_0^\infty \frac{t^{\alpha+1}}{1+t}\, dt.$$

On the other hand, if $\alpha := (2m+1)/(2n)$, $n, m \in \mathbb{N}$, we find by changing variables $t = x^{2n}$

$$\int_0^\infty \frac{t^{\frac{2m+1}{2n}-1}}{(1+t)}\, dt = 2n \int_0^\infty \frac{x^{2m}}{1+x^{2n}}\, dx = \frac{\pi}{\sin\left(\frac{2m+1}{2n}\pi\right)},$$

see, e.g., [GM2, 5.36]. This yields (2.34) when $\alpha = (2m+1)/2n$ for some $n, m \in \mathbb{N}$. The claim now follows for all $\alpha \in]0, 1[$ since the numbers of type $(2m+1)/(2n)$ are dense in $[0, 1]$ and both functions on the left and on the right side of (2.34) are continuous.

2.5 Measure and Area

We are interested in computing not only volumes of n-dimensional objects in $\mathbb{R}^n$ but also the "k-dimensional *area*" of "k-dimensional surfaces" in $\mathbb{R}^n$, $k < n$, as for example the two-dimensional area of the graph of a function $f : \mathbb{R}^2 \to \mathbb{R}_+$.

This is a question that can be treated at various levels of difficulty finding formulas that apply to more or less general objects, or using measure theory to define the k-measure of a subset of $\mathbb{R}^n$, $k < n$. In fact, in contrast with the n-dimensional measure that is *essentially* unique (one can show that the $\mathcal{L}^n$ Lebesgue measure is the only measure that is invariant under rotations and translations, is homogeneous of degree n, and for which the unit cube has measure 1), there are several k-measures suited to measure subsets of $\mathbb{R}^n$, $k < n$: they are different on *nonregular* subsets but agree on "regular surfaces". Among these measures, the *Hausdorff k-dimensional measure* appears as the most suited in many contexts.

2.5.1 Hausdorff's measures

It is convenient to define Hausdorff s-dimensional measure $\mathcal{H}^s(E)$ of a set $E \subset \mathbb{R}^n$ also for noninteger $s \geq 0$. For $s \in \mathbb{R}$, $s \geq 0$, we set

$$\omega_s := \frac{\pi^{s/2}}{\Gamma(1 + s/2)},$$

recalling that, if s is an integer, then ω_s is the $\mathcal{L}^s$ measure of the s-dimensional ball $B(0,1) \subset \mathbb{R}^s$, $\omega_s = \mathcal{L}^s(B(0,1))$. For $E \subset \mathbb{R}^n$ and $\delta > 0$, we define

$$\mathcal{H}^s_\delta(E) := \inf\left\{\omega_s \sum_{j=1}^\infty \left(\frac{\operatorname{diam} E_j}{2}\right)^s \,\middle|\, E \subset \cup_j E_j, \ \operatorname{diam}(E_j) \leq \delta\right\}$$

and, since $\mathcal{H}^s_\delta$ is nondecreasing in $\delta > 0$, we set

$$\mathcal{H}^s(E) := \lim_{\delta \to 0} \mathcal{H}^s_\delta(E).$$

The set-function $\mathcal{H}^s : \mathcal{P}(X) \to \overline{\mathbb{R}}_+$ is by definition the (*exterior*) s-dimensional Hausdorff measure in $\mathbb{R}^n$. We then say that a set $E \subset \mathbb{R}^n$ is $\mathcal{H}^s$-measurable if E satisfies the *Carathéodory criterion* for measurability: for any set $A \subset \mathbb{R}^n$ we have

$$\mathcal{H}^s(A) = \mathcal{H}^s(A \cap E) + \mathcal{H}^s(A \setminus E).$$

Methods of *measure theory*, see [GM5], allow us to prove the following.

  (i) The class $\mathcal{M}$ of $\mathcal{H}^s$-measurable sets is a σ-algebra of sets and $\mathcal{H}^s$ is σ-addditive on $\mathcal{M}$.

(ii) Open and closed sets of $\mathbb{R}^n$ are $\mathcal{H}^s$-measurable.
(iii) In $\mathbb{R}^n$, $\mathcal{H}^n$ and the Lebesgue n-dimensional measure $\mathcal{L}^n$ agree.

Moreover, it is not difficult to prove the following.

(i) For $\delta > 0, \mathcal{H}^s_\delta(E) < +\infty$ for all bounded sets.
(ii) $\mathcal{H}^s$ is not necessarily finite on compact sets. For example, if $E \subset \mathbb{R}^n$ has a nonempty interior and $s < n$, then $\mathcal{H}^s(E) = +\infty$.
(iii) In the definition of $\mathcal{H}^s_\delta(E)$ we may replace the generic sets E_j with closed, or closed and convex sets, or with open sets without changing the definition of $\mathcal{H}^s$. However, we cannot replace the E_j's by balls. If we do it, for the new measure $\mathcal{H}^s_{\mathrm{sph}}(E)$ we have $\mathcal{H}^s_{\mathrm{sph}}(E) > \mathcal{H}^s(E)$ for some subsets $E \subset \mathbb{R}^n$.
(iv) $\mathcal{H}^0$ is the *counting measure*, $\mathcal{H}^0(E) = \#$ points of E.
(v) $\mathcal{H}^s$ is invariant under orthogonal transformations: if $E \subset \mathbb{R}^n$ and $\mathbf{R}^T\mathbf{R} = \mathrm{Id}$, then $\mathcal{H}^s(\mathbf{R}(E)) = \mathcal{H}^s(E)$.
(vi) $\mathcal{H}^s$ is positively homogeneous of degree s, i.e., for all $\lambda > 0$ and $E \subset \mathbb{R}^n$, we have $\mathcal{H}^s(\lambda E) = \lambda^s \mathcal{H}^s(E)$.
(vii) $\mathcal{H}^s = 0$ if $s > n$.
(viii) If $0 \le t < s \le n$, then $\mathcal{H}^s \le \mathcal{H}^t$. Moreover, $\mathcal{H}^s(E) > 0$ implies $\mathcal{H}^t(E) = +\infty$ and $\mathcal{H}^t(E) < \infty$ implies $\mathcal{H}^s(E) = 0$.
(ix) If $f : \mathbb{R}^n \to \mathbb{R}^k$ is Lipschitz-continuous, then $\forall 0 \le s \le n$ we have

$$\mathcal{H}^s(f(E)) \le (\mathrm{Lip}\, f)^s\, \mathcal{H}^s(E).$$

2.76 Remark. Notice the following:

(i) The claim in (vii) shows that $\mathcal{H}^s(E)$, $E \subset \mathbb{R}^n$, is finite and nonzero for at most one value of s, $0 \le s \le n$, which is called the *Hausdorff dimension of E*, defined in general as

$$\dim_{\mathcal{H}}(E) := \sup\left\{s \,\Big|\, \mathcal{H}^s(E) > 0\right\} = \sup\left\{s \,\Big|\, \mathcal{H}^s(E) = +\infty\right\}$$
$$= \inf\left\{s \,\Big|\, \mathcal{H}^s(E) < \infty\right\} = \inf\left\{s \,\Big|\, \mathcal{H}^s(E) = 0\right\}$$

where the equalities follow from (viii).
(ii) The estimate (ix) is useful to estimate from below the Hausdorff measure of a set. Estimates from above are usually obtained by estimating from above $\mathcal{H}^s_\delta$ by choosing suitable coverings of E with sets of diameter less than δ.

Finally, observe that we may construct an integral with respect to the Haussdorff measure with the same procedure we used to define the Lebesgue integral from Lebesgue's measure, compare [GM5]. From now on, for a given $\mathcal{H}^s$-measurable set of $\mathbb{R}^n$ and an $\mathcal{H}^s$-integrable map, the symbol

$$\int_E f(x)\, d\mathcal{H}^s(x)$$

is well-understood.

2.5.2 Area formula

We did not need any measure theory to define the length of a curve in $\mathbb{R}^n$. If $\gamma : [a, b] \to \mathbb{R}^n$ is a curve, for any subdivision $\sigma := \{t_j\}$, $a = t_0 < t_1 < \cdots < t_N = b$, of $[0, 1]$ we compute

$$\sum_{i=1}^{N} |\gamma(t_{i-1}) - \gamma(t_i)|$$

and define the length of γ as

$$L(\gamma) := \sup\left\{ \sum_{i=1}^{n} |\gamma(t_{i-1}) - \gamma(t_i)| \,\Big|\, a = t_0 < t_1 < \cdots < t_N = b \right\},$$

the supremum being taken over all possible subdivisions, see [GM2].

If we want to imitate the previous procedure to define the area of a C^1-image of an open set of $\mathbb{R}^2$ into $\mathbb{R}^3$, we may think of triangularizing the space of parameters and, associated to it, considering the polyhedral surface in $\mathbb{R}^3$ with triangular faces whose vertices are the images of the vertices of the triangulation of the space of parameters. Then, we may compute the area of these approximating polyhedral surfaces and define the area of the surface as the supremum of the areas of the inscribed polyhedral surfaces when the triangulation of the parameters varies. The following example due to Hermann Schwarz (1843–1921) shows how illusory it is to imagine being able to come to a reasonable definition of the area in this way.

2.77 Example (Schwarz). Consider the map $\varphi : [0, 2\pi[\times[0, 1] \to \mathbb{R}^3$ $\varphi(\theta, z) := (\cos\theta, \sin\theta, z)$ that maps one-to-one the square $[0, 2\pi[\times[0, 1]$ onto a portion of a cylinder $S := \{(x, y, z) \mid x^2 + y^2 = 1, 0 \le z \le 1\}$. Trivially the elementary area of S is $A(S) = 2\pi$.

We divide the side of the square $[0, 2\pi] \times [0, 1]$ in n and m parts, respectively, then we divide each rectangle obtained in this way in four triangles by means of its diagonals, obtaining a triangulation of $[0, 2\pi] \times [0, 1]$ in $4nm$ triangles. We construct a polyhedral surface S_{mn} with triangular faces inscribed to the cylinder using the images of the vertices of the triangulation as vertices. A tedious computation yields the area A_{mn} of the inscribed polyhedral surface S_{mn},

$$A_{mn} = 2n \sin\frac{\pi}{2n} + \left[\frac{1}{4} + \frac{4m^2}{n^4}\left(n\sin\frac{\pi}{2n}\right)^4\right]^{1/2} \cdot 2n \sin\frac{\pi}{n}.$$

Now, if we choose $m = n$, then $A_{mn} \to 2\pi$ as suggested by intuition; but, if $m = n^3$, then $A_{mn} \to +\infty$. The supremum of the areas of all polyhedral surfaces obtained by triangulations on the space of parameters is therefore $+\infty$ and not the area of S. The intuitive reason for this behavior is the following: If m, the number of subdivisions of the z-axis, is large with respect to the number of subdivisions of the angle, then the triangles of the inscribed polyhedral surface to the cylinder tend to become closer to the orthogonal to the surface of the cylinder. Consequently, the area of the polyhedral surface is large.

This example motivated a flourishing of possible definitions (and, consequently, of treatises) for the area of two-dimensional surfaces in $\mathbb{R}^3$. Among those definitions, the most effective, at least for elementary purposes, has proved to be the one based on Hausdorff measure.

Let $\Omega \subset \mathbb{R}^n$ and $f : \Omega \to \mathbb{R}^N$, $n \leq N$, be a map of class C^1. For any measurable set $A \subset \mathbb{R}^n$, we say that the image $f(A) \subset \mathbb{R}^N$ *is parameterized by f*, and we think of the area of $f(A)$ as of $\mathcal{H}^n(f(A))$ when f is injective. An important formula, called the *area formula*, allows us to compute $\mathcal{H}^n(f(A))$.

2.78 Theorem. *Let $\Omega \subset \mathbb{R}^n$ be an open set and let $f \in C^1(\Omega, \mathbb{R}^N)$, $N \geq n$. If $A \subset \Omega$ is a $\mathcal{L}^n$-measurable set and f is injective on A, then $f(A)$ is $\mathcal{H}^n$-measurable and*

$$\int_A J(\mathbf{D}f(x))\, dx = \mathcal{H}^n(f(A)) \tag{2.35}$$

where $J_f(x) := J(\mathbf{D}f(x)) = \sqrt{\det \mathbf{D}f^T(x)\mathbf{D}f(x)}$ is the Jacobian of f.

2.79 Remark. Notice the following:

(i) If $n = 1$, then $\mathbf{D}f = f'$ and $J(\mathbf{D}f) = |f'|$: the area formula (2.35) says that the length of a curve agrees with the one-dimensional Hausdorff measure of the trajectory.

(ii) If f is linear, $f(x) = \mathbf{L}x$, (2.35) simply reads as

$$\mathcal{H}^n(\mathbf{L}(A)) = J(\mathbf{L})\mathcal{L}^n(A).$$

Actually, this is the starting point for the proof and follows from the invariance of the Hausdorff measure under rotations. In fact, using the polar decomposition of $\mathbf{L}$, and identifying $\mathbb{R}^n$ with the n coordinate plane of the first n coordinates of $\mathbb{R}^N$, $\mathbf{L}$ writes as $\mathbf{L} = \mathbf{U}\Delta\mathbf{S}$ where $\mathbf{S} \in M_{n,n}$ is symmetric, $\mathbf{U} \in M_{N,N}$ is orthogonal, and

$$\Delta = \begin{pmatrix} \mathrm{Id} \\ 0 \end{pmatrix}.$$

Using the invariance of $\mathcal{H}^n$ and the change of variable formula for $\mathcal{L}^n$, we find

$$\mathcal{H}^n(\mathbf{U}\Delta\mathbf{S}(A)) = \mathcal{H}^n(\Delta\mathbf{S}(A)) = \mathcal{H}^n(\mathbf{S}(A)) = \mathcal{L}^n(\mathbf{S}(A))$$
$$= |\det \Delta\mathbf{S}|\mathcal{L}^n(A) = |\det \mathbf{S}|\mathcal{L}^n(A) = J(\mathbf{L})\mathcal{L}^n(A).$$

(iii) The area formula implies that the *image of the points at which the linear tangent map is not injective has zero $\mathcal{H}^n$-measure*. In fact, for any $\mathbf{T} \in M_{N,n}$ $N \geq n$, $\ker \mathbf{T} \neq \{0\}$ if and only if $J(\mathbf{T}) = 0$. Hence from (2.35) we get $\mathcal{H}^n(f(A)) = 0$ if

$$A := \Big\{ x \in \Omega \,\Big|\, \ker \mathbf{D}f(x) \neq 0 \Big\} = \Big\{ x \in \Omega \,\Big|\, J(\mathbf{D}f(x)) = 0 \Big\}.$$

(iv) From the area formula we also get the following: *Let Ω, Δ be open sets of $\mathbb{R}^n$ and $\phi : \Omega \to \mathbb{R}^N$ and $\psi : \Delta \to \mathbb{R}^N$ be two maps of class C^1 that are injective, respectively in $A \subset \Omega$ and $B \subset \Delta$, A and B being $\mathcal{L}^n$-measurable. If $\phi(A) = \psi(B)$, then*

$$\int_A J(\mathbf{D}\phi(x)) \, dx = \int_B J(\mathbf{D}\psi(y)) \, dy.$$

(v) *If $n = N$, then (2.35) is simply the change of variable formula for $\mathcal{L}^n$.*

The area formula extends in several ways. First, we can drop the injectivity hypothesis by introducing the *multiplicity function* or *Banach's indicatrix*

$$y \to N(f, A, y) := \mathcal{H}^0(A \cap f^{-1}(y))$$

which counts the points of A in the inverse image of y. Under the hypotheses that $f \in C^1(\Omega)$, one shows that the multiplicity function is $\mathcal{H}^N$-measurable and

$$\int_A J(\mathbf{D}f) \, dx = \int_{\mathbb{R}^N} N(f, A, y) \, d\mathcal{H}^n(y). \tag{2.36}$$

Moreover, we can also relax the regularity of the map f: One can in fact prove that (2.36) holds also if f is Lipschitz-continuous (recall that, if f is Lipschitz-continuous, then $J(\mathbf{D}f)$ is defined $\mathcal{L}^n$-a.e. since f is differentiable $\mathcal{L}^n$-almost-everywhere by the Rademacher theorem).

Starting from (2.36), by approximating $\mathcal{L}^n$-measurable functions u by simple functions and then passing to the limit by means of the monotone convergence theorem of Beppo Levi, we also get the following.

2.80 Theorem (Change of variables formula). *Let $\Omega \subset \mathbb{R}^n$ be open, let $f : \Omega \subset \mathbb{R}^n \to \mathbb{R}^N$, $n \leq N$, be of class $C^1(\Omega)$ (or, more generally, locally Lipschitz-continuous in Ω), and let $u : \Omega \to \mathbb{R}$ be $\mathcal{L}^n$-measurable and nonnegative, or such that $|u| \, J(\mathbf{D}f)$ is $\mathcal{L}^n$-summable. Then the function*

$$y \to \sum_{x \in f^{-1}(y)} u(x)$$

is $\mathcal{H}^n$-measurable and

$$\int_{\mathbb{R}^n} u(x) \, J(\mathbf{D}f)(x) \, dx = \int_{\mathbb{R}^N} \Big(\sum_{x \in f^{-1}(y)} u(x) \Big) \, d\mathcal{H}^n(y). \tag{2.37}$$

In particular, if $v : \mathbb{R}^N \to \mathbb{R}$ is $\mathcal{H}^n$-measurable and nonnegative, then

$$\int_A v(f(x)) \, J(\mathbf{D}f)(x) \, dx = \int_{\mathbb{R}^N} v(y) \, N(f, A, y) \, d\mathcal{H}^n(y) \, . \tag{2.38}$$

a. Calculus of the area of a surface

Parameterizing, at least locally, a k-dimensional surface in $\mathbb{R}^n$ by a C^1 map, we can easily compute its area by means of the area formula (2.35).

In this procedure we need to compute the Jacobian of the parameterization, and the following information may be useful.

(i) Let $\mathbf{A} \in M_{N,n}$. The alternative theorem yields

$$\operatorname{Rank}(\mathbf{A}^T\mathbf{A}) = \operatorname{Rank}\mathbf{A}^T = \operatorname{Rank}\mathbf{A} = \operatorname{Rank}\mathbf{A}\mathbf{A}^T \leq \min(n, N). \tag{2.39}$$

It follows that $\det \mathbf{A}\mathbf{A}^T = 0$ if $N > n$.

(ii) We have $\ker \mathbf{A}^T\mathbf{A} = \ker \mathbf{A}$, consequently the three claims
  (a) $J(\mathbf{A}) = (\det \mathbf{A}^T\mathbf{A})^{1/2} = 0$,
  (b) $\ker \mathbf{A} \neq \{0\}$,
  (c) $\operatorname{Rank}\mathbf{A}$ is not maximal,
  are equivalent.

(iii) (AREA AND METRIC TENSOR) Let $\{A_1, A_2, \ldots, A_n\}$ denote the columns of $\mathbf{A}$, $\mathbf{A} = [A_1|A_2|\ldots|A_n]$. Then

$$\mathbf{A}^T\mathbf{A} = \mathbf{G}, \qquad \text{where} \qquad \mathbf{G} = (g_{ij}),\ g_{ij} := A_i \bullet A_j\,.$$

Consequently, if $f : \Omega \subset \mathbb{R}^n \to \mathbb{R}^N$, $N \geq n$, is of class C^1, then

$$J(\mathbf{D}f) = \sqrt{\det \mathbf{G}}, \qquad \mathbf{G} = (g_{ij}),\ g_{ij} = f_{x^i} \bullet f_{x^j}\,,$$

and the area formula becomes

$$\mathcal{H}^n(f(\Omega)) = \int_\Omega \sqrt{g(x)}\,dx, \qquad g(x) := \det \mathbf{G}(x).$$

(iv) (THE CAUCHY–BINET FORMULA) Let $\mathbf{A} \in M_{N,n}$, $N \geq n$. For every multiindex $\alpha = (\alpha_1, \alpha_2, \ldots, \alpha_n)$, $1 \leq \alpha_1 < \alpha_2 < \cdots < \alpha_n \leq N$, let $\mathbf{A}^\alpha$ be the $n \times n$-submatrix of $\mathbf{A}$ made of the rows $\alpha_1, \alpha_2, \ldots, \alpha_n$ of $\mathbf{A}$. Then, see, e.g., [GM5], the following Cauchy–Binet formula holds

$$J(\mathbf{A})^2 = \sum_{\alpha \in I(n,N)} (\det(\mathbf{A}^\alpha))^2.$$

2.81 Example (Two-dimensional parameterized surfaces in $\mathbb{R}^3$). Let $n = 2$ and $N = 3$. Then

$$\mathbf{D}f := \begin{pmatrix} a & d \\ b & e \\ c & f \end{pmatrix}$$

where the columns are the vectors with components the partial derivatives of f with respect to x and y, $f_x := (a, b, c)^T$ and $f_y := (d, e, f)^T$. If we set

$$E := |f_x|^2, \qquad F := f_x \bullet f_y, \qquad G := |f_y|^2,$$

we find

$$\mathbf{D}f^T\mathbf{D}f = \begin{pmatrix} E & F \\ F & G \end{pmatrix}$$

and

$$\mathcal{H}^2(f(\Omega)) = \int_\Omega J(\mathbf{D}f)\,dxdy = \int_\Omega \sqrt{EG - F^2}\,dxdy.$$

Alternatively, we can compute $J(\mathbf{D}f)$ and the area of $u(\Omega)$ by means of the Cauchy–Binet formula. If

$$\mathbf{A}^{12} := \begin{pmatrix} a & d \\ b & e \end{pmatrix}, \qquad \mathbf{A}^{23} := \begin{pmatrix} b & e \\ c & f \end{pmatrix}, \qquad \mathbf{A}^{13} := \begin{pmatrix} a & d \\ c & f \end{pmatrix},$$

then

$$J(\mathbf{D}f)^2 = (ae - bd)^2 + (bf - ec)^2 + (af - dc)^2.$$

Notice that the three numbers $ae - bd$, $-(bf - ec)$, and $af - dc$ are the three components of the *vector product*

$$f_x \times f_y$$

of the columns of $\mathbf{D}f$, hence

$$\mathcal{H}^2(f(\Omega)) = \int_\Omega |f_x \times f_y|\,dxdy.$$

2.82 Example (Graphs of codimension 1). Let $u : \Omega \subset \mathbb{R}^n \to \mathbb{R}$ be a function of class C^1 and

$$\mathcal{G}_{u,\Omega} := \left\{ (x,y) \in \Omega \times \mathbb{R} \,\middle|\, y = u(x) \right\}$$

be its graph. $\mathcal{G}_{u,\Omega}$ is the image of the injective map $f(x) = (x, u(x))$ from Ω into $\mathbb{R}^{n+1}$. Since

$$\mathbf{D}f(x) = \left(\boxed{\begin{array}{c} \text{Id} \\[1em] \mathbf{D}u(x) \end{array}} \right),$$

the Cauchy–Binet formula gives $J(\mathbf{D}f(x)) = \sqrt{1 + |\mathbf{D}u(x)|^2}$, hence

$$\mathcal{H}^n(\mathcal{G}_{u,\Omega}) = \int_\Omega \sqrt{1 + |\mathbf{D}u(x)|^2}\,dx.$$

2.83 Example (Parameterized hypersurfaces). Let $u : \Omega \subset \mathbb{R}^n \to \mathbb{R}^{n+1}$ be an injective map of class C^1. The Jacobian matrix of u has $n+1$ rows and n columns, and its $n \times n$ submatrices can be indexed by the missing row. If

$$\frac{\partial(u^1, \ldots, u^{i-1}, u^{i+1}, \ldots, u^n)}{\partial(x^1, x^2, \ldots, x^n)}$$

denote the determinant of the submatrix obtained by removing the ith row, we then get

$$\mathcal{H}^n(u(\Omega)) = \int_\Omega \left(\sum_{i=1}^n \left(\frac{\partial(u^1, \ldots, u^{i-1}, u^{i+1}, \ldots, u^n)}{\partial(x^1, x^2, \ldots, x^n)} \right)^2 \right)^{1/2} dx.$$

2.84 Example (Rotational surfaces). A rotational surface around an axis is well described by its perpendicular sections to its axis that are circles. We can describe its points P by means of two parameters: the orthogonal projection of P on the rotational axis and a parameter describing the points on the circle in the perpendicular plane to the axis through P. This way, if S is a rotational surface around the axis z, S is the one-to-one image of

$$A := [a,b] \times [0, 2\pi[$$

by a map $\phi : [a,b] \times \mathbb{R} \to \mathbb{R}^3$ of the type

$$\phi(z,\theta) := \begin{cases} x = \rho(z)\cos\theta, \\ y = \rho(z)\sin\theta, \\ z = z \end{cases}$$

where $\rho(z)$ is the radius of the section at level z. Assuming $\rho(z) \in C^1([a,b])$, we have

$$\mathbf{D}\phi(z,\theta) := \begin{pmatrix} \rho'\cos\theta & -\rho\sin\theta \\ \rho'\sin t & -\rho\cos\theta \\ 1 & 0 \end{pmatrix}, \qquad (z,t) \in A := [a,b] \times [0, 2\pi[$$

hence, by the Cauchy–Binet formula,

$$J(\mathbf{D}\phi)(z,\theta) = \rho(z)\sqrt{1 + {\rho'}^2(z)};$$

therefore

$$\mathcal{H}^2(S) = \mathcal{H}^2(\phi(A)) = \int_0^{2\pi}\int_a^b \rho(z)\sqrt{1 + (\rho'(z))^2}\, dz = 2\pi \int_a^b \rho(z)\sqrt{1 + (\rho'(z))^2}\, dz.$$

2.5.3 The coarea formula

Consider a function $f : \mathbb{R}^n \to \mathbb{R}^N$, $N \leq n$ and, for $y \in \mathbb{R}^N$ its inverse image $f^{-1}(y)$. When y varies, the family $\{f^{-1}(y)\}$ yields a sort of *foliation* of $\mathbb{R}^n$, for example think of $f : \mathbb{R}^2 \to \mathbb{R}$, $f(x,y) = x^2 + y^2$, for which $f^{-1}(t) := \{(x,y) = x^2 + y^2 = t\}$. As we shall see in Chapter 5, if $\mathbf{D}f(x)$ is of maximal rank N, then the *leaf* $f^{-1}(y)$ is an $(n - N)$-dimensional submanifold of $\mathbb{R}^n$. The coarea formula provides a formula that allows us to express the $\mathcal{L}^n$-integration on a set $A \subset \mathbb{R}^n$ as the integration with respect to y of an $\mathcal{H}^{n-N}$-integration over $f^{-1}(y)$.

2.85 Theorem (Coarea formula). *Let Ω be an open set and $A \subset \Omega$ be $\mathcal{L}^n$-measurable, let $f : \Omega \to \mathbb{R}^N$ be a map of class C^1, and assume $N \leq n$. For $\mathcal{L}^N$-a.e. y the set $A \cap f^{-1}(y)$ is $\mathcal{H}^{n-N}$-measurable, the function $y \to \mathcal{H}^{n-N}(A \cap f^{-1}(y))$ is $\mathcal{L}^N$-measurable and*

$$\int_A J(\mathbf{D}f(x))\, d\mathcal{L}^n(x) = \int_{\mathbb{R}^N} \mathcal{H}^{n-N}(A \cap f^{-1}(y))\, d\mathcal{L}^N(y); \qquad (2.40)$$

here

$$J_f(x) := J(\mathbf{D}f(x)) = \sqrt{\det(\mathbf{D}f(x)\mathbf{D}f(x)^T)}$$

denotes the Jacobian of f.

Actually the previous theorem can be generalized in several ways. First, it suffices to assume that f be locally Lipschitz-continuous. Moreover, by approximating measurable maps u with simple functions, one shows the following.

2.86 Theorem. *Let $\Omega \subset \mathbb{R}^n$ be open, let $f : \Omega \subset \mathbb{R}^n \to \mathbb{R}^N$, $N \le n$, be a map of class $C^1(\Omega)$ (or merely locally Lipschitz-continuous) and let $u : \Omega \to \mathbb{R}$ be a measurable function on Ω such that $|u| J(\mathbf{D}f)$ is $\mathcal{L}^n$-integrable. Then*

$$\int_\Omega u(x) J(\mathbf{D}f)(x) \, dx = \int_{\mathbb{R}^N} \left(\int_{f^{-1}(y)} u(x) \, d\mathcal{H}^{n-N}(x) \right) d\mathcal{L}^N(y). \quad (2.41)$$

2.87 Remark. We notice the following:

(i) If we split $\mathbb{R}^n$ as $\mathbb{R}^n = \mathbb{R}^{n-N} \times \mathbb{R}^N$ denoting its coordinates by (x, y), $x \in \mathbb{R}^{n-N}$, $y \in \mathbb{R}^N$, and we choose $f(x, y) := y$, then $J(\mathbf{D}f)(x, y) = |\mathbf{D}f(x, y)| = 1$, $A \cap f^{-1}(y) = \{(x, z) \in A \,|\, z = y\} = A_y$. Therefore, Theorem 2.85 simply reduces to Fubini's theorem.

(ii) Let $f : \mathbb{R}^n \to \mathbb{R}$. Since

$$J(\mathbf{D}f) = \sqrt{\det(\mathbf{D}f)(\mathbf{D}f)^T} = |\mathbf{D}f|,$$

we then obtain

$$\int_A |Df| \, dx = \int_{-\infty}^{+\infty} \mathcal{H}^{n-1}(A \cap f^{-1}(t)) \, dt$$

and

$$\int_A g(x)|Df(x)| \, dx = \int_{-\infty}^{+\infty} \left(\int_{A \cap f^{-1}(t)} g(z) \, d\mathcal{H}^{n-1}(z) \right) dt$$

for any measurable $A \subset \mathbb{R}^n$ and any measurable $g : A \to \mathbb{R}$ such that $g(x)|Df(x)|$ is $\mathcal{L}^n$-integrable.

2.88 Example (Measure of the unit sphere in $\mathbb{R}^n$). The volume of the ball of radius r in $\mathbb{R}^n$ is $\omega_n r^n$ where $\omega_n = \mathcal{L}^n(B(0, 1))$, and the measure of the sphere of radius t, $\mathcal{H}^{n-1}(\partial B(0, t))$ is positively homogeneous of degree $n - 1$, in particular it is continuous in t. If we choose $f(x) = |x|$, then $J(\mathbf{D}f(x)) = |\mathbf{D}f(x)| = 1$: from the coarea formula

$$\int_{B(0,r+h) \setminus B(0,r)} dx = \int_r^{r+h} \mathcal{H}^{n-1}(\partial B(0, t)) \, dt$$

and on account of the fundamental theorem of calculus we infer

$$\mathcal{H}^{n-1}(\partial B(0, r)) = \lim_{h \to 0} \frac{1}{h} \int_r^{r+h} \mathcal{H}^{n-1}(\partial B(0, t)) \, dt = \lim_{h \to 0} \frac{1}{h} \int_{B(0,r+h) \setminus B(0,r)} dx$$

$$= \frac{d\mathcal{L}^n(B(0, r))}{dr}(r) = n\omega_n r^{n-1}.$$

2.89 Example. Notice that if $f(x) := |x|$, then $f^{-1}(y) = \partial B(0, y)$ and Theorem 2.86 yields the well-known formula of integration on polar coordinates

$$\int_{\{s < |x| < t\}} u \, dx = \int_r^s \left(\int_{\partial B(0,\rho)} u(y) \, d\mathcal{H}^{n-1}(y) \right) d\rho.$$

In particular, for $h \neq 0$ we have

$$\frac{1}{h} \int_{\{r < |x| < r+h\}} u \, dx = \frac{1}{h} \int_r^{r+h} \left(\int_{\partial B(0,\rho)} u(y) \, d\mathcal{H}^{n-1}(y) \right) d\rho,$$

hence, by the theorem of differentiation of the integral, for $\mathcal{L}^1$-a.e. r (for all r if, for instance, u is continuous),

$$\frac{d}{dr} \left(\int_{B(0,r)} u \, dx \right) = \int_{\partial B(0,r)} u \, d\mathcal{H}^{n-1}.$$

For $u = 1$ we therefore find again

$$\mathcal{H}^{n-1}(\partial B(0,r)) = \frac{d}{dr}(\omega_n r^n) = n\omega_n r^{n-1}.$$

2.90 Example (Measure of the unit ball in $\mathbb{R}^n$). The coarea formula yields also an alternative way to compute the measure ω_n of the unit ball $B(0,1)$ of $\mathbb{R}^n$. Using Fubini's theorem we find

$$\int_{\mathbb{R}^n} e^{-|x|^2} \, dx = \int_{-\infty}^{+\infty} e^{-x_1^2} \, dx_1 \int_{-\infty}^{+\infty} e^{-x_2^2} \, dx_2 \dots \int_{-\infty}^{+\infty} e^{-x_n^2} \, dx_n = \pi^{n/2}.$$

On the other hand, the coarea formula yields

$$\int_{\mathbb{R}^n} e^{-|x|^2} \, dx = \int_0^{+\infty} \left(\int_{\partial B(0,t)} e^{-t^2} \, d\mathcal{H}^{n-1}(t) \right) dt = \frac{n}{2} |B(0,1)| \int_0^{+\infty} e^{-t^2} t^{n-1} \, dt$$

$$= \frac{n}{2} |B(0,1)| \int_0^{+\infty} s^{\frac{n}{2}-1} e^{-s} \, ds = \frac{n}{2} |B(0,1)| \Gamma\left(\frac{n}{2}\right)$$

in terms of Euler's Γ-function. If follows

$$\omega_n = |B(0,1)| = \frac{2}{n} \frac{\pi^{n/2}}{\Gamma\left(\frac{n}{2}\right)} = \frac{\pi^{n/2}}{\Gamma\left(\frac{n}{2} + 1\right)}.$$

2.91 Example. Let $f : \mathbb{R}^n \to \mathbb{R}$ be a function of class C^1 (or merely locally Lipschitz-continuous). From the coarea formula, for any positive h we have

$$\frac{1}{h} \int_{\{t < f < t+h\}} |Df| \, dx = \frac{1}{h} \int_t^{t+h} \mathcal{H}^{n-1}(\{f = t\}) \, dt$$

$$\frac{1}{h} \int_{\{t-h < f < t\}} |Df| \, dx = \frac{1}{h} \int_{t-h}^t \mathcal{H}^{n-1}(\{f = t\}) \, dt$$

hence, by the theorem of differentiation of integral, for $\mathcal{L}^1$-a.e. $t \in \mathbb{R}$

$$\frac{d}{dt} \int_{\{f < t\}} |Df| \, dx = -\frac{d}{dt} \int_{\{f > t\}} |Df| \, dx = \mathcal{H}^{n-1}(\{f = t\}).$$

2.6 Gauss–Green Formulas

In this section, we state the *Gauss–Green theorem* and discuss the *divergence theorem*. These topics are of fundamental relevance for the development of the calculus for functions of several variables. In particular, Gauss–Green formulas extend the fundamental theorem of calculus

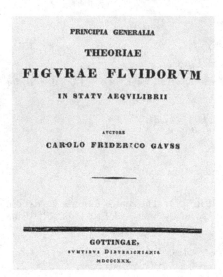

Figure 2.16. Two pages from the paper by Carl Friedrich Gauss (1777–1855), in which the Gauss–Green formula appears.

$$f(b) - f(a) = \int_a^b f'(t)\, dt$$

to functions of several variables.

a. Two simple situations

We begin with a very simple situation.

2.92 Proposition. *Let $A \subset \mathbb{R}^n$ be open and $f \in C_c^1(A)$. Then*

$$\int_A D_i f(x)\, dx = 0 = \int_{\partial A} f\, d\mathcal{H}^{n-1} \qquad \forall i = 1, \ldots, n.$$

Proof. We extend f to all $\mathbb{R}^n$ as zero outside A; we call it f. It is not restrictive to assume that its support is contained in the unit cube. Since for every $x_1, \ldots, x_{i-1}, x_{i+1}, \ldots, x_n$ we have

$$\int_{-1}^1 D_i f(x_1, \ldots, x_{i-1}, x_i, x_{i+1}, \ldots, x_n)\, dx_i =$$

$$= f(x_1, \ldots, x_{i-1}, 1, x_{i+1}, \ldots, x_n) - f(x_1, \ldots, x_{i-1}, -1, x_{i+1}, \ldots, x_n) = 0,$$

integrating with respect to the remaining variables we get

$$\int_A D_i f(x)\, dx = \int_Q D_i f(x)\, dx = 0.$$

$\square$

Split $\mathbb{R}^n$ as $\mathbb{R}^{n-1} \times \mathbb{R}$ and let $x = (x', x_n)$, $x' = (x_1, x_2, \ldots, x_{n-1}) \in \mathbb{R}^{n-1}$, $x_n \in \mathbb{R}$, be its coordinates. Let Q be a bounded open set in $\mathbb{R}^{n-1}$ and let $\alpha : Q \to\,]a, b[$ be a function of class C^1; set

$$A := \left\{ x = (x', x_n) \in Q \times [a, b] \,\middle|\, a < x_n < \alpha(x') \right\}.$$

Since the vector $(-\nabla\alpha(x'), 1)$ is perpendicular to the plane tangent to the graph of α at $(x', \alpha(x'))$, the exterior normal vector to A at $(x', \alpha(x'))$ has components $\nu = (\nu_1, \nu_2, \ldots, \nu_n)$ given by

$$\begin{cases} \nu_i := \dfrac{-D_i\alpha}{\sqrt{1 + |D\alpha(x')|^2}} & \forall i = 1, \ldots, n-1, \\[4mm] \nu_n := \dfrac{1}{\sqrt{1 + |D\alpha(x')|^2}}. \end{cases}$$

2.93 Proposition. *Let $f \in C^1(A) \cap C^0(\overline{A})$ with $|Df| \in \mathcal{L}^1(A)$. Suppose that f vanishes near $\partial(Q \times [a, b]) \cap \overline{A}$, trivially*

$$\int_A D_i f(x)\, dx = \int_{\partial A} f\nu_i \mathcal{H}^{n-1}.$$

Proof. Since f vanishes on $\partial(Q \times [a, b]) \cap \overline{A}$, trivially

$$\int_{\partial A} f\nu_i d\mathcal{H}^{n-1} = \int_{\mathcal{G}_{\alpha, Q}} f\nu_i d\mathcal{H}^{n-1}$$

and, since the element of area on $\mathcal{G}_{\alpha, Q}$ is $d\mathcal{H}^{n-1} = \sqrt{1 + |D\alpha(x')|^2}\, dx'$, we have

$$\int_{\partial A} f\nu_i \mathcal{H}^{n-1} = \begin{cases} -\displaystyle\int_Q f(x', \alpha(x')) \dfrac{\partial\alpha}{\partial x_i}(x')\, dx' & \text{if } i - 1, \ldots, n-1, \\[4mm] \displaystyle\int_Q f(x', \alpha(x'))\, dx' & \text{if } i = n. \end{cases} \tag{2.42}$$

From the fundamental theorem of calculus, since $f = 0$ near $\partial(Q \times [a, b]) \cap \overline{A}$ and $|Df| \in \mathcal{L}^1(A)$, we infer

$$\int_A D_n f(x', x_n)\, dx' dx_n = \int_Q \int_a^{\alpha(x')} D_n f(x', x_n)\, dx_n dx'$$

$$= \int_Q (f(x', \alpha(x')) - f(x', a))\, dx' = \int_Q f(x', \alpha(x'))\, dx'. \tag{2.43}$$

A comparison of (2.42) and (2.43) yields the result for $i = n$.

Now set

$$F(x') := \int_a^{\alpha(x')} f(x', x_n)\, dx_n, \qquad x' \in Q.$$

Differentiating under the integral sign with respect to x_i, we infer

$$D_i F(x') = \int_a^{\alpha(x')} D_i f(x', x_n)\, dx_n + f(x', \alpha(x'))D_i\alpha(x');$$

on the other hand

$$\int_A D_i f(x', x_n)\, dx'\, dx_n = \int_Q \left(\int_a^{\alpha(x')} D_i f(x', x_n)\, dx_n \right) dx'$$

$$= \int_Q D_i F(x')\, dx' - \int_Q f(x', \alpha(x')) D_i \alpha(x')\, dx',$$ (2.44)

and, since f vanishes if $x \in \partial(Q \times [a, b]) \cap \overline{A}$, $F(x') = 0$ if $x' \in \partial Q$ and

$$\int_Q D_i F(x')\, dx' = \int_Q D_i F(x')\, dx_1 \ldots dx_{n-1}$$

$$= \int \int \left(\int_{-1}^1 D_i F(x') dx_i \right) dx_1 \ldots dx_{i-1}\, dx_{i+1} \ldots dx_{n-1}$$

$$= \int \int \Big(F(x_1, \ldots, x_{i-1}, 1, x_{i+1}, \ldots, x_{n-1}) - F(x_1, \ldots, x_{i-1}, -1, x_{i+1}, \ldots, x_{n-1}) \Big)$$

$$dx_1 \ldots dx_{i-1}\, dx_{i+1} \ldots dx_{n-1}$$

$$= \int \int 0\, dx_1 \ldots dx_{i-1}\, dx_{i+1} \ldots dx_{n-1} = 0.$$

Therefore, (2.44) becomes

$$\int_A D_i f(x', x_n)\, dx'\, dx_n = - \int_Q f(x', \alpha(x')) D_i \alpha(x')\, dx'.$$ (2.45)

From (2.45) and (2.42), we infer the result for $i = 1, \ldots, n-1$. $\square$

b. Admissible sets

In the sequel we shall limit ourselves to prove Gauss–Green formulas for a class of sets, which we now introduce, sufficiently large for the applications. Actually, measure theory would allow us to prove them for a much larger class.

Let $A \subset \mathbb{R}^n$ be an open set. In this context, we say that $x \in \partial A$ is a *regular point* for ∂A if there exists an open cube with center at x and sides parallel to the axes (that we write as $Q \times [a, b]$ where Q is a cube on $\mathbb{R}^{n-1}$) and a function $\alpha : Q \to]a, b[$ of class $C^1(Q)$ such that

(i) $U_x \cap A = \{(x', x_n) \,|\, a < x_n < \alpha(x'),\ x' \in Q\}$,
(ii) $U_x \cap \partial A = \{(x', x_n) \,|\, x_n = \alpha(x'),\ x' \in Q\}$.

The set $r(A) \subset \partial A$ of regular points for ∂A is open relatively to ∂A, and for every $x \in \partial A$, the *exterior unit vector* to A at x is given by the vector $\nu = (\nu_1, \nu_2, \ldots, \nu_n)$ given by

$$\nu_i := \frac{-D_i \alpha}{\sqrt{1 + |D\alpha(x')|^2}} \quad \forall i = 1, \ldots, n-1, \qquad \nu_n := \frac{1}{\sqrt{1 + |D\alpha(x')|^2}}.$$

Obviously $|\nu| = 1$, ν is perpendicular to the tangent plane to the graph of α at $x = (x', \alpha(x'))$ and $x - t\nu(x) \in A$ for all $t > 0$ sufficiently small positive t. Of course, neither the cube nor the function α are uniquely defined by A; however, it is not difficult to show, compare Chapter 5, that the exterior unit normal is uniquely defined at the points $x \in r(A)$. Moreover, one sees, compare Chapter 5, that $x \in \partial A$ is regular if and only if there exist an open cube Q centered at x and a function $\varphi : U_x \to \mathbb{R}$ of class C^1 such that

(i) $U_x \cap A = \{y \in U_x \mid \varphi(y) < 0\}$,
(ii) $U_x \cap \partial A = \{y \in U_x \mid \varphi(y) = 0\}$,
(iii) $\nabla \varphi \neq 0$ in $U_x \cap \partial A$.

In this case the exterior normal vector at $x \in r(A)$ at $x \in r(A)$ is

$$\nu(x) = \frac{\nabla \varphi(x)}{|\nabla \varphi(x)|}.$$

2.94 Definition. *We shall say that an open set $A \subset \mathbb{R}^n$ is admissible if A is open, $\mathcal{H}^{n-1}(\partial A) < +\infty$ and $\mathcal{H}^{n-1}(\partial A \setminus r(A)) = 0$.*

For example, an open set in $\mathbb{R}^2$ whose boundary is the union of a finite number of closed and disjoint piecewise regular curves is an admissible set of $\mathbb{R}^2$. Also a bounded set whose boundary is a polyhedron with a finite number of faces is an admissible set. Actually, it is easily seen that A is admissible if A is bounded and ∂A is a finite disjoint union $\partial A = \cup_{i=0}^{N} \Gamma_i$ where Γ_0 is closed with $\mathcal{H}^{n-1}(\Gamma_0) = 0$, and, for $i = 1, \ldots, N$, Γ_i is a $(n-1)$-submanifold of $\mathbb{R}^n$, see Chapter 5.

c. Decomposition of unity

The *decomposition* (or *partition*) *of unity* is a useful tool when we want to transfer local information to global ones.

2.95 Theorem. *Let $\{V_\alpha\}$ be a family of open sets in $\mathbb{R}^n$ and $\Omega := \cup_\alpha V_\alpha$. There exists a locally finite covering of Ω with balls $B_j \subset\subset \Omega$ such that for every j we have $\overline{B_j} \subset V_\alpha$ for some α.*

Proof. For all $j = 1, 2, \ldots$, we choose a sequence $\{H_j\}$ of compact sets contained in Ω with $H_j \subset\subset \operatorname{int}(H_{j+1})$ and $\Omega = \cup_j H_j$; we also set $H_{-1} = H_{-2} := \emptyset$. For $j := 0, 1, \ldots$ we consider the compact sets $K_j := H_j \setminus \operatorname{int}(H_{j-1})$ and the open sets $A_j := \operatorname{int} H_{j+1} \setminus H_{j-2}$. We have $K_j \subset\subset A_j$, $\Omega = \cup_j A_j$ and $A_i \cap A_j = \emptyset$ except for $i = j-1, j$ o $j+1$. Now, for every $x \in K_j$ choose $\lambda = \lambda(x)$ such that $x \in V_{\lambda(x)}$ and a ball $B(x, r(x))$ with closure in $A_j \cap V_{\lambda(x)}$. The family $\{B(x, r(x))\}_{x \in K_j}$ is clearly an open covering of the compact set K_j from which we can choose a finite covering $\{B_{j,1}, B_{j,2}, \ldots B_{j,N_j}\}$. The family

$$\mathcal{B} := \Big\{ B_{j,i} \,\Big|\, j = 0, 1, \ldots, i = 1, \ldots N_j \Big\}$$

has the required properties. $\square$

2.96 Lemma. *The function*

$$\varphi(x) := \begin{cases} \exp\left(\frac{1}{1-|x|^2}\right) & \text{if } |x| < 1, \\ 0 & \text{if } |x| \geq 1 \end{cases}$$

is of class C^∞ and nonzero exactly on $B(0,1)$.

2.97 Theorem. *Let $\{B_j\}$ be a locally finite covering of $\Omega = \cup_j B_j$, B_j being balls. There exists functions $w_j : \mathbb{R}^n \to \mathbb{R}$ of class C^∞ such that*

(i) $0 \leq \alpha_j(x) \leq 1 \; \forall x \in \mathbb{R}^n$,
(ii) $\alpha_j(x) > 0$ *if and only if* $x \in B_j$,
(iii) $\sum_{j=1}^\infty \alpha_j(x) = 1 \; \forall x \in \Omega$.

Proof. For $j = 1, 2, \ldots$, we choose $\varphi_j \in C^\infty(\mathbb{R}^n)$ with $\varphi_j > 0$ on B_j and $\varphi_j = 0$ outside B_j. The function $\sum_{j=1}^\infty \varphi_j(x)$ is well defined on $\mathbb{R}^n$ since locally it is a finite sum ($\{B_j\}$ being locally finite) and positive in Ω, since $\{B_j\}$ is a covering of Ω. Thus, we readily see that the functions

$$\alpha_j(x) := \frac{\varphi_j(x)}{\sum_{j=0}^\infty \varphi_j(x)}$$

have the desired properties.   □

We notice that the number of functions α_j of the decomposition of unity that are nonzero at each x is finite and that they are exactly the nonzero functions of the decomposition of unity that are nonzero at y if y is sufficiently close to x. Consequently, we also have

$$\sum_{j=1}^\infty D\alpha_j(x) = D\Big(\sum_{j=1}^\infty \alpha_j(x) \Big) = 0 \qquad \forall x \in \Omega$$

and

$$\int \sum_{j=1}^\infty \alpha_j(x) \, d\mu = \sum_{j=1}^\infty \int \alpha_j(x) \, d\mu$$

for $\mu = \mathcal{L}^n$ or $\mathcal{H}^{n-1}$.

d. Gauss–Green formulas

2.98 Theorem (Gauss–Green formulas). *Let $A \subset \mathbb{R}^n$ be an admissible open set and let f be a function of class C^1 in a neighborhood of A with $|Df| \in \mathcal{L}^1(A)$. Denote by $\nu : r(A) \to \mathbb{R}^n$ the field of exterior unit normal vectors to A. Then ν is defined $\mathcal{H}^{n-1}$-a.e. on ∂A. We have*

$$\int_A D_i f(x) \, dx = \int_{\partial A} f \nu_i \, d\mathcal{H}^{n-1} \qquad \forall i = 1, \ldots, n.$$

Proof. Recall that $r(A)$ is the set of regular points for ∂A. We set $s(A) := \partial A \setminus r(A)$, Δ to be an open set so that $\Delta \supset \overline{A}$, and, finally, $\Omega := \Delta \setminus s(A)$. Since $s(A)$ is closed, Ω is open. Now, for $x \in \Omega$ we can choose an open neighborhood U_x of x so that

(i) if $x \in \Omega \setminus \overline{A}$, then U_x is a cube centered at x and contained in $\Omega \setminus \overline{A}$,
(ii) if $x \in A \cap \Omega$, then U_x is a cube centered at x and contained in $\Omega \cap A$,
(iii) if $x \in \partial A \cap \Omega$, i.e., $x \in r(\partial A)$, then we choose U_x as in the definition of regular points and, without loss of generality, we assume that U_x is small enough so that $U_x \subset \Omega$.

The family $\{U_x\}$ covers Ω. Therefore, there exists a denumerable locally finite refinement $\{B_j\}$ of $\{U_x\}$, Theorem 2.95, with the associated decomposition of unity $\{\alpha_j\}$, Theorem 2.97, and we distinguish the following three cases:

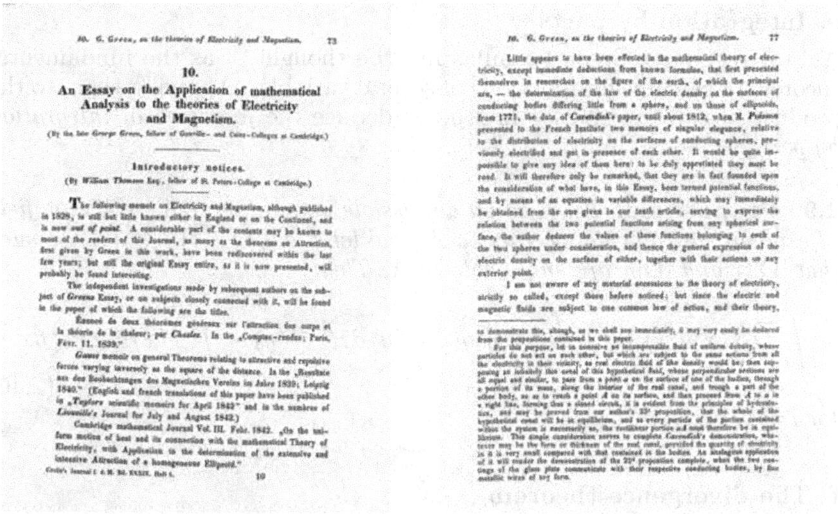

10. G. Green, on the theories of Electricity and Magnetism. 73

10.

An Essay on the Application of mathematical Analysis to the theories of Electricity and Magnetism.

(By the late George Green, fellow of Gonville- and Caius-Colleges at Cambridge.)

Introductory notices.

(By William Thomson Esq., fellow of St. Peters-College at Cambridge.)

The following memoir on Electricity and Magnetism, although published in 1828, is still but little known either in England or on the Continent, and is now out of print. A considerable part of the contents may be known to most of the readers of this Journal, as many as the theorems on Attraction, first given by Green in this work, have been rediscovered within the last few years; but still the original Essay entire, as it is now presented, will probably be found interesting.

The independent investigations made by subsequent authors on the subject of Greens Essay, or on subjects closely connected with it, will be found in the paper of which the following are the titles.

Énoncé de deux théorèmes généraux sur l'attraction des corps et la théorie de la chaleur; par Chasles. In the „Comptes-rendus; Paris, Fevr. 11. 1839."

Gauss memoir on general Theorems relating to attractive and repulsive forces varying inversely as the square of the distance. In the „Resultate aus den Beobachtungen des Magnetischen Vereins im Jahre 1839; Leipzig 1840." (English and french translations of this paper have been published in „Taylors scientific memoirs for April 1842" and in the numbers of Liouville's Journal for July and August 1842.)

Cambridge mathematical Journal Vol. III. Febr. 1842. „On the uniform motion of heat and its connection with the mathematical Theory of Electricity, with Application to the determination of the extensive and intensive Attraction of a homogeneous Ellipsoid."

Crelle's Journal f. d. M. Bd. XXXIX. Heft 1. 10

10. G. Green, on the theories of Electricity and Magnetism. 77

Little appears to have been effected in the mathematical theory of electricity, except immediate deductions from known formulae, that first presented themselves in researches on the figure of the earth, of which the principal are, — the determination of the law of the electric density on the surfaces of conducting bodies differing little from a sphere, and on those of ellipsoids, from 1771, the date of Cavendish's paper, until about 1812, when M. Poisson presented to the French Institute two memoirs of singular elegance, relative to the distribution of electricity on the surfaces of conducting spheres, previously electrified and put in presence of each other. It would be quite impossible to give any idea of them here: to be duly appretiated they must be read. It will therefore only be remarked, that they are in fact founded upon the consideration of what have, in this Essay, been termed potential functions, and by means of an equation in variable differences, which may immediately be obtained from the one given in our tenth article, serving to express the relation between the two potential functions arising from any spherical surface, the author deduces the values of these functions belonging to each of the two spheres under consideration, and thence the general expression of the electric density on the surface of either, together with their actions on any exterior point.

I am not aware of any material accessions to the theory of electricity, strictly so called, except those before noticed; but since the electric and magnetic fluids are subject to one common law of action, and their theory,

to demonstrate this, although, as we shall see immediately, it may very easily be deduced from the propositions contained in this paper.

For this purpose, let us conceive an incompressible fluid of uniform density, whose particles do not act on each other, but which are subject to the same actions from all the electricity in their vicinity, as real electric fluid of like density would be; then supposing an infinitely thin canal of this hypothetical fluid, whose perpendicular sections are all equal and similar, to pass from a point a on the surface of one of the bodies, through a portion of its mass, along the interior of the real canal, and through a part of the other body, so as to reach a point A on its surface, and then proceed from A to a in ...

Figure 2.17. Two pages from the *Essay* by George Green (1793–1841), which appeared in 1828 and was reprinted in 1850 in *Crelle's Journal* where Gauss–Green formulas appear.

- B_j is exterior to $\overline{A}$. Then $\alpha_j = 0$ in $\overline{A}$ hence

$$\int_A D_i(f\alpha_j)\,dx = 0 = \int_{\partial A} f\nu_i\alpha_j\,d\mathcal{H}^{n-1}.$$

- B_j is interior to A. Then from Proposition 2.92 and $\alpha_j = 0$ in ∂A

$$\int_A D_i(f\alpha_j)\,dx = 0 = \int_{\partial A} f\nu_i\alpha_j\,d\mathcal{H}^{n-1}.$$

- $B_j \cap \partial A \neq \emptyset$, then B_j is contained in some U_x of type (iii) and $f\alpha_j : U_x \to \mathbb{R}$ satisfies the assumptions of Proposition 2.93. If follows

$$\int_A D_i(f\alpha_j)\,dx = \int_{U_x} D_i(f\alpha_j)\,dx = \int_{\partial A\cap U_x} f\nu_i\alpha_j\,d\mathcal{H}^{n-1} = \int_{\partial A} f\nu_i\alpha_j\,d\mathcal{H}^{n-1}.$$

Summing on $j = 1, \dots,$ since $\sum_{j=1}^{\infty}\alpha_j = 1$ in Ω, $\{B_j\}$ is locally finite and

$$\mathcal{H}^{n-1}(\partial A\cap\Omega) = \mathcal{H}^{n-1}(r(A)) = \mathcal{H}^{n-1}(\partial A),$$

we conclude

$$\begin{aligned}
\int_A D_i f\,dx &= \int_{A\cap\Omega} D_i f\,dx = \int_A \sum_{j=1}^{\infty}(D_i f)\alpha_j\,dx = \sum_{j=1}^{\infty}\int_A D_i(f\alpha_j)\,dx \\
&= \sum_{j=1}^{\infty}\int_{\partial A} f\nu_i\alpha_j\,d\mathcal{H}^{n-1} = \int_{\partial A} f\nu_i\sum_{j=1}^{\infty}\alpha_j\,d\mathcal{H}^{n-1} = \int_{\partial A\cap\Omega} f\nu_i\,d\mathcal{H}^{n-1} \\
&= \int_{\partial A} f\nu_i\,d\mathcal{H}^{n-1}.
\end{aligned}$$

□

e. Integration by parts

As stated, Gauss–Green formulas may be thought of as the fundamental theorem of calculus for functions of several variables. Applying them to the product of two functions f and g, we deduce the *formulas of integration by parts*.

2.99 Proposition. *Let A be an admissible domain, $\nu : \partial A \to \mathbb{R}^n$ the field of exterior unit normal vectors to A, and let $f, g \in C^0(\overline{A}) \cap C^1(A)$ be such that $|\mathbf{D}f|$ and $|\mathbf{D}g|$ are summable in A. Then*

$$\int_A D_i f(x) g(x)\, dx = \int_{\partial A} f(y) g(y) \nu_i(y)\, d\mathcal{H}^{n-1}(y) - \int_A f(x) D_i g(x)\, dx$$

$$(2.46)$$

for $i = 1, 2, \ldots, n$.

f. The divergence theorem

Let A be an admissible domain and $E : \overline{A} \to \mathbb{R}^n$, $E = (E^1, E^2, \ldots, E^n)$ a field of class $C^0(\overline{A}) \cap C^1(A)$ with summable Jacobian matrix $\mathbf{D}E$. The *divergence* of E at $x \in A$ is the number

$$\operatorname{div} E(x) := \operatorname{tr} \mathbf{D}E(x) = \sum_{i=1}^n \frac{\partial E^i}{\partial x^i}(x) = \sum_{i=1}^n D_i E^i(x).$$

Since the functions $D_i E^i : A \to \mathbb{R}$, $i = 1, \ldots, n$, are summable, if we apply the Gauss–Green formulas to them, we find in particular

$$\int_A D_i E^i\, dx = \int_{\partial A} E^i \nu_i\, d\mathcal{H}^{n-1} \qquad \forall i = 1, \ldots, n$$

and, summing over i, the *divergence theorem*

$$\int_A \operatorname{div} E(x)\, dx = \int_{\partial A} E \bullet \nu\, d\mathcal{H}^{n-1}.$$

$$(2.47)$$

The quantity

$$\phi(E, A) := \int_{\partial A} E \bullet \nu\, d\mathcal{H}^{n-1}$$

is called the *flux* of E outgoing from A.

g. Geometrical meaning of the divergence

Let $E : A \to \mathbb{R}^n$ be a field that we assume of class $C^1(A)$. For every ball $B(x, r) \subset\subset A$ we denote by $\phi(E, r)$ the flux of E outgoing from $B(x, r)$,

$$\phi(E, r) := \int_{\partial B(x,r)} E \bullet \nu\, d\mathcal{H}^{n-1}, \qquad \nu(x) := x/|x|.$$

The divergence theorem yields

$$\phi(E, r) = \int_{B(x,r)} \operatorname{div} E(x)\, dx$$

hence, if we divide by $|B(x,r)| = \omega_n r^n$ and let $r \to 0$, we infer

$$\lim_{r \to 0} \frac{\phi(E, r)}{\omega_n r^n} = \lim_{r \to 0} \frac{1}{|B(x,r)|} \int_{B(x,r)} \operatorname{div} E(y)\, dy = \operatorname{div} E(x),$$

because of the continuity of $\operatorname{div} E(x)$, or

$$\phi(E, r) = \omega_n \operatorname{div} E(x) r^n + o(r^n) \qquad \text{as } r \to 0.$$

In other words, $\operatorname{div} E(x)$ represents the (rescaled) flow outgoing from an infinitesimal ball centered at x.

h. Divergence and transport of volume

Let $A \subset \mathbb{R}^n$ be open and $F : \mathbb{R} \times A \to \mathbb{R}^n$ be smooth. A curve $\gamma(t) :$ $I \to A$ satisfying the differential equation $\gamma'(t) = F(t, \gamma(t))$, i.e., a curve $t \to (t, \gamma(t))$ with velocity $(1, F(t, \gamma(t)))$, is called a *flux line* or an *integral line* of F. As we shall see in Chapter 6, for every $x \in A$ there exists a unique flux line defined for small times that at time $t = 0$ is at x. If we denote by $\phi(x, t)$ these flux lines, i.e.,

$$\begin{cases} \dfrac{\partial}{\partial t}\phi(t, x) = F(t, \phi(t, x)), \\[2mm] \phi(0, x) = x, \end{cases}$$

and set $\phi_t(x) := \phi(t, x)$, then $D\phi_0(x) = \operatorname{Id}$, and, for $K \subset\subset \Omega$ there exists ϵ_0 such that $\phi(t, x)$ is defined on $]-\epsilon_0, \epsilon_0[\times K$ with $\det \mathbf{D}\phi_t(x) > 0$. From (1.28) with $A(t) := \mathbf{D}\phi_t(x)$, we then infer

$$\begin{aligned} \frac{\partial}{\partial t}[\det \mathbf{D}\phi_t(x)](t) &= \det \mathbf{D}\phi_t(x)\operatorname{tr}\left(\mathbf{D}\phi_t(x)^{-1}\frac{\partial}{\partial t}\mathbf{D}\phi_t(x)\right) \\ &= \det \mathbf{D}\phi_t(x)\operatorname{tr}\left(\mathbf{D}\phi_t(x)^{-1}\mathbf{D}\frac{\partial}{\partial t}\phi(t, x)\right) \\ &= \det \mathbf{D}\phi_t(x)\operatorname{tr}\left(\mathbf{D}\phi_t(x)^{-1}\mathbf{D}F(t, \phi(t, x))\,\mathbf{D}\phi_t(x)\right) \\ &= \det \mathbf{D}\phi_t(x)\operatorname{tr}\mathbf{D}F(t, \phi(t, x)) \\ &= \det \mathbf{D}\phi_t(x)\operatorname{div}F(t, \phi(t, x)). \end{aligned}$$

If $\Omega \subset\subset A$ and $\Omega_t := \phi_t(\Omega)$ is the image of Ω at time t transported by the flow, then the area formula says

$$\mathcal{L}^n(\Omega_t) = \int_\Omega |\det \mathbf{D}\phi_t(x)|\, dx = \int_\Omega \det \mathbf{D}\phi_t(x)\, dx$$

and, differentiating under the integral sign,

$$\frac{d\mathcal{L}^n(\Omega_t)}{dt}(t) = \int_\Omega \frac{\partial}{\partial t} \det \mathbf{D}\phi_t(x) dx$$

$$= \int_\Omega \det \mathbf{D}\phi_t(x) \operatorname{div} F(t, \phi(t, x)) \, dx = \int_{\Omega_t} \operatorname{div} F(t, x) \, dx.$$

In the so-called *autonomous case*, $F = F(x)$, and for $t = 0$, we get

$$\frac{1}{\mathcal{L}^n(\Omega)} \frac{d\mathcal{L}^n(\Omega_t)}{dt}(0) = \frac{1}{\mathcal{L}^n(\Omega)} \int_\Omega \operatorname{div} F(x) \, dx,$$

i.e., div $E(x)$ is the percentage variation of the infinitesimal volume when transported by the flow at time $t = 0$.

2.7 Exercises

2.100 ¶. Let $C(A)$ be the cone of basis $A = \{(x, y) \in \mathbb{R}^2 \,|\, x^2 < y < 1\}$, and vertex $(0, 0, 1)$. Compute the volume of $C \setminus B((0, 0, 1), 1/2)$.

2.101 ¶. Prove Schwarz's theorem, Theorem 1.34, for functions of class $C^2(\Omega)$ by using the theorem of differentiation under the integral sign. [*Hint:* Differentiate at (t_0, x_0) the identity

$$f(t, x_0 + h) - f(t, x_0) = \int_0^h \frac{\partial f}{\partial x}(t, s) \, dt$$

for $|t - t_0|, |h|$ small enough, and then use the fundamental theorem of calculus.]

2.102 ¶. Show that *Airy's function* $\phi(t) := \frac{1}{\sqrt{\pi}} \int_0^\infty \cos\left(tx + \frac{x^3}{3}\right) dx$ solves the equation

$$\varphi''(t) - t\varphi(t) = 0.$$

2.103 ¶. Show a sequence $\{f_n\}$ of nonnegative summable functions on $[0, 1]$ such that

$$\lim_{n \to \infty} \int_0^1 f_n(x) \, dx = 0 \qquad \text{and} \qquad \limsup_{n \to \infty} f_n(x) = +\infty \quad \forall x \in [0, 1].$$

2.104 ¶. Show that $f' : [0, 1] \to \mathbb{R}$ is measurable if $f : [0, 1] \to \mathbb{R}$ is differentiable.

2.105 ¶. Show that

$$\int_0^1 \frac{x^{1/3}}{1 - x} \log \frac{1}{x} \, dx = \sum_{n=0}^\infty \frac{9}{(3n + 4)^2},$$

$$\int_0^\infty e^{-x} \cos \sqrt{x} \, dx = \sum_{n=0}^\infty (-1)^n \frac{n!}{(2n)!},$$

$$\int_0^\pi \sum_{n=1}^\infty \frac{n^2 \sin nx}{a^n} = \frac{2a(1 + a^2)}{(a^2 - 1)^2} \qquad \forall a > 1.$$

2.106 ¶. Show that for $p, q > 0$ we have

$$\int_0^1 \frac{x^{p-1}}{1+x^q} = \sum_{n=0}^{\infty} \frac{(-1)^n}{p+nq};$$

infer that $\frac{\pi}{4} = \sum_{n=0}^{\infty} \frac{(-1)^n}{2n+1}$.

2.107 ¶. Show that for $|a| < 1$,

$$\int_0^1 \frac{1-t}{1-at^3}\, dt = \sum_{n=0}^{\infty} \frac{a^n}{(3n+1)(3n+2)};$$

infer that

$$\frac{\pi}{3\sqrt{3}} = \sum_{n=0}^{\infty} \frac{1}{(3n+1)(3n+2)}.$$

2.108 ¶. Compute $\int_D \frac{e^{-xy}}{y}\, dxdy$ where $D := \{(x,y) \in \mathbb{R}^2 \mid x \geq 0,\ x^2 \leq y, 0 \leq y \leq 2\}$.

2.109 ¶. Show that

$$\int_0^\infty \frac{x^{\alpha-1}}{e^{at}-1}\, dt = \frac{\Gamma(\alpha)}{a^\alpha} \sum_{j=0}^{\infty} \frac{1}{(n+1)^\alpha},$$

and

$$\int_0^1 \frac{\arctan t}{\sqrt{1-t^2}}\, dt = \frac{\pi}{2} \log(1 + \sqrt{2}).$$

2.110 ¶. Let $\mathbf{A}$ be a positive $n \times n$ symmetric matrix. Show that

$$\int_{\mathbb{R}^n} \exp\left(-\mathbf{A}x \bullet x\right) dx = \sqrt{\frac{\pi^n}{\det \mathbf{A}}}.$$

2.111 ¶. Let $f : [0,1] \to \mathbb{R}$ be a continuous function. Show that $\mathcal{L}^2(\mathcal{G}_{f,[0,1]}) = 0$.

2.112 ¶. Let $E \subset \mathbb{R}^n$. Show that E is measurable if $\mathcal{L}^{n*}(\partial E) = 0$.

2.113 ¶. Compute

$$\lim_{t \to +\infty} \int_{-1}^4 \frac{t^2 + \sqrt{|x|}}{1+t^2 x^2}\, dx, \qquad \lim_{t \to 0+} \int_0^1 \frac{x + \sqrt{tx}}{t+x}\, dx.$$

2.114 ¶. Show that for $\alpha > 0$

$$\lim_{n \to \infty} \int_0^n \left(1 - \frac{x}{n}\right)^n x^{\alpha-1}\, dx = \int_0^\infty e^{-x} x^{\alpha-1}\, dx.$$

2.115 ¶ Astroid. Compute the area and the length of the boundary of the astroid

$$A := \left\{(x,y) \in \mathbb{R}^2 \,\middle|\, x^{2/3} + y^{2/3} \leq 1\right\}.$$

2.116 ¶. If $S^2 := \{(x,y,z) \mid x^2 + y^2 + z^2 = 1\}$, compute

$$\int_{S^2} x^2\, d\mathcal{H}^2.$$

2.117 ¶. Let T be the triangle in $\mathbb{R}^3$ with vertices $(1,0,0)$, $(0,1,0)$ and $(0,0,1)$. Compute

$$\int_T x\, d\mathcal{H}^2.$$

2.118 ¶. If $G \subset \mathbb{R}^3$ is the graph of the function $f : [0,1] \times [-1,1] \to \mathbb{R}$, $f(x,y) = x^2 + y$, compute

$$\int_G x\, d\mathcal{H}^2.$$

2.119 ¶. For $a, L > 0$, let $C \subset \mathbb{R}^3$ be the truncated cone

$$C := \Big\{ (x,y,z) \in \mathbb{R}^3 \,\Big|\, z^2 = a\,(x^2 + y^2),\ 0 \le z \le L \Big\}.$$

Compute the volume of C and the area of the boundary of C.

2.120 ¶ The Viviani solid. Let

$$V := \Big\{ (x,y,z) \,\Big|\, x^2 + y^2 + z^2 \le 1,\ x^2 + y^2 \le x \Big\}$$

be the intersection of the unit ball in $\mathbb{R}^3$ with the vertical cylinder $\{(x,y,z) \in \mathbb{R}^3 \,|\, x^2 + y^2 - x \le 0\}$.

(i) Compute the colume of V.
(ii) Show that $S := \partial V = S_1 \cup S_2$ where

$$S_1 := \Big\{ (x,y,z) \in \mathbb{R}^3 \,\Big|\, x^2 + y^2 + z^2 = 1,\ x^2 + y^2 \le x \Big\},$$

$$S_2 := \Big\{ (x,y,z) \in \mathbb{R}^3 \,\Big|\, x^2 + y^2 + z^2 \le 1,\ x^2 + y^2 = x \Big\},$$

and compute the area of S_1 and S_2.
(iii) Show that the curve $s(\alpha) := (\cos^2 \alpha, \cos \alpha \sin \alpha, \sin \alpha)$ maps the interval $]-\pi, \pi[$ onto $S_1 \cap S_2$, and compute the length of $S_1 \cap S_2$.

2.121 ¶. Compute

$$\int_{\mathbb{R}^{n-1}} \frac{1}{(1+|x|^2)^n}\, dx.$$

2.122 ¶. Compute $\mathcal{H}^{n-1}(\Sigma_{n-1})$ where

$$\Sigma_{n-1} := \Big\{ x \in \mathbb{R}^n,\ \Big|\ \sum_{i=1}^n x_i = 1,\ 0 \le x_i \le 1\ \forall i \Big\}.$$

2.123 ¶ Feynman's formula. Let $a \in \mathbb{R}^n$ be a point with positive coordinates. Show that

$$\int_{S_+^{n-1}} \frac{1}{a \bullet x}\, d\mathcal{H}^{n-1}(x) = \frac{1}{(n-1)!\, \prod_{1 \le j \le n} a_j}.$$

2.124 ¶. Let $f : \mathbb{R}^n \setminus \{0\} \to \mathbb{R}$ be positively homogeneous of degree d, $f(tx) = t^d f(x)$ $\forall x \in \mathbb{R}^n \setminus \{0\}$ $\forall t > 0$. Prove that

$$\int_{B(0,1)} \Delta f(x)\, dx = d \int_{\partial B(0,1)} f(x)\, d\mathcal{H}^{n-1}(x).$$

In particular, if $x = (x_1, \ldots, x_n)$, show that $\forall j = 1, \ldots, n$

$$\mathcal{L}^n(B(0,1)) = \int_{S^{n-1}} x_j^2\, d\mathcal{H}^{n-1}.$$

2.125 ¶. If $B_R \subset \mathbb{R}^n$ denotes the ball of radius R around 0 in $\mathbb{R}^n$, show that for every $f \in C^1(\overline{B_R})$ we have

$$\int_{B_R} \left(\sum_{i=1}^n x^i D_i f(x) + n f(x) \right) dx = R \int_{\partial B_R} f(x) \, d\mathcal{H}^{n-1}(x).$$

2.126 ¶. If $f \in C^3(\overline{\Omega})$ and $\nabla f = 0$ on $\partial\Omega$, show that

$$\int_\Omega (\Delta f)^2 \, dx = \int_\Omega \sum_{1 \le i,j \le n} (D_i D_j f)^2 \, dx.$$

2.127 ¶. Compute the outgoing flux from the unit ball in $\mathbb{R}^3$ centered at 0 of the field $E := (2x, y^2, z^2)$.

2.128 ¶. Compute the outgoing flux from the lateral surface of the cylinder

$$C := \left\{ (x,y,z) \in \mathbb{R}^3 \,\middle|\, x^2 + y^2 \le 1, \; -1 \le z \le 1 \right\}$$

of the field $E = (xy^2, xy, y)$.

2.129 ¶. Let Ω be an open admissible set. Then

$$\mathcal{L}^n(\Omega) = \frac{1}{n} \int_{\partial\Omega} x \bullet \nu_\Omega \, d\mathcal{H}^{n-1}.$$

3. Curves and Differential Forms

In this chapter we discuss notions such as force, work, vector field, differential form, conservative vector field and its potential, and the solvability in an open set $\Omega \subset \mathbb{R}^n$ of the equation

$$\operatorname{grad} U = F.$$

We shall see that the vector field F is conservative, i.e., the equation $\operatorname{grad} U = F$ is solvable, if and only if the work along closed curves in Ω is zero, and we shall discuss how to compute a solution, a potential.

When $n = 3$, every function U of class C^2 satisfies the equation $\operatorname{rot} \operatorname{grad} U = 0$. Therefore, $\operatorname{rot} F = 0$ in Ω is a necessary condition in order for the vector field $F \in C^1$ to be conservative in Ω. In terms of differential forms, we shall also see that $\operatorname{rot} F = 0$ suffices for F to be conservative in simply connected domains.

Though Lebesgue's theory of integration would allow us more general results, here we prefer to limit ourselves to the use of Riemann integral.

3.1 Differential Forms, Vector Fields, and Work

a. Vector fields and differential forms

We recall that with respect to a basis $(e_1, e_2, \ldots, e_n)$ in $\mathbb{R}^n$ with coordinates $(x^1, x^2, \ldots, x^n)$, every linear map $\ell : \mathbb{R}^n \to \mathbb{R}$ writes as

$$\ell = \sum_{i=1}^{n} \ell_i dx^i$$

where, for $i = 1, \ldots, n$, $\ell_i := \ell(e_i)$ and for $h = \sum_{i=1}^{n} h^i e_i \in \mathbb{R}^n$ $dx^i(h) := h^i$. From now on we shall denote the action of a linear map over the vector h by

$$< \ell, h > := \ell(h).$$

M. Giaquinta and G. Modica, *Mathematical Analysis: An Introduction to Functions of Several Variables*, DOI: 10.1007/978-0-8176-4612-7_3, © Birkhäuser Boston, a part of Springer Science + Business Media, LLC 2010

We shall also recall that, if $\bullet$ is an inner product in $\mathbb{R}^n$, then for every linear map $\ell : \mathbb{R}^n \to \mathbb{R}$, there is by Riesz theorem a unique vector $F \in \mathbb{R}^n$ such that

$$< \ell, h > = \ell(h) = F \bullet h \qquad \forall h \in \mathbb{R}^n. \tag{3.1}$$

The coordinates of F can be easily computed writing (3.1) for $h = e_1, e_2, \ldots, e_n$, to get $F = \mathbf{G}^{-1}\mathbf{L}^T$, where $\mathbf{G}$ is the metric tensor $\mathbf{G} := [G_{ij}]$, $G_{ij} = e_i \bullet e_j$, and $\mathbf{L}$ is the row vector $\mathbf{L} := (\ell(e_1), \ell(e_2), \ldots, \ell(e_n))$.

The structure we have just described is used for instance in mechanics to model the relationship between force and work. If we think of a force as a primitive notion, (3.1) defines $(F|h)$ as the elementary work done by F in the direction $h \in \mathbb{R}^n$, whereas, if the work $h \to \ell(h)$ is our primitive notion (notice that, in fact, we measure the work and not the force), then (3.1) provides us with the force F that does the work, see [GM3].

3.1 Definition. *Let Ω be an open set of $\mathbb{R}^n$.*

(i) *A* vector field *in Ω is a map $F : \Omega \to \mathbb{R}^n$.*
(ii) *A* differential form *ω in Ω is a map $\omega : \Omega \to \mathcal{L}(\mathbb{R}^n, \mathbb{R})$ that associates to every $x \in \Omega$ a linear map $\omega(x) : \mathbb{R}^n \to \mathbb{R}$.*

Hence, in coordinates a differential form can be written as

$$\omega(x) = \sum_{i=1}^{n} \omega_i(x)\, dx^i, \qquad \omega_i(x) :=< \omega(x), e_i >, \; x \in \Omega,$$

and a vector field as $F(x) = (F(x)^1, F(x)^2, \ldots, F(x)^n)$. If ω is a differential form on Ω, and F is the (unique) vector field on Ω such that

$$< \omega(x), h > = F(x) \bullet h \qquad \forall h \in \mathbb{R}^n \; \forall x \in \Omega, \tag{3.2}$$

we say that F is the vector field *associated* to ω or that ω is the differential form *associated* to F. We say that a differential form is of class C^k if its components in a basis are of class C^k. Notice that a differential form is of class C^k if and only if its associated vector field is of class C^k.

b. Curves

We briefly recall a few facts about curves, see, e.g., [GM3]. Let Ω be an open set of $\mathbb{R}^n$. A curve in Ω is a map $\gamma : [a, b] \to \Omega$. Its image $\gamma([a, b])$ is the trajectory of the curve, $\gamma(a)$ is its initial point, $\gamma(b)$ is its final point. A curve $\gamma : [a, b] \to \Omega$ is said to be *regular* if $\gamma'(t) \neq 0$ for all $t \in [a, b]$, *closed* if $\gamma(a) = \gamma(b)$, and *simple* if $\gamma(t) \neq \gamma(s) \; t \neq s \in]a, b]$. The length of γ is given by

$$L(\gamma) := \int_a^b |\gamma'(s)|\, ds.$$

At a point t where $|\gamma'(t)| \neq 0$, the *unit tangent vector* to γ is defined by

$$\mathbf{t}(t) := \frac{\gamma'(t)}{|\gamma'(t)|}. \tag{3.3}$$

Notice that $\operatorname{Span} \mathbf{t}(t)$ is the tangent line to $\gamma([a, b])$ at $\gamma(t)$.

A reparameterization of γ is a new curve $\delta \in C^1([c, d], \mathbb{R}^n)$ such that there exists $h : [c, d] \to [a, b]$ of class C^1 that is one-to-one and such that $\delta(s) = \gamma(h(s)) \ \forall s \in [c, d]$. Since either $h' > 0$ or $h' < 0$ on $[c, d]$, we say in the former case that δ is an *increasing reparameterization* of γ or that h is an *orientation preserving* reparameterization, whereas in the latter case we say that δ is a *decreasing reparameterization* of γ or that h is a *reversing orientation* reparameterization. Trivially

$$\frac{\delta'(s)}{|\delta'(s)|} = \operatorname{sgn}(h') \frac{\gamma'(h(s))}{|\gamma'(h(s))|}.$$

Recall that the length of a curve does not change under reparameterization of it and also that two *simple* curves of class C^1 are reparameterizations of the same curve if and only if they have the same trajectory.

A natural reparameterization of a regular curve $\gamma \in C^1([a, b], \mathbb{R}^n)$ is obtained in terms of the *curvilinear abscissa*, or *arc length*, i.e., the parameterization in terms of the traveled space

$$s(t) := \int_a^t |\gamma'(\tau)| \, d\tau.$$

In fact, the function $s(t) : [a, b] \to [0, L(\gamma)]$ is strictly increasing and of class C^1 with $s'(t) = |\gamma'(t)| > 0$. Therefore, its inverse $t(s) : [0, L(\gamma)] \to [a, b]$ is also of class C^1 and increasing with $t'(s) = 1/|\gamma'(t(s))|$ for all $s \in [0, L]$. Consequently, the parameterization of γ by arc length

$$\delta(s) := \gamma(t(s)), \qquad t \in [0, L(\gamma)]$$

has the same orientation of γ and $|\delta'(s)| = 1 \ \forall s \in [0, L(\gamma)]$.

3.2 Remark. Let $\gamma : [a, b] \to \mathbb{R}^n$ and $\delta : [c, d] \to \mathbb{R}^n$ be two simple curves with the same trajectory, $\gamma([a, b]) = \delta([c, d])$. Though in general they are not one the reparameterization of the other, we have $\delta = \gamma \circ h$ where $h : [c, d] \to [a, b]$ is a homeomorphism. This suffices to show that γ and δ have the same length, see [GM3]. This follows also from the area formula, Theorems 2.78 and 2.80, since for a simple curve γ of class C^1 we have

$$L(\gamma) = \int_a^b |\gamma'(t)| \, dt = \mathcal{H}^1(\gamma([a, b])).$$

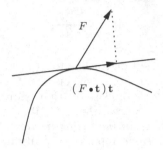

Figure 3.1. The tangent component of a vector field at a point.

c. Integration along a curve and work

3.3 Definition. *Let $\Omega \subset \mathbb{R}^n$ be open and let $\gamma : [a,b] \to \Omega$ be a curve of class C^1. The* work, *or integral of a differential form ω along γ is the number, denoted by $\int_\gamma \omega$ or $\mathcal{L}(\gamma, \omega)$, given by*

$$\int_\gamma \omega = \mathcal{L}(\gamma, \omega) := \int_a^b <\omega(\gamma(s)), \gamma'(s)> \, ds. \qquad (3.4)$$

Similarly, the work *of a continuous vectorfield F in Ω is defined as*

$$\int_\gamma (F|ds) = \mathcal{L}(\gamma, F) := \int_a^b F(\gamma(s)) \bullet \gamma'(s) \, ds.$$

It is easily seen that

○ the work functional is linear on forms,

$$\int_\gamma (\omega + \eta) = \int_\gamma \omega + \int_\gamma \eta \quad \text{and} \quad \int_\gamma \lambda\omega = \lambda \int_\gamma \omega$$

for all couple of forms ω, η and every $\lambda \in \mathbb{R}$,

○ if $<\omega(x), h> = F(x) \bullet h$ $\forall h$ $\forall x \in \Omega$, then $\int_\gamma \omega = \int_\gamma F \, ds$,

○ the work of ω along γ does not change if we reparameterize γ preserving the orientation and changes sign if we reverse the orientation,

○ the definition of work extends to continuous curves that are piecewise regular.

3.4 ¶. The use of continuous and piecewise regular curves is useful, but not necessary. In fact, every such curve admits a reparameterization of class C^1 with $h' \geq 0$, therefore with $\mathcal{L}(\delta, \omega) = \mathcal{L}(\gamma, \omega)$.

If $\gamma_i : [a,b] \to \Omega$, $i = 1, \ldots, k$, are k curves of class $C^1([a,b])$ with $\gamma_{i+1}(a) = \gamma_i(b)$ on an open set $\Omega \subset \mathbb{R}^n$, we may and do define a new curve, traveling successively on $\gamma_1, \gamma_2, \ldots, \gamma_n$, called the *join* of $\gamma_1, \gamma_2, \ldots, \gamma_n$ as follows. We choose $(n+1)$ points $\{t_j\}$ in $[a,b]$, for instance $t_j := j(b-a)/k + a$, $j = 0, \ldots, k$, and for $i = 1, \ldots, k$ we reparameterize γ_i on $[t_{i-1}, t_i]$ as

$$\delta_i := \gamma \circ \phi_i. \qquad \phi_i(s) := a + \frac{b-a}{t_i - t_{i-1}}(s - t_{i-1}).$$

The new curve γ, defined by $\gamma(s) := \delta_i(s)$ if $s \in [t_{i-1}, t_i[$, is also denoted by

$$\gamma = \sum_{i=1}^{n} \gamma_i$$

and, for any continuous differential form on Ω, we have

$$\mathcal{L}\left(\sum_{i=1}^{k} \gamma_i, \omega\right) = \sum_{i=1}^{k} \mathcal{L}(\omega, \delta_i) = \sum_{i=1}^{k} \mathcal{L}(\omega, \gamma_i)$$

since $\mathcal{L}(\gamma_i, \omega) = \mathcal{L}(\delta_i, \omega) \ \forall i$.

Finally, if $-\gamma(s) := \gamma((1-s)b + sa)$, $s \in [0,1]$, then $-\gamma$ is a decreasing reparametrization of γ and

$$\mathcal{L}(-\gamma, \omega) = -\mathcal{L}(\gamma, \omega).$$

3.5 Remark. Let $\gamma : [a, b] \to \mathbb{R}^n$ be a simple curve of class C^1 in an open set Ω of $\mathbb{R}^n$. For every t for which $\gamma'(t) \neq 0$, the unit tangent vector

$$\mathbf{t}_\gamma(y) := \frac{\gamma'(t)}{|\gamma'(t)|}, \qquad \gamma(t) = y,$$

is well defined in y. If ω is a continuous differentiable form, we wish to write

$$L(\gamma, \omega) = \int_a^b < \omega(\gamma(t)), \gamma'(t) > dt = \int_R < \omega(\gamma(t)), \mathbf{t}_\gamma(\gamma(t)) > |\gamma'(t)| \, dt \tag{3.5}$$

where $R := \{t \in [a, b] \,|\, \gamma'(t) \neq 0\}$. However, one can show nonnegative scalar continuous functions for which the characteristic function of $\{x \,|\, f(x) \neq 0\}$ is not Riemann integrable, hence the right-hand side in (3.5) is meaningless, in general. However, the difficulty is overcome if we interpret the integrals as Lebesgue's integral. In fact, R is open in $[a, b]$, thus Lebesgue-measurable. Moreover, by the area formula, Theorems 2.78 and 2.80, we have

$$0 = \int_{[a,b]\setminus R} |\gamma'(t)| \, dt = \mathcal{H}^1(\gamma([a, b] \setminus R)).$$

It follows that $\mathbf{t}_\gamma(y)$ exists for $\mathcal{H}^1$-a.e. $y \in \gamma([a, b])$ and, again by the area formula,

$$\int_a^b <\omega(\gamma(t)), \gamma'(t)> dt = \int_R <\omega(\gamma(t)), \mathbf{t}_\gamma(y)> |\gamma'(t)| \, dt$$

$$= \int_{\gamma(R)} <\omega(y), \mathbf{t}_\gamma(y)> d\mathcal{H}^1(y) \qquad (3.6)$$

$$= \int_{\gamma([a,b])} <\omega(y), \mathbf{t}_\gamma(y)> d\mathcal{H}^1(y).$$

As a consequence of (3.6) we can state: if $\gamma : [a, b]$ and $\delta : [c, d] \to \mathbb{R}^n$ are two simple curves of class C^1 with the same trajectory $\Delta = \gamma([a, b]) = \delta([c, d])$ (not necessarily one the reparameterization of the other), then either $\mathbf{t}_\gamma(y) = \mathbf{t}_\delta(y)$ for $\mathcal{H}^1$-q.o. $y \in \Delta$ and $\mathcal{L}(\omega, \gamma) = \mathcal{L}(\omega, \delta)$, or $\mathbf{t}_\gamma(y) = -\mathbf{t}_\delta(y)$ for $\mathcal{H}^1$-a.e. $y \in \Delta$ and $\mathcal{L}(\gamma, \omega) = -\mathcal{L}(\delta, \omega)$.

3.2 Conservative Fields and Potentials

a. Exact differential forms

3.6 Definition. *Let Ω be an open set of $\mathbb{R}^n$.*

(i) *A continuous differential form ω in Ω is said to be* exact *in Ω if there exists $f \in C^1(\Omega)$ such that $\omega(x) = df(x)$ for all $x \in \Omega$. The function f is called a* potential *of ω in Ω.*

(ii) *A continuous vector field, F in Ω is* conservative *in Ω if there exists $f \in C^1(\Omega)$ such that $F(x) = \nabla f(x)$ for all $x \in \Omega$. The function f is called a* potential *for F in Ω.*

Trivially, a differential form is exact if and only if its associated vector field is conservative, and a potential of ω is a potential of F and vice versa. We notice that two potentials of ω (or of F) in a connected open set differ by a constant, see Corollary 1.45.

3.7 Example. According to Hooke's law, the pulling force exerted by an ideal spring fixed at the origin on a point $x \in \mathbb{R}^n$ is given by $F(x) = -kx$, where $k > 0$ is called the elastic constant of the spring. The vector field $F(x) := -kx$, $x \in \mathbb{R}^n$, is conservative, since $F(x) = \nabla f(x)$ where $f(x) = -\frac{k}{2}|x|^2 \ \forall x \in \mathbb{R}^n$. In general, a *radial* vector field $F : \mathbb{R}^n \setminus \{0\} \to \mathbb{R}^n$,

$$F(x) = \varphi(|x|)x, \qquad \text{where } \varphi : \mathbb{R}_+ \to \mathbb{R}, \ \varphi \in C^0(]0, \infty[),$$

is conservative in $\mathbb{R}^n \setminus \{0\}$ since the function

$$f(x) := \int_1^{|x|} s\varphi(s) \, ds, \qquad , \ x \in \mathbb{R}^n \setminus \{0\},$$

is a potential of F in $\mathbb{R}^n \setminus \{0\}$, as it is easily seen by differentiating f.

A potential of the *gravitational field* $F(x) := -\frac{G}{|x|^3}x$, $x \in \mathbb{R}^3$, is

$$V(x) = \int_{+\infty}^{|x|} \frac{G}{s^2}\,ds = \frac{G}{|x|}.$$

We have chosen as an extremal point $+\infty$ since $1/s^2$ is summable at ∞. This way the potential V vanishes at ∞.

Let Ω be a connected open set in $\mathbb{R}^n$, let $f \in C^1(\Omega)$, and let $\gamma : [a,b] \to \Omega$ be a curve in Ω of class C^1. By definition

$$
\begin{aligned}
\int_\gamma df &= \int_a^b <df(\gamma(s)), \gamma'(s)>\,ds \\
&= \int_a^b \frac{d}{ds} f(\gamma(s))\,ds = f(\gamma(b)) - f(\gamma(a))
\end{aligned}
\tag{3.7}
$$

or, in other words, *the work of an exact differential form of class C^1 depends only on the difference of the potential at the extreme points of the curve.*

3.8 ¶. Show that (3.7) also holds for continuous and piecewise smooth curves.

Actually, the following holds.

3.9 Theorem (Fundamental theorem of calculus). *Let ω be a continuous differential form in an open and connected set $\Omega \subset \mathbb{R}^n$ and let $f : \Omega \to \mathbb{R}$. The following claims are equivalent.*

(i) *f is of class C^1 and is a potential for ω in Ω.*
(ii) *For any couple of points $x_0, x \in \Omega$ and for any piecewise smooth curve $\gamma_{x_0,x} : [u,b] \to \Omega$ joining x_0 to x we have*

$$f(x) - f(x_0) = \int_{\gamma_{x_0,x}} \omega.$$

(iii) *For any $x_0 \in \Omega$ there exists $\delta > 0$ such that for every $x \in B(x_0,\delta)$ we have*

$$f(x) - f(x_0) = \int_{r_{x_0,x}} \omega$$

where $r(t) := (1-t)x_0 + tx$, $t \in [0,1]$, is the segment curve going from x_0 to x.

Proof. As we have seen in (3.7), (i) implies (ii) and trivially (ii) implies (iii). Let us show that (iii) implies (i). for $x \in \Omega$, $i = 1,\dots,n$ and $h \in \mathbb{R}$, $|h|$ small, consider the segment $\delta(s) := x + she_i$, $0 \le s \le 1$, joining x to $x + he_i$. On account of the mean value theorem and of the continuity of ω_i, we find

$$
\begin{aligned}
\frac{f(x+he_i) - f(x)}{h} &= \frac{1}{h}\int_0^1 <\omega(x + she_i), he_i>\,ds \\
&= \int_0^1 \omega_i(x + she_i)\,ds \\
&= \frac{1}{h}\int_0^h \omega_i(x + te_i)\,dt \longrightarrow \omega_i(x),
\end{aligned}
$$

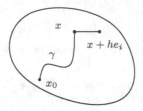

Figure 3.2. The proof of Theorem 3.9.

hence the partial derivatives of f at x_0 exist and

$$\frac{\partial f}{\partial x^i}(x) = \omega_i(x) \qquad \forall x \in \Omega.$$

Since ω is continuous, the partial derivatives are continuous in Ω, hence $f \in C^1(\Omega)$ and $df(x) = \omega(x)$ in Ω. □

3.10 Theorem. *Let ω be a continuous differential form in an open and connected set $\Omega \subset \mathbb{R}^n$. Then ω is exact if and only if the work of ω along any closed piecewise smooth curve is zero. In this case, a potential for ω is obtained as follows: fix $x_0 \in \Omega$ and for any $x \in \Omega$ choose a curve $\gamma_{x_0,x} : [a,b] \to \Omega$ joining x_0 to x with $\gamma(a) = x_0$ and $\gamma(b) = x$; then the work function*

$$f(x) := \int_{\gamma_{x_0,x}} \omega, \qquad x \in \Omega, \tag{3.8}$$

is a potential of ω in Ω.

Proof. Trivially (3.7) yields $\int_\gamma \omega = 0$ for any closed piecewise continuous curve, if ω has a potential. Conversely, assume that $\int_\gamma \omega = 0$ for all closed γ's. If $x, y \in \Omega$ and $\delta : [a,b] \to \Omega$ is a piecewise smooth curve with $\delta(a) = x$ and $\delta(b) = y$, then $\gamma := \gamma_{x_0,x} + \delta - \gamma_{x_0,y}$ is a closed curve in Ω, hence

$$0 = \mathcal{L}(\gamma, \omega) = \int_{\gamma_{x_0,x}} \omega + \int_\delta \omega - \int_{\gamma_{x,x_0}} \omega = \int_\delta \omega + f(x) - f(y).$$

□

3.11 ¶. We ask the reader to state the analogy of Theorem 3.9 for vector fields.

3.12 Exact forms in practice. Theorem 3.9 gives us a necessary and sufficient condition for a form ω to be exact. The necessary condition is useful to show that a differential form is *not exact*: if we find a closed curve along which ω does work, then ω is *not* exact. Instead, it is difficult to use the sufficient condition, as it requires us to verify that the work done on *every* closed curve is zero.

To prove that a given differential form is exact, in practice, it is easier to exhibit a potential, and, to do it, we essentially have two alternatives.

(i) Guess a potential.

(ii) Suppose Ω is an open connected subset of $\mathbb{R}^n$. Fix $x_0 \in \Omega$, and for $x \in \Omega$ choose a curve γ_x from x_0 to x, for example traveling parallel to the coordinates axes going from x_0 to x and compute the work $f(x) := \mathcal{L}(\omega, \gamma_x)$ along that curve. If $f \in C^1(\Omega)$ and $\frac{\partial f}{\partial x^i}(x) = \omega_i(x)$ for all $x \in \Omega$, then f is a potential for ω in Ω.

Notice that any two potentials of an (exact) differential form on a connected open set differ by a constant by Theorem 3.9.

3.13 Example. The differential form $w(x,y) := \sqrt{y/x}\, dx + \sqrt{x/y}\, dy$ is exact in $\Omega = \{(x,y) \in \mathbb{R}^2 \,|\, x > 0,\ y > 0\}$.

In fact, the function $f(x,y) = 2\sqrt{xy}$ is a potential of ω in Ω as

$$\frac{\partial \sqrt{xy}}{\partial x} = \sqrt{\frac{y}{x}}, \qquad \frac{\partial \sqrt{xy}}{\partial y} = \sqrt{\frac{x}{y}}, \qquad x > 0,\ y > 0.$$

3.14 Example. The differential form

$$\omega(x,y) = e^{x/y}\, dx + \left(1 - \frac{x}{y}\right)e^{x/y}\, dy,$$

is exact in $\Omega = \{(x,y) \in \mathbb{R}^2 \,|\, y > 0\}$. In fact, if we fix the point $(0,1)$ and join it to (x,y) first traveling from $(0,1)$ to $(0,y)$ vertically and then from $(0,y)$ to (x,y) horizontally, the work done is

$$f(x,y) = \int_1^y \left(1 - \frac{0}{y}\right)e^{0/y}\, dy + \int_0^x e^{x/y}\, dx = (y-1) + y\int_0^{x/y} e^t\, dt = y - 1 + y(e^{x/y} - 1).$$

It is easy to show that $f \in C^1(\Omega)$ and that $df(x,y) = \omega(x,y)$ in Ω, therefore f is a potential of ω in Ω.

3.3 Closed Forms and Irrotational Fields

a. Closed forms

There is another necessary condition for a differential form of class C^1 to be exact: If $\omega = df$ in an open set Ω, then $f \in C^2(\Omega)$, hence

$$\frac{\partial \omega_i}{\partial x^j} = \frac{\partial}{\partial x^j}\frac{\partial f}{\partial x^i} = \frac{\partial}{\partial x^i}\frac{\partial f}{\partial x^j} = \frac{\partial \omega_j}{\partial x^i}$$

for $i, j = 1, n$, by Schwarz's theorem. Motivated by this remark, we set the following.

3.15 Definition. *Let Ω be an open set of $\mathbb{R}^n$. We say that a differential form ω in Ω of class C^1 is closed in Ω if*

$$\frac{\partial \omega_i}{\partial x^j}(x) = \frac{\partial \omega_j}{\partial x^i}(x) \qquad \forall i,j = 1, n,\ \forall x \in \Omega.$$

A vector field $F : \Omega \to \mathbb{R}^n$ of class C^1 is said to be irrotational *in Ω if*

$$\frac{\partial F^i}{\partial x^j}(x) = \frac{\partial F^j}{\partial x^i}(x) \qquad \forall i,j = 1, n, \ \forall x \in \Omega.$$

The reason for naming it irrotational is that for $n = 3$, the *curl* of F given by

$$\operatorname{curl} F := \left(\frac{\partial F^3}{\partial y} - \frac{\partial F^2}{\partial z}, \frac{\partial F^1}{\partial z} - \frac{\partial F^3}{\partial x}, \frac{\partial F^2}{\partial x} - \frac{\partial F^1}{\partial y} \right)$$

is also called the *rotor* of F and denoted rot F: F is irrotational if rot $F =$ curl $F = 0$.

Notice that F is irrotational if and only if its associated differential form is closed.

3.16 Proposition. *Let $\Omega \subset \mathbb{R}^n$ be an open set. An exact differential form of class $C^1(\Omega)$ is closed in Ω. A conservative vector field of class $C^1(\Omega)$ is irrotational.*

3.17 Example. There are closed forms that are not exact. For instance, the *angle form*

$$\omega(x,y) := -\frac{y}{x^2 + y^2} \, dx + \frac{x}{x^2 + y^2} \, dy$$

is closed in $\Omega := \mathbb{R}^2 \setminus \{0\}$, as one sees differentiating. However, ω is not exact since the work of ω along the curve $t \to \gamma(t) = (\cos t, \sin t)$, $t \in [0, 2\pi]$ that travels along the boundary of the unit circle, is nonzero,

$$\int_\gamma \omega = \int_0^{2\pi} (\sin^2 t + \cos^2 t) \, dt = 2\pi \neq 0.$$

Notice that ω is exact in $\Omega := \{(x,y) \mid y > 0\}$: The function $f(x,y) := -\arctan(x/y)$, $(x,y) \in \Omega$, is a potential for ω in Ω.

3.18 ¶. Let Γ be a halfline in $\mathbb{R}^2$ from the origin. Show that the angle form in Example 3.17 is exact on $\mathbb{R}^2 \setminus \Gamma$.

Every matrix $\mathbf{A} \in M_{n,n}(\mathbb{R})$ splits into the sum of its symmetric part and antisymmetric part

$$\mathbf{A} = \frac{1}{2}(\mathbf{A} + \mathbf{A}^T) + \frac{1}{2}(\mathbf{A} - \mathbf{A}^T).$$

The symmetric part of the Jacobian matrix of a vector field F is called the *deformation gradient* of F and its antisymmetric part the *rotational gradient* of F that coincides with $\frac{1}{2}$rot F,

$$\begin{cases} \mathbf{D}F = \epsilon(F) + W(F) \\ \epsilon(F) = [\epsilon(F)^i_j], \qquad \epsilon(F)^i_j := \frac{1}{2}\left(\frac{\partial F^i}{\partial x^j} + \frac{\partial F^j}{\partial x^i} \right), \\ \mathbf{W}(F) = [\mathbf{W}(F)^i_j], \qquad \mathbf{W}(F)^i_j := \frac{1}{2}\left(\frac{\partial F^i}{\partial x^j} - \frac{\partial F^j}{\partial x^i} \right). \end{cases}$$

EINFÜHRUNG IN DIE THEORIE
DER SYSTEME VON
DIFFERENTIALGLEICHUNGEN

VON

E. KÄHLER

1934
LEIPZIG UND BERLIN
VERLAG UND DRUCK VON B. G. TEUBNER

Figure 3.3. George Gabriel Stokes (1819–1903) and an introduction to differential forms.

Similarly, if ω is a differential form, we call the matrix

$$\mathbf{W}(\omega) := [\mathbf{W}(\omega)_{ij}], \qquad \mathbf{W}(\omega)_{ij} := \frac{\partial \omega_i}{\partial x^j} - \frac{\partial \omega_j}{\partial x^i}$$

the *rotation matrix* of ω. Thus, ω is closed if and only if its *rotation matrix* is zero.

b. Poincaré lemma

We shall see that the obstruction to the exactness of every closed form in Ω is in the form of Ω. First we consider a class of special domains.

3.19 Definition. *We say that an open set $\Omega \subset \mathbb{R}^n$ is star-shaped with respect to one of its points x_0 if the segment joining it to any other point $x \in \Omega$ is contained in Ω. We simply say that Ω is star-shaped if Ω is star-shaped with respect to one of its points.*

3.20 Theorem (Poincaré lemma). *Let Ω be a star-shaped domain of $\mathbb{R}^n$. Every closed differential form ω in Ω of class C^1 is exact. Equivalently, every irrotational vector field F in Ω of class C^1 is conservative.*

Proof. After a translation we may assume that Ω is star-shaped with respect to the origin. If we parameterize the segment joining the origin to x as $\gamma(t) := tx/|x|$, $t \in [0, |x|]$, we need to prove that

$$f(x) := \mathcal{L}(\gamma, \omega) = \int_0^{|x|} < \omega(th), h > dt = \int_0^1 \sum_{i=1}^n \omega_i(th) h^i \, dt, \qquad x \in \Omega,$$

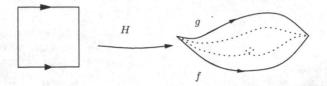

Figure 3.4. A homotopy with fixed end points between f and g.

is a potential for ω in Ω. Differentiating under the integral sign, we find

$$\frac{\partial f}{\partial x^j}(x) = \frac{\partial}{\partial x^j} \int_0^1 \sum_{i=1}^n \omega_i(tx)x^i \, dt = \int_0^1 \frac{\partial}{\partial x^j} \Big[\sum_{i=1}^n \omega_i(tx)x^i \Big] dt$$

and, using the closedness of ω,

$$\frac{\partial f}{\partial x^j}(x) = \int_0^1 \Big(\omega_j(tx) + t\frac{\partial \omega_i}{\partial x^j}(tx)x^i \Big) dt = \int_0^1 \Big(\omega_j(tx) + t\frac{\partial \omega_j}{\partial x^i}(tx)x^i \Big) dt$$

$$= \int_0^1 \Big(\omega_j(tx) + t\frac{d}{dt}\omega_j(tx) \Big) dt = \int_0^1 \frac{d}{dt}(t\omega_j(tx)) \, dt = \omega_j(x).$$

$\square$

Of course, the balls of $\mathbb{R}^n$ are connected and star-shaped with respect to its center, consequently we observe the following.

3.21 Proposition. *Every closed form of class C^1 is locally exact. Every irrotational vector field of class C^1 is locally conservative.*

c. Homotopic curves and work

Let Ω be an open connected set in $\mathbb{R}^n$. We recall, see [GM3], that two continuous curves $f, g : [0, 1] \to \Omega$ with $f(0) = g(0)$, $f(1) = g(1)$, are said to be *homotopic* in Ω if there exists a *continuous* map

$$H : R = [0, 1] \times [0, 1] \to \Omega$$

such that

$$\begin{cases} H(0, t) = f(t), \ \forall t \in [0, 1], \\ H(1, t) = g(t), \ \forall t \in [0, 1], \\ H(s, a) = f(0) = g(0), \ \forall s \in [0, 1], \\ H(s, b) = f(1) = g(1), \ \forall s \in [0, 1], \end{cases}$$

see Figure 3.4. In this case, and with the previous notations, $f(t) = H \circ \delta_0(t)$ and $g(t) = H \circ \delta_1(t)$; moreover, $\gamma_2 := H \circ \delta_2$ and $\gamma_3 := H \circ \delta_3$ are constant.

3.22 Theorem. *Let $\gamma, \delta : [0, 1] \to \Omega$ be two piecewise curves of class C^1 that are homotopic with fixed extreme points and let ω be a continuous locally exact form in Ω. Then we have*

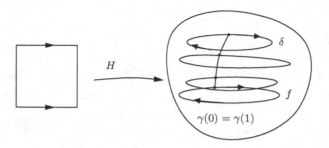

Figure 3.5. A homotopy between γ and δ as closed curves.

$$\mathcal{L}(\gamma, \omega) = \mathcal{L}(\delta, \omega).$$

Similarly, if F is a locally conservative continuous field in Ω, we have

$$\mathcal{L}(\gamma, F) = \mathcal{L}(\delta, F).$$

Proof. Let $h : [0,1] \times [0,1] \to \Omega$ be a homotopy from γ to δ and, for every $x \in \Omega$, let $F_x : B(x, \epsilon(x)) \to \mathbb{R}$ be a potential of ω in $B(x, \epsilon(x))$. Since $h([0,1] \times [0,1])$ is compact, one can suppose that $\epsilon(x) \geq \epsilon_0 > 0$ if $x \in h([0,1] \times [0,1])$. Since h is uniformly continuous on $[0,1] \times [0,1]$, we decompose $[0,1] \times [0,1]$ in adjacent squares $\{R_{i,j}\}$, $i, j = 1, \ldots, N$ with side length of $1/N$, N large, such that $h(R_{i,j}) \subset B(x, \epsilon_0)$ for some $x = x_{ij} \in h([0,1] \times [0,1])$. For every $i, j = 1, \ldots, N$, let

$$\phi_{i,j}(s, t) := F_{x(i,j)}(h(s, t)).$$

Since two potentials differ by a constant on a connected set, we can add, sequentially by rows, a constant $c_{i,j}$ to $\phi_{i,j}$ in such a way that

$$\phi(t, s) := \phi_{i,j}(t, s) + c_{i,j} \qquad \text{if } (t, s) \in R_{i,j}$$

is well defined on $[0,1] \times [0,1]$. Clearly $\phi(t, s)$ is continuous on $[0,1] \times [0,1]$.

Now, let $t_0 = 0 < t_1 < \cdots < t_{N-1} < t_N = 1$ be the subdivision of $[0,1]$ such that $R_{i,j} = [t_{i-1}, t_i] \times [t_{j-1}, t_j]$. Since we have

$$\int_\gamma \omega = \sum_{i=1}^N \int_{\gamma|[t_{i-1}, t_i]} \omega = \sum_{i=1}^N \Big(\phi_{i,j}(0, t_i) - \phi_{i,j}(0, t_{i-1})\Big)$$

$$= \sum_{i=1}^N \Big(\phi(0, t_i) - \phi(0, t_{i-1})\Big) = \phi(0, 1) - \phi(0, 0)$$

and, similarly,

$$\int_\delta \omega = \phi(1, 1) - \phi(1, 0),$$

the claim follows being also

$$\phi(0, 0) = \phi(1, 0) \qquad \text{and} \qquad \phi(1, 0) = \phi(1, 1). \tag{3.9}$$

$\square$

We also recall, see [GM3], that two continuous *closed* curves γ and $\delta : [0,1] \to \Omega$ are said to be *homotopic* as closed curves if there exists a continuous map $H : R \to \Omega$ such that

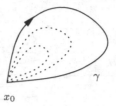

x_0

Figure 3.6. γ is homotopic to x_0.

$$\begin{cases} H(0,t) = \gamma(t), \ \forall t \in [0,1], \\ H(1,t) = \delta(t), \ \forall t \in [0,1], \\ H(s,0) = H(s,1), \ \forall s \in [0,1], \end{cases}$$

see Figure 3.5. We have

3.23 Theorem. *Let* $\gamma, \delta : [0,1] \to \Omega$ *be two piecewise closed curves of class* C^1 *that are homotopic as closed curves. If* ω *is a locally exact form in* Ω, *then*

$$\mathcal{L}(\gamma, \omega) = \mathcal{L}(\delta, \omega),$$

and, similarly, if F *is a locally conservative continuous field in* Ω, *then*

$$\mathcal{L}(\gamma, F) = \mathcal{L}(\delta, F).$$

Proof. We proceed as in the proof of Theorem 3.23. The conclusion then follows since this time we have $\phi(1,1) - \phi(1,0) = \phi(1,0) - \phi(0,0)$ instead of (3.9). □

d. Simply connected subsets and closed forms

3.24 Definition. *An open connected set is said to be* simply connected *if every closed curve in* Ω *is homotopic to a constant curve in* Ω.

Since every open and connected set is also arc-connected, Ω is simply connected if and only if for $x_0 \in \Omega$ every closed curve with initial and final end points at x_0 is homotopic to the constant curve x_0 with initial and final points x_0. In other words, an open and connected set Ω is simply connected if and only if its first homotopy group $\pi_1(\Omega)$ vanishes.

As consequence of Theorem 3.23 the following holds.

3.25 Corollary. *Let* Ω *be an open, connected and simply connected domain. Every closed form in* Ω *is exact in* Ω, *or, equivalently, every irrotational field* F *in* Ω *is conservative in* Ω, *i.e., if* rot $F = 0$ *in* Ω, *then there exists* $f \in C^2(\Omega)$ *such that* $F = \nabla f$ *in* Ω.

We now illustrate a few situations in which Corollary 3.25 applies.

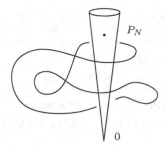

Figure 3.7. $\mathbb{R}^3 \setminus \{x_0\}$ is simply connected.

For example $\mathbb{R}^n$ and balls in $\mathbb{R}^n$ are star-shaped with respect to any of their points, as well as open convex sets. Instead, $\mathbb{R}^n \setminus \{0\}$ or the annulus, $\{x \in \mathbb{R}^n \mid 1 < |x| < 2\}$ are not star-shaped. Notice that star-shaped domains are necessarily connected.

If Ω is star-shaped with respect to x_0, then every closed curve $f :$ $[0,1] \to \Omega$ with $f(1) = f(0) = x_0$ is homotopic to x_0 as the homotopy $H(t,x) = (1-s)x_0 + sf(t)$, $s \in [0,1]$, $t \in [0,1]$, shows. Consequently every star-shaped domain is connected and simply connected.

3.26 Example. Trivially $\mathbb{R}^2 \setminus \{0\}$ is not simply connected at least because the angle form of Example 3.17 is not exact though closed. Instead, $\mathbb{R}^3 \setminus \{0\}$ *is* simply connected. In fact, if γ is a closed curve in $\mathbb{R}^3 \setminus \{0\}$, there exists $\delta > 0$ such that the trajectory of γ lies outside $B(0,\delta)$, since $\gamma([0,1])$ is compact. The curve $\delta(t) := \gamma(t)/|\gamma(t)|$, $t \in [0,1]$ is then a well-defined curve of class C^1, its trajectory is on S^2 and necessarily has to avoid a point, say the North Pole, and consequently, a neighborhood of the North Pole, again by Weierstrass theorem. In conclusion, γ lies outside a cone with axis through the origin and the North Pole and, therefore, it can be homotoped to the South Pole, see Figure 3.7.

3.3.1 Pull back of a differential form

Let Δ and Ω be open sets respectively in $\mathbb{R}^r$ and $\mathbb{R}^n$. If $\phi : \Delta \to \Omega$ is continuous and $f \in C^0(\Omega)$, then $f \circ \phi$ is continuous on Δ and, if ϕ is of class C^1 and $f \in C^1(\Delta)$, then $f \circ \phi \in C^1(\Delta)$; therefore, if we think of ϕ as an operator $\phi^\#$, $\phi^\#(f) := f \circ \phi$, $\phi^\#$ maps $C^0(\Omega)$ into $C^0(\Delta)$, $C^1(\Omega)$ into $C^1(\Delta)$. Moreover, if ϕ is of class C^k, then $\phi^\#$ maps $C^k(\Omega)$ into $C^k(\Delta)$.

Similarly, we may pull back differential forms in Ω to differential forms in Δ.

3.27 Definition. *Let $\phi \in C^1(\Delta, \Omega)$ be a map of class C^1 from an open set $\Delta \subset \mathbb{R}^r$ into an open set $\Omega \subset \mathbb{R}^n$. The* pull back *of a continuous differential form ω in Ω is the continuous differential form in Δ defined by*

$$< \phi^\# \omega(x), h > := < \omega(\phi(x)), \mathbf{D}\phi(x)h > \qquad \forall h \in \mathbb{R}^r. \tag{3.10}$$

3.28 ¶ Inverse image of a vector field. It is less intuitive to pull back a vector field. Show that if $\phi : \Delta \to \Omega$ and $F : \Omega \subset \mathbb{R}^r \to \mathbb{R}^n$, then the unique definition of pull back that is compatible with the definition of pull back of the associated form is

$$\phi^{\#} F(x) := \mathbf{D}\phi(x)^* F(\phi(x)),$$

where $\mathbf{D}\phi(x)^*$ is the adjoint operator of $\mathbf{D}\phi(x)$.

3.29 $\phi^{\#}\omega$. We can write $\phi^{\#}\omega$ in several ways.

○ Let $\omega = \sum_{i=1}^{n} \omega_i(y)\, dy^i$ and let $y = \phi(x) = (\phi(x)^1, \phi(x)^2, \ldots, \phi(x)^n)^T$, then we have

$$< \phi^{\#}\omega(x)\,,h> \; = <\omega(\phi(x))\,,\mathbf{D}\phi(x)h> = \sum_{i=1}^{n} \omega_i(\phi(x))\mathbf{D}\phi^i(x)h$$

$$= \sum_{i=1}^{n} \omega_i(\phi(x)) <d\phi^i(x)\,,h>$$

for every $h \in \mathbb{R}^n$ hence

$$\phi^{\#}\omega(x) = \sum_{i=1}^{n} \omega_i(\phi(x))d\phi^i(x)$$

and

$$\phi^{\#}\omega(x) = \sum_{i=1}^{n} \omega_i(\phi(x))d\phi^i(x) = \sum_{j=1}^{r} \Big(\sum_{i=1}^{n} \omega_i(\phi(x)) \frac{\partial \phi_i}{\partial x^j} \Big) dx^j.$$

○ If we write the components of ω as a row vector, $\omega = (\omega_1, \omega_2, \ldots, \omega_n)$, with n components, we can write $\phi^{\#}\omega$ as the row vector with r components

$$\phi^{\#}\omega(x) = ((\phi^{\#}\omega(x))_1, (\phi^{\#}\omega(x))_2, \ldots, (\phi^{\#}\omega(x))_r)$$
$$= (\omega(\phi(x))_1, \omega(\phi(x))_2, \ldots, \omega(\phi(x))_n)\, \mathbf{D}\phi(x).$$

3.30 Invariance of work. From the definition (3.10) we see that

$$d(\phi^{\#}f)(x) = \phi^{\#}df(x) \qquad\qquad (3.11)$$

for all $f \in C^1(\Omega)$, which implies in particular that $\phi^{\#}\omega$ is exact in Δ if ω is exact in $\omega := \phi(\Delta)$. Moreover, we have

$$< \phi^{\#}\omega(\gamma(s)), \gamma'(s) >: = <\omega(\phi(\gamma(s))), \mathbf{D}\phi(\gamma(s))\gamma'(s) >$$
$$= <\omega(\phi \circ \gamma(s)), (\phi \circ \gamma)'(s) >$$

for every curve $\gamma : [a, b] \to \Delta$ of class C^1. Integrating we get

$$\mathcal{L}(\gamma, \phi^{\#}\omega) = \mathcal{L}(\phi \circ \gamma, \omega). \qquad\qquad (3.12)$$

In particular, (3.11) (3.12) state respectively *the invariance of the differential of a function* and *the invariance of work of a differential form* under changes of references in $\mathbb{R}^n$.

3.31 ¶. Write the angle form of Example 3.17 in polar coordinates, i.e., compute the pull back via the transformation $x = \rho \cos\theta$, $y = \rho \sin\theta$.

3.32 Proposition (Differential of the pull back of forms). *Let ω be a differential form of class C^1 in an open set $\Omega \subset \mathbb{R}^n$, let $H : \Delta \subset \mathbb{R}^r \to \Omega$ be a map of class C^2 from the open set Δ into Ω and denote by $(u^1, u^2, \ldots, u^r)$ the coordinates in $\mathbb{R}^r$. Then $H^\#\omega$ writes as*

$$H^\#\omega = \sum_{i=1}^r P_i(u)\, du^i,$$

where $P_i(u)$ are of class C^1 and we have

$$\frac{\partial P^h}{\partial u_k} - \frac{\partial P_k}{\partial u^h} = \sum_{i,j=1}^n \left(\frac{\partial \omega_i}{\partial x^j} - \frac{\partial \omega_j}{\partial x^i}\right)\frac{\partial H^i}{\partial u^h}\frac{\partial H^j}{\partial u^k}; \tag{3.13}$$

here $\frac{\partial P_h}{\partial u^k}, \frac{\partial P_k}{\partial u^h}, \frac{\partial H^i}{\partial u^h}, \frac{\partial H^j}{\partial u^k}$, are evaluated at $u \in \Delta$ and $\frac{\partial \omega_i}{\partial x^j}$ and $\frac{\partial \omega_j}{\partial x^i}$ in $H(u)$.

In particular, $H^\#\omega$ is closed in Δ if ω is closed in Ω.

Proof. In fact, computing the derivatives of $P_h := \sum_{i=1}^n \omega_i \frac{\partial H^i}{\partial u^h}$, we find

$$\begin{cases} \dfrac{\partial P_h}{\partial u^k} = \displaystyle\sum_{i,j=1}^n \dfrac{\partial \omega_i}{\partial x^j}\dfrac{\partial H^j}{\partial u^k}\dfrac{\partial H^i}{\partial u^h} + \displaystyle\sum_{i=1}^n \omega_i \dfrac{\partial^2 H^i}{\partial u^k \partial u^h} \\ \dfrac{\partial P_k}{\partial u^h} = \displaystyle\sum_{i,j=1}^n \dfrac{\partial \omega_i}{\partial x^j}\dfrac{\partial H^j}{\partial u^h}\dfrac{\partial H^i}{\partial u^k} + \displaystyle\sum_{i=1}^n \omega_i \dfrac{\partial^2 H^i}{\partial u^h \partial u^k}, \end{cases}$$

and (3.13) follows subtracting the two equations, since the Hessian matrices of the components of H are symmetric.   □

3.33 ¶. Since the rotation matrix of a form is antisymmetric, (3.13) can also be written as

$$\frac{\partial P^h}{\partial u^k} - \frac{\partial P^k}{\partial u^h} = \frac{1}{2}\sum_{i,j=1}^n \left(\frac{\partial \omega_i}{\partial x^j} - \frac{\partial \omega_j}{\partial x^i}\right)\left(\frac{\partial H^i}{\partial u^h}\frac{\partial H^j}{\partial u^k} - \frac{\partial H^j}{\partial u^h}\frac{\partial H^i}{\partial u^k}\right). \tag{3.14}$$

3.3.2 Homotopy formula

a. Stokes's theorem in a square

If $R = [0,1] \times [0,1] \subset \mathbb{R}^2$, we denote by $\delta_0, \delta_1, \delta_2, \delta_3 : [0,1] \to \mathbb{R}^2$ the curves defined by

$$\begin{cases} \delta_0(t) := (0,t), \\ \delta_1(t) := (1,t), \\ \delta_2(t) := (t,1), \\ \delta_3(t) := (t,0), \end{cases} \quad t \in [0,1],$$

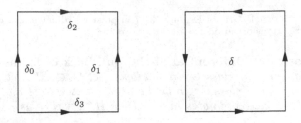

Figure 3.8. δ travels along ∂R anticlockwise.

in such a way that the curve

$$\delta := \delta_3 + \delta_1 - \delta_2 - \delta_0, \tag{3.15}$$

goes anticlockwise along ∂R, see Figure 3.8.

As we know, if $f : R \subset \mathbb{R}^2 \to \mathbb{R}$ is continuous (or piecewise continuous) and bounded on $R = [a, b] \times [c, d]$, then we can change the order of integration,

$$\int_a^b \left(\int_c^d f(x, y) \, dy \right) dx = \int_c^d \left(\int_a^b f(x, y) \, dx \right) dy =: \int_c^d \int_a^b f(x, y) \, dx \, dy.$$

3.34 ¶. The reader is invited to prove this directly, first showing that the formula holds for constant functions $f(x, y) = \lambda$ (the value of the two integrals is $\lambda(b - a)(d - c)$), then for a piecewise constant function on a squaring of $[a, b] \times [c, d]$ and finally for a continuous function (approximating it with piecewise constant functions).

3.35 Proposition (Stokes). *Let $R := [0, 1] \times [0, 1]$ and let $\delta : [0, 1] \to \mathbb{R}^2$ be the curve in (3.15) with trajectory ∂R oriented anticlockwise. Suppose that $\eta(s, t) := P(s, t) \, ds + Q(s, t) \, dt$ is a differential form of class C^1 in a neighborhood of R. Then*

$$\mathcal{L}(\delta, \eta) = \int_0^1 \int_0^1 \left(\frac{\partial Q}{\partial s} - \frac{\partial P}{\partial t} \right) ds \, dt.$$

Proof. In fact, if δ_i, $i = 0, 3$ are the parameterizations in (3.15), we have

$$\int_\delta \eta = \int_{\delta_1} \eta - \int_{\delta_0} \eta + \int_{\delta_3} \eta - \int_{\delta_2} \eta$$

$$= \int_0^1 Q(1, t) \, dt - \int_0^1 Q(0, t) \, dt + \int_0^1 P(s, 0) \, ds - \int_0^1 P(s, 1) \, ds$$

$$= \int_0^1 \left(\int_0^1 \frac{\partial Q}{\partial s}(s, t) \, ds \right) dt - \int_0^1 \left(\int_0^1 \frac{\partial P}{\partial t} \, dt \right) ds$$

$$= \int_0^1 \int_0^1 \left(\frac{\partial Q}{\partial s} - \frac{\partial P}{\partial t} \right) ds \, dt.$$

Here the first two equalities are by definition of work, the third follows from the fundamental theorem of calculus and the last by interchanging the integral signs.   □

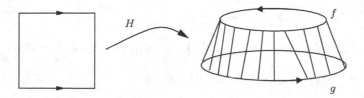

Figure 3.9. A linear homotpy between f and g.

b. Homotopy formula

3.36 Proposition (Homotopy formula). *Let $f, g : [0,1] \to \mathbb{R}^n$ be two curves of class C^1 and let $H : R \to \mathbb{R}^2$, $H(s,t) := sf(t) + (1-s)g(t)$ be the linear homotopy. Set $K := H(R)$, and denote by $L(f)$ and $L(g)$ respectively, the lengths of f and g. Then for every differential form ω of class C^1 in a neighborhood of K we have*

$$\left| \mathcal{L}(f,\omega) - \mathcal{L}(g,\omega) \right| \leq \frac{1}{2} \|W(\omega)\|_{\infty,K} (L(f) + L(g)) \|f - g\|_{\infty,[0,1]},$$

where $W(\omega)$ is the rotation matrix of ω, see (3.14). In particular, $\mathcal{L}(f,\omega) = \mathcal{L}(g,\omega)$ if ω is closed.

Proof. If δ is the curve in (3.15), by (3.12) we have

$$\mathcal{L}(f,\omega) - \mathcal{L}(g,\omega) = \mathcal{L}(\delta, H^{\#}\omega);$$

thus, we need to estimate $\mathcal{L}(\delta, H^{\#}\omega)$. Writing $H^{\#}\omega =: P \, ds + Q \, dt$, the functions $s \to Q(s,t)$ and $t \to P(s,t)$ are piecewise continuous of class C^1, therefore by Stokes's theorem

$$\mathcal{L}(\delta, H^{\#}\omega) = \int_0^1 \int_0^1 \left(\frac{\partial P}{\partial t} - \frac{\partial Q}{\partial s} \right) ds \, dt. \tag{3.16}$$

From (3.13) we infer

$$\frac{\partial P}{\partial t} - \frac{\partial Q}{\partial s} = \sum_{i,j} W_{ij}(\omega) \left(s f^{i'}(t) + (1-s)g^{i'}(t) \right) (f^j(t) - g^j(t)),$$

thus

$$\left| \frac{\partial P}{\partial t} - \frac{\partial Q}{\partial s} \right| \leq \|W(\omega)\|_{\infty,K} \|f - g\|_{\infty,[0,1]} \left(s|f'(t)| + (1-s)|g'(t)| \right).$$

We therefore conclude from (3.16)

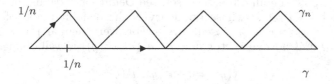

Figure 3.10. For every C^1 form, $\mathcal{L}(\gamma_n, \omega) \to \mathcal{L}(\gamma, \omega)$ though the lengths of γ_n do not converge to the length of γ.

$$\left| \mathcal{L}(\delta, H^{\#}\omega) \right| = \left| \int_0^1 \int_0^1 \left(\frac{\partial P}{\partial t} - \frac{\partial Q}{\partial s} \right) dt\, ds \right| = \int_0^1 \int_0^1 \left| \frac{\partial P}{\partial t} - \frac{\partial Q}{\partial s} \right| dt\, ds$$

$$\leq \|W(\omega)\|_{\infty, K} \|f - g\|_{\infty, [0,1]} \int_0^1 \left(s \int_0^1 |\phi'(t)|\, dt + (1 - s) \int_0^1 |\gamma'(t)|\, dt \right) ds$$

$$= \frac{1}{2} \|W(\omega)\|_{\infty, K} \left(L(f) + L(g) \right) \|f - g\|_{\infty, [0,1]}.$$

This proves the first part of the claim. If ω is a closed form, then $W(\omega) = 0$ everywhere, and the second part follows at once.   □

As a consequence we find that the work of a differential form is continuous, contrary to the length, with respect to the uniform convergence.

3.37 Proposition. *Let Ω be an open set in $\mathbb{R}^n$ and let $\gamma : [0, 1] \to \Omega$ and $\gamma_k : [0, 1] \to \mathbb{R}^n$, $k = 1, 2, \ldots$, be piecewise smooth curves, say of class C^1. It the lengths of the γ_k's are equibounded and $\gamma_k \to \gamma$ uniformly, i.e.,*

$$\sup_k L(\gamma_k) \leq M < +\infty, \qquad \|\gamma_k - \gamma\|_{\infty, [0,1]} \to 0,$$

then for large k's both γ_k and the linear homotopy $H_k(s, t) := s\gamma_k(t) + (1 - s)\gamma(t)$ have image in Ω and

$$\mathcal{L}(\gamma_k, \omega) \to \mathcal{L}(\gamma, \omega) \qquad \text{as } k \to \infty.$$

Proof. Let us show that for k large, γ_k is a curve in Ω and the image $K_k := H_k(R)$ of H_k is contained in Ω. Let K_0 be the trajectory of γ and $\delta_0 := \text{dist}(K_0, \partial\Omega)$. Since K_0 is compact, we have $\delta_0 > 0$. Since $\{\gamma_k\}$ converges uniformly to γ, there exists k_0 such that $|\gamma_k(t) - \gamma(t)| < \delta_0$ for $k \geq k_0$ and therefore

$$|H_k(t, s) - \gamma(t)| \leq |\gamma_k(t) - \gamma(t)| < \delta_0,$$

i.e., the images of the K_k's and of the trajectories of the γ_k's are in Ω for $k \geq k_0$. From the homotopy formula we also infer for $k \geq k_0$

$$\left| \mathcal{L}(\gamma_k, \omega) - \mathcal{L}(\gamma, \omega) \right| \leq \|W(\omega)\|_{\infty, K_{k_0}} M \|\gamma_k - \gamma\|_{\infty, [0,1]}$$

which yields the result when $k \to \infty$.   □

3.4 Stokes's Formula in the Plane

Let $A \subset \Omega \subset \mathbb{R}^2$ be an admissible set, see Definition 2.94, as for instance a bounded open set Ω whose boundary is the disjoint union of a finite number of trajectories of closed curves. For every regular point $y \in \partial A$, an exterior unit normal vector $\mathbf{n}(y) = (n^1(y), n^2(y))$ is well defined, and

$$\mathbf{t}(y) := (-n^2(y), n^1(y))$$

is tangent to ∂A at y. We define the work of a continuous differential form $\omega \in C^0(\partial A)$ along the anticlockwise oriented boundary of A as

$$\int_{\partial^+ A} \omega := \int_{\partial^+ A} <\omega(y), \mathbf{t}(y)> d\mathcal{H}^1(y).$$

Notice, see Remark 3.5, that $\int_{\partial^+ A} \omega = \mathcal{L}(\omega, \gamma)$ if $\gamma : [0, 1] \to \mathbb{R}^2$ is a simple closed curve piecewise of class C^1 whose trajectory is the boundary of A and such that $\det[\gamma'(t)|\mathbf{n}(\gamma(t))] < 0$.

3.38 Proposition (Stokes). *Let A be an admissible open domain of $\mathbb{R}^2$ and let ω, $\omega(x, y) = P(x, y)\,dx + Q(x, y)\,dy$, be a differential form of class C^1 in a neighborhood of A. We have*

$$\int_{\partial^+ A} \omega = \iint_A \left(\frac{\partial Q}{\partial x} - \frac{\partial P}{\partial y} \right) dx\,dy. \tag{3.17}$$

Proof. In fact, Gauss–Green formulas yield

$$\int_{\partial^+ A} \omega = \int_{\partial A} (P(y)t^1(y) + Q(y)t^2(y))\,d\mathcal{H}^1(y)$$

$$= \int_{\partial A} (-P(y)n^2(y) + Q(y)n^1(y))\,d\mathcal{H}^1(y)$$

$$= \iint_A \left(\frac{\partial Q}{\partial x} - \frac{\partial P}{\partial y} \right) dx\,dy.$$

$\square$

Formula (3.17) allows us to compute the area of a plane figure as a boundary integral. In fact, if ω is one of the differential forms $x\,dy$, $-y\,dx$, $\frac{1}{2}(x\,dy - y\,dx)$ or $\alpha x\,dy - \beta y\,dx$, $\alpha + \beta = 1$, we get

$$\int_{\partial^+ A} \omega = \iint_A 1\,dx\,dy.$$

3.39 Example. Suppose we want to compute the area of the cardioid

$$C := \Big\{ (\rho\cos\theta, \rho\sin\theta) \,\Big|\, 0 \le \theta \le 2\pi, 0 \le \rho \le (1 - \cos\theta) \Big\}.$$

Its anticlockwise oriented boundary is the trajectory of the curve $\gamma(t) := ((1 - \cos\theta)\cos\theta, (1 - \cos\theta)\sin\theta)$, $t \in [0, 2\pi]$, thus

$$\mathcal{L}^2(C) = \int_C 1\,dx\,dy = \frac{1}{2} \int_{\partial^+ C} (x\,dy - y\,dx) = \frac{1}{2} \int_0^{2\pi} (1 - \cos\theta)^2\,dt = \frac{3}{2}\pi.$$

3.40 Example. In general, if $\varphi(\theta)$, $\theta \in [0, 2\pi]$, is piecewise of class C^1 and nonnegative, the area of the figure defined in polar coordinates by

$$A := \Big\{ (\rho, \theta) \,\Big|\, 0 \le \rho < \varphi(\theta), \ \theta \in [0, 2\pi] \Big\}$$

is

$$\mathcal{L}^2(A) = \iint_A dx\,dy = \frac{1}{2} \int_{\partial^+ A} (x\,dy - y\,dx)$$

$$= \frac{1}{2} \int_0^{2\pi} \Big(\varphi(\theta)\cos\theta(\varphi(\theta)\sin\theta)' - \varphi(\theta)\sin\theta(\varphi(\theta)\cos\theta)' \Big) d\theta$$

$$= \frac{1}{2} \int_0^{2\pi} \varphi^2(\theta)\,d\theta.$$

3.5 Exercises

3.41 ¶. Decide whether the following differential forms are exact:

$$xy\,dx + \frac{1}{3}x^2\,dy$$

$$yz\,dx + xz\,dy + xy\,dz$$

$$\left(\frac{1}{x^2} + \frac{1}{y^2}\right)(y\,dx - x\,dy), \qquad x \neq 0,\ y \neq 0.$$

3.42 ¶. A vector field is said to be *central* if $F(x) := f(x)\frac{x}{|x|}$, $x \in \mathbb{R}^n \setminus \{0\}$, where $f : \mathbb{R}^n \setminus \{0\} \to \mathbb{R}$. Prove that F is conservative if and only if F is a *radial vector field*, i.e., $F(x) := \varphi(|x|)\frac{x}{|x|}$ where $\varphi : \mathbb{R} \to \mathbb{R}$.

3.43 ¶. Let $\Omega = \mathbb{R}^2 \setminus \{0\}$, let $\gamma(t) := (\cos t, \sin t)$, $t \in [0, 2\pi]$ and let ω be a closed differential form on Ω with $\mathcal{L}(\omega, \gamma) = 0$. Show that ω is exact on $\mathbb{R}^2 \setminus \{0\}$. Show that every closed form ω decomposes as

$$\omega = \lambda\omega_0 + \alpha$$

where $\lambda \in \mathbb{R}$, α is an exact form and ω_0 is the angle form, see Example 3.17.

3.44 ¶. Let ω be a closed differential form in $\mathbb{R}^n \setminus \{0\}$. Prove that ω is exact in $\mathbb{R}^n \setminus \{0\}$ if

$$\lim_{x \to 0} \omega(x)|x| = 0.$$

3.45 ¶. Let Ω be an open set in $\mathbb{R}^3$. For every $u \in C^2(\Omega)$ we have div rot $u = 0$ in Ω, therefore a necessary condition for the solvability of rot $u = f$ is that div $f = 0$ in Ω. Suppose Ω star-shaped with respect to the origin and prove the following.

(i) Two solutions of rot $u = f$ differ by the gradient of an arbitrary function.
(ii) If div $f = 0$, then

$$\text{rot}\,(tf(x) \times x) = \frac{d}{dt}(t^2 f(tx)),$$

thus integrating, infer that

$$u(x) := \int_0^1 tf(tx) \times x\,dt$$

is a solution of rot $u = f$. Here $a \times b$ denotes the *vector product* defined by

$$a \times b := (a^2 b^3 - a^3 b^2, -(a^1 b^3 - a^3 b^1), a^1 b^2 - a^2 b^1)$$

if $a = (a^1, a^2, a^3)$ and $b = (b^1, b^2, b^3)$, see [GM5, Chapter 4].

4. Holomorphic Functions

The theory of functions of one complex variable is one of the most central and fascinating chapters of mathematics. It has its prehistory with the works of Leonhard Euler (1707–1783), Joseph-Louis Lagrange (1736–1813), and Carl Friedrich Gauss (1777–1855), its gold period with Augustin-Louis Cauchy (1789–1857), G. F. Bernhard Riemann (1826–1866), Hermann Schwarz (1843–1921), and Karl Weierstrass (1815–1897), and it is the result of the contributions of many mathematicians in the period 1800–1950. The ideas, the methods, and the results of the theory of holomorphic functions play a fundamental role in several fields of mathematics both pure and applied, beyond their essential beauty. Here we shall limit ourselves to an elementary introduction.

4.1 Functions from $\mathbb{C}$ to $\mathbb{C}$

a. Complex numbers

Recall that the correspondence $z = a + ib \in \mathbb{C}$ to $(a, b) \in \mathbb{R}^2$ identifies $\mathbb{C}$ and $\mathbb{R}^2$ as vector spaces, and the product in $\mathbb{C}$ gives a simple way of describing oriented rotations in the plane. In fact, for $z = a + ib$ and $w = c + id$ we have

$$z\overline{w} = (a + ib)(c - id) = (ac + bd) + i(bc - ad) = (z|w)_{\mathbb{R}^2} + i\det(w, z) \quad (4.1)$$

hence, if θ is the angle between the two vectors $z = (a, b)$ and $w = (c, d)$ measured from w to z, then

$$w\overline{z} = |z||w|(\cos\theta + i\sin\theta).$$

In particular, multiplying z by i means rotating anticlockwise the vector z by $\pi/2$: $iz\overline{z} = i|z|^2 = |z|^2(0 + i1)$.

b. Complex derivative

4.1 Definition. *Let* $f : \Omega \subset \mathbb{C} \to \mathbb{C}$ *be a complex-valued function where* Ω *is open in* $\mathbb{C}$ *and let* $z_0 \in \Omega$. *We say that* f *is* differentiable in the complex

M. Giaquinta and G. Modica, *Mathematical Analysis: An Introduction to Functions of Several Variables*, DOI: 10.1007/978-0-8176-4612-7_4,
© Birkhäuser Boston, a part of Springer Science + Business Media, LLC 2010

sense at z_0, *in short f is* $\mathbb{C}$*-differentiable with complex derivative $f'(z_0)$, if the following limit*

$$f'(z_0) = \lim_{z \to z_0} \frac{f(z) - f(z_0)}{z - z_0}$$

exists in $\mathbb{C}$*. If f has complex derivative at every point of Ω, we say that f is* holomorphic *(or* analytic*) in Ω. The class of holomorphic functions on Ω is denoted $\mathcal{H}(\Omega)$.*

4.2 ¶. Show that the following facts hold:
 (i) If f is $\mathbb{C}$-differentiable at z_0, then f is continuous at z_0.
 (ii) If f and g are $\mathbb{C}$-differentiable at z_0, then $f + g$ and fg are $\mathbb{C}$-differentiable at z_0 and

$$(f + g)'(z_0) = f'(z_0) + g'(z_0), \qquad (fg)'(z_0) = f'(z_0)g(z_0) + f(z_0)g'(z_0).$$

 (iii) If f and g are $\mathbb{C}$-differentiable at z_0 and $g(z_0) \neq 0$, then f/g is $\mathbb{C}$-differentiable at z_0 and

$$\left(\frac{f}{g}\right)'(z_0) = \frac{f'(z_0)g(z_0) - g'(z_0)f(z_0)}{g^2(z_0)}.$$

 (iv) Let f be a map from an open set $\Omega \subset \mathbb{C}$ into $\mathbb{C}$ and $z_0 \in \Omega$; let $g : A \subset \mathbb{C} \to \mathbb{C}$ be a map with $f(z_0) \in A$. If f is $\mathbb{C}$-differentiable at z_0 and g is $\mathbb{C}$-differentiable at $f(z_0)$, then $g \circ f$ is $\mathbb{C}$-differentiable at z_0 and $(g \circ f)'(z_0) = g'(f(z_0))f'(z_0)$.
 (v) Let $F \in \mathcal{H}(\Omega)$ and $\gamma : [0, 1] \to \Omega$ be of class C^1, then $t \to F(\gamma(t))$ is differentiable on $[0, 1]$ and

$$\frac{d}{dt}F(\gamma(t)) = F'(\gamma(t))\gamma'(t) \qquad \forall t \in [0, 1].$$

4.3 ¶. Let $f \in \mathcal{H}(\Omega)$ and $g \in \mathcal{H}(\Delta)$ for Ω and Δ open sets, and let $f = g$ on $\Omega \cap \Delta$. Prove that the function

$$F(z) := \begin{cases} f(z) & \text{if } z \in \Omega, \\ g(z) & \text{if } z \in \Delta \end{cases}$$

is holomorphic in $\Omega \cup \Delta$.

4.4 ¶. Show that
 (i) Polynomials in one complex variable are holomorphic functions on $\mathbb{C}$.
 (ii) A rational function $R(z) := P(z)/Q(z)$ for P and Q polynomials is holomorphic on $\Omega := \{z \in \mathbb{C} \mid Q(z) \neq 0\}$.

c. Cauchy–Riemann equations

Let us identify complex numbers and vectors of $\mathbb{R}^2$, i.e., $z = x + iy$ with (x, y), and let $f : \Omega \subset \mathbb{C} \to \mathbb{C}$ be a function. Clearly, f is $\mathbb{R}$-differentiable at $z_0 = x_0 + iy_0$ if

$$f(z_0 + w) - f(z_0) = \mathbf{D}f(z_0)\,w + o(|w|) \qquad \text{as } w \to 0, \qquad (4.2)$$

and f is $\mathbb{C}$-differentiable at z_0 if

$$f(z_0 + w) - f(z_0) = f'(z_0)\,w + o(|w|) \qquad \text{as } w \to 0. \qquad (4.3)$$

A comparison between (4.2) and (4.3) yields at once the following.

Figure 4.1. Augustin-Louis Cauchy (1789–1857), Karl Weierstrass (1815–1897), and G. F. Bernhard Riemann (1826–1866).

4.5 Proposition. *The function $f : \Omega \subset \mathbb{C} \to \mathbb{C}$ has complex derivative at $z_0 \in \Omega$ if and only if f is $\mathbb{R}$-differentiable at z_0 and for some $\lambda \in \mathbb{C}$*

$$\frac{\partial f}{\partial w}(z_0) = \mathbf{D}f(z_0)(w) = \lambda w. \tag{4.4}$$

If so, then $\lambda = f'(z_0)$.

Condition (4.4) states that the $\mathbb{R}$-differential of a $\mathbb{C}$-differentiable function f at z_0 exists and acts on $\mathbb{C}$ as a complex multiplication

$$w \;\longrightarrow\; \frac{\partial f}{\partial w}(z_0) = df(z_0)(w) = \lambda w.$$

This is quite a restrictive condition. In fact, two vectors $w_1, w_2 \in \mathbb{C}$ are mapped by the differential into the vectors $\lambda w_1, \lambda w_2$, i.e., into vectors rotated of the same angle and scaled by $|\lambda|$. In particular, perpendicular vectors w_1, w_2 of the same length have as images the perpendicular vectors λw_1 and λw_2 of the same length.

Write now f as $f(x,y) = u(x,y) + iv(x,y)$ for $z := x + iy$, and denote by f_x and f_y the first and the second column of the Jacobian matrix of $f(x,y)$

$$\mathbf{D}f(z) = [f_x | f_y] = \begin{pmatrix} u_x & u_y \\ v_x & v_y \end{pmatrix}.$$

One easily shows that (4.4) is equivalent to

$$f_y(z_0) = i f_x(z_0) \tag{4.5}$$

which says that the vector f_y is obtained by rotating anticlockwise by $\pi/2$ the vector f_x. In fact, if (4.4) holds, then

$$f_x(z_0) = \frac{\partial f}{\partial(1,0)}(z_0) = \lambda(1 + i0),$$

$$f_y(z_0) = \frac{\partial f}{\partial(0,1)}(z_0) = \lambda(0 + i1),$$

and, conversely, if $w = a + ib \in \mathbb{C}$, (4.5) yields

$$\mathbf{D}f(z_0)(w) = f_x(z_0)a + f_y(z_0)b = f_x(z_0)(a + ib) = f_x(z_0)w.$$

Equation (4.5) can also be written as

$$\begin{cases} \dfrac{\partial u}{\partial x}(x_0, y_0) = \dfrac{\partial v}{\partial y}(x_0, y_0), \\[2mm] \dfrac{\partial u}{\partial y}(x_0, y_0) = -\dfrac{\partial v}{\partial x}(x_0, y_0) \end{cases} \tag{4.6}$$

in terms of the components of f, $f =: u + iv$. Moreover, (4.5) is equivalent to

$$\begin{cases} |f_x(z_0)| = |f_y(z_0)|, \\ (f_x(z_0)|f_y(z_0)) = 0, \\ \det \mathbf{D}f(z_0) \geq 0. \end{cases} \tag{4.7}$$

Finally, if one defines the partial derivatives of f with respect to z and $\bar{z}$ as

$$\frac{\partial f}{\partial z} = f_z := \frac{1}{2}(f_x - if_y), \qquad \frac{\partial f}{\partial \bar{z}} = f_{\bar{z}} := \frac{1}{2}(f_x + if_y),$$

(4.5) writes also as

$$\frac{\partial f}{\partial \bar{z}}(z_0) = 0. \tag{4.8}$$

Summarizing, we can state the following

4.6 Proposition. *Let Ω be an open set of $\mathbb{C}$. Then $f \in \mathcal{H}(\Omega)$ if and only if it is $\mathbb{R}$-differentiable and one of the following equivalent conditions holds.*

(i) $f_y(z) = if_x(z) \ \forall z \in \Omega,$
(ii) $\frac{\partial f}{\partial \bar{z}}(z) = 0 \ \forall z \in \Omega,$
(iii) $f := u + iv$ *satisfies the* Cauchy–Riemann equations

$$\begin{cases} \dfrac{\partial u}{\partial x}(x, y) = \dfrac{\partial v}{\partial y}(x, y), \\[2mm] \dfrac{\partial u}{\partial y}(x, y) = -\dfrac{\partial v}{\partial x}(x, y) \end{cases} \qquad \forall z = x + iy \in \Omega,$$

(iv) *f fulfills the* conformality relations

$$\begin{cases} |f_x(z)| = |f_y(z)|, \\ (f_x(z)|f_y(z)) = 0, \end{cases}$$

and preserves the orientation,

$$\det \mathbf{D}f(z) \geq 0 \qquad \forall z \in \Omega.$$

If f is $\mathbb{C}$-differentiable, then

$$f'(z) = \frac{\partial f}{\partial z}(z) = \frac{\partial f}{\partial x}(z) = f_x(z) \qquad \forall z \in \Omega.$$

4.7 Remark. Notice that currently holomorphic functions *a priori* may not be of class C^1 and we are not allowed to use theorems that require the continuity of the derivatives. We shall see in the following that in fact holomorphic functions are of class C^∞ and even more.

4.2 The Fundamental Theorem of Calculus on $\mathbb{C}$

a. Line integrals

Let $\Omega \subset \mathbb{C}$ be an open domain, let $f : \Omega \to \mathbb{C}$ be a continuous function, and let $\gamma : [a,b] \to \Omega \subset \mathbb{C}$ be a C^1 curve in Ω. The *integral* $\int_\gamma f(z)\,dz$ of f along γ is defined as

$$\int_\gamma f(z)\,dz := \int_a^b f(\gamma(t))\gamma'(t)\,dt.$$

If $f := u + iv$ and $\gamma(t) = x(t) + iy(t)$, then $f(\gamma)\gamma' = (ux' - vy') + i(uy' + vx')$ hence

$$\int_\gamma f(z)\,dz = \int_a^b (ux' - vy')\,dt + i \int_a^b (uy' + vx')\,dt,$$

is the line integral along γ of the differential 1-form $(u\,dx - v\,dy) + i(v\,dx + u\,dy)$. Notice that if γ is the segment joining $(x_0, 0)$ to $(x, 0)$, then

$$\int_\gamma f(z)\,dz = \int_{x_0}^x f(s)\,ds,$$

i.e., the usual oriented integral of f on the interval $[x_0, x]$ of the real line.

By the change of variable formula, one easily sees that, if $\delta : [a,b] \to \Omega$ is a *reparameterization* of γ, i.e., $\delta = \gamma \circ h$ where $h : [a,b] \to [0,1]$ is of class C^1, then

$$\int_\delta f(z)\,dz = \int_\gamma f(z)\,dz,$$

if h is orientation preserving ($h' \geq 0$) and

$$\int_\delta f(z)\,dz = -\int_\gamma f(z)\,dz,$$

if h reverses the orientation. Recall also that if $\gamma : [0,1] \to \Omega$ is a simple curve of class C^1,

$$\int_\gamma f(z)\,dz$$

depends only on the trajectory of γ and on its orientation.

Finally, from the definition we easily get

$$\left| \int_\gamma f(z)\,dz \right| \le \int_\gamma |f(z)|\,|dz| \le \|f\|_{\infty,\Gamma} L(\gamma)$$

where $|dz|$ denotes the element of length, $L(\gamma)$ is the length of γ, $\Gamma := \gamma([a,b])$, and

$$\|f\|_{\infty,\Gamma} := \sup_{z \in \Gamma} |f(z)|.$$

b. Holomorphic primitives and line integrals

4.8 Definition. *Let $\Omega \subset \mathbb{C}$ be an open set and let $f, F : \Omega \to \mathbb{C}$ be two functions. We say that F is a* holomorphic primitive *of f in Ω if $F \in \mathcal{H}(\Omega)$ and $F'(z) = f(z)\ \forall z \in \Omega$.*

Assuming $f \in C^0(\Omega)$, we shall now give neccessary and sufficient conditions in order for a function $F : \Omega \to \mathbb{C}$ to be a holomorphic primitive of f.

Let $z_0 = x_0 + iy_0$, $z = x + iy$ be two points in $\mathbb{C}$. Denote by $\delta_{z_0,z}(t)$ the polygonal line that joins linearly z_0 to $x + iy_0$ and then $x + iy_0$ to $z = x + iy$. Notice that if Ω is a rectangle with sides parallel to the axis, the curve $\delta_{z_0,z}(t)$ is inside the rectangle for every couple $z_0, z \in \Omega$, while in general, if Ω is open, we have $\delta_{z_0,z} \subset \Omega$ if z_0 and z are sufficiently close.

4.9 Theorem (Fundamental theorem of calculus). *Let $f \in C^0(\Omega, \mathbb{C})$ and let $F : \Omega \to \mathbb{C}$. The following claims are equivalent.*

(i) *F is a holomorphic primitive of f in Ω, $F \in \mathcal{H}(\Omega)$ and $F'(z) = f(z)$ $\forall z \in \Omega$.*

(ii) *For every couple of points $z, w \in \Omega$ and every curve $\gamma : [0,1] \to \Omega$ of class C^1 with $\gamma(0) = w$ and $\gamma(1) = z$ we have*

$$F(z) - F(w) = \int_\gamma f(z)\,dz. \tag{4.9}$$

(iii) *For every $z \in \Omega$ there exists $\delta > 0$ such that for all $w \in B(z, \delta)$ we have*

$$F(z) - F(w) = \int_{\delta_{w,z}} f(\zeta)\,d\zeta.$$

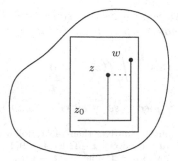

Figure 4.2. Illustration of the proof of the fundamental theorem of calculus.

Proof. (i) $\Rightarrow$ (ii). We have $f(\gamma(t))\gamma'(t) = F'(\gamma(t))\gamma'(t) = \frac{d}{dt}(F(\gamma(t)))$, see Exercise 4.2. From the fundamental theorem of calculus for functions of one real variable, we deduce

$$\int_\gamma f(z)\,dz = \int_0^1 F'(\gamma(t))\gamma'(t)\,dt = \int_0^1 \frac{d}{dt}(F(\gamma(t)))\,dt = F(\gamma(1)) - F(\gamma(0))$$

(ii) $\Rightarrow$ (iii) is trivial.

(iii) $\Rightarrow$ (i) Fix $z \in \Omega$. For every $h \in \mathbb{C}$, $|h| < \delta$, from the assumption we have

$$F(z+h) - F(z) = \int_{\delta_{z,z+h}} f(w)\,dw$$

where $\delta_{z,z+h}$ is the polygonal line that joins z to $z+h$ first moving horizontally and then vertically, see Figure 4.2. Of course, the length of $\delta_{z,z+h}$ is not greater than $\sqrt{2}\,|h|$ and its image is contained in $\overline{B(z,|h|)}$. Since

$$\int_{\delta_{z,z+h}} dw = h,$$

we have

$$\left| F(z+h) - F(z) - hf(z) \right| = \left| \int_{\delta_{z,z+h}} (f(\zeta) - f(z))\,d\zeta \right| \leq \sup_{\zeta \in \overline{B(z,|h|)}} \left| f(\zeta) - f(z) \right| \sqrt{2}\,|h|$$

hence

$$\left| \frac{F(z+h) - F(z)}{h} - f(z) \right| \leq \sqrt{2} \sup_{\zeta \in \overline{B(z,|h|)}} |f(\zeta) - f(z)|.$$

For $h \to 0$, we find $F'(z) = f(z)$, f being continuous in z. $\qquad\square$

A continuous, or even holomorphic function in Ω does not need to have a holomorphic primitive in Ω.

4.10 Example. Let $f(z) = \frac{1}{z}$, $z \neq 0$ and let $\gamma(t) := e^{it}$, $t \in [0, 2\pi]$. Trivially $f \in \mathcal{H}(\Omega)$, $\Omega = \mathbb{C} \setminus \{0\}$. However, if F were a primitive of f on Ω, then by (4.9)

$$0 = F(1) - F(1) = F(\gamma(2\pi)) - F(\gamma(0)) = \int_\gamma \frac{dz}{z} = \int_0^{2\pi} \frac{ie^{it}}{e^{it}}\,dt = 2\pi i, \qquad (4.10)$$

an absurdity.

4.11 Theorem. *Let Ω be an open connected set of $\mathbb{C}$. A continuous function $f : \Omega \to \mathbb{C}$ has a holomorphic primitive in Ω if and only if*

$$\int_\gamma f(z)\, dz = 0 \qquad (4.11)$$

for every closed curve γ piecewise of class C^1 with image in Ω.

Proof. If f has a holomorphic primitive, then (4.11) follows from (4.9). Conversely, suppose (4.11). Fix $z_0 \in \Omega$ and for every $z \in \Omega$ let $\delta_z : [0,1] \to \Omega$ a curve piecewise of class C^1 with $\delta(0) = z_0$ and $\delta(1) = z$. Define $F : \Omega \to \mathbb{C}$ by

$$F(z) := \int_{\delta_z} f(\zeta)\, d\zeta, \qquad z \in \mathbb{C}.$$

We then compute for every $z, w \in \Omega$ and every curve $\gamma : [0,1] \to \Omega$ piecewise of class C^1 with $\gamma(0) = w$ and $\gamma(1) = z$

$$F(z) - F(w) = \int_{\delta_z} f(\zeta)\, d\zeta - \int_{\delta_w} f(\zeta)\, d\zeta = \int_\gamma f(\zeta)\, d\zeta$$

on account of (4.11). Theorem 4.9 says that F is a holomorphic primitive of f on Ω. $\square$

When Ω is a rectangle, a condition weaker than (4.11) suffices. For a rectangle R we denote by $\partial^+ R$ a simple, regular, piecewise smooth, closed curve whose oriented trajectory is the boundary of R oriented counterclockwise.

4.12 Theorem. *Let Ω be a rectangle with sides parallel to the axis. A continuous function $f : \Omega \to \mathbb{C}$ has a holomorphic primitive in Ω if and only if for every rectangle $R \subset\subset \Omega$ with sides parallel to the sides of Ω we have*

$$\int_{\partial^+ R} f(z)\, dz = 0. \qquad (4.12)$$

Proof. Fix $z_0 \in \Omega$ and for every $z \in \Omega$, denote by $\delta_{z_0, z}$ the polygonal line that travels from z_0 to z first horizontally and then vertically. Then define $F : \Omega \to \mathbb{C}$ by

$$F(z) := \int_{\delta_z} f(\zeta)\, d\zeta, \qquad z \in \Omega.$$

Using (4.12), we infer for every $z, w \in \Omega$ that

$$F(z) - F(w) = \int_{\delta_{z_0, z}} f(\zeta)\, d\zeta - \int_{\delta_{z_0, w}} f(\zeta)\, d\zeta = \int_{\delta_{w, z}} f(\zeta)\, d\zeta,$$

thus F is a holomorphic primitive of f on Ω, again by Theorem 4.9. $\square$

It is worth emphasizing the differences between Theorems 4.11 and 4.12.

4.13 Corollary. *Let Ω be a connected open set in $\mathbb{C}$ and let $f : \Omega \to \mathbb{C}$ be a continuous function. Then*

(i) $f : \Omega \to \mathbb{C}$ *has locally holomorphic primitives if and only if for every rectangle* $R \subset\subset \Omega$ *with sides parallel to the axis we have*

$$\int_{\partial^+ R} f(z)\, dz = 0.$$

(ii) f *has a holomorphic primitive in* Ω *if and only if*

$$\int_{\gamma} f(z)\, dz = 0$$

for any curve $\gamma : [0,1] \to \Omega$ *piecewise of class* C^1.

Later we shall prove the following.

4.14 Theorem. *Let* Ω *be a connected open set in* $\mathbb{C}$ *and let* $f : \Omega \to \mathbb{C}$ *be a function.*

(i) f *has locally holomorphic primitives if and only if* f *is holomorphic.*
(ii) f *has a holomorphic primitive in* Ω *if* f *holomorphic and* Ω *is simply connected.*

4.3 Fundamental Theorems about Holomorphic Functions

In this section we prove some basic theorems about holomorphic functions, in particular we shall prove that holomorphic functions are exactly the functions that locally admit a power series development.

4.3.1 Goursat and Cauchy theorems

a. Goursat lemma

4.15 Lemma. *Let* Ω *be an open set of* $\mathbb{C}$ *and* $f \in \mathcal{H}(\Omega)$. *Then*

$$\int_{\partial^+ R} f(z)\, dz = 0$$

for every rectangle $R \subset\subset \Omega$. *In particular,* f *has locally holomorphic primitives.*

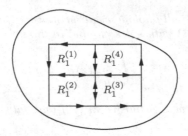

Figure 4.3. Illustration of the proof of the Goursat lemma.

Proof. Following Edouard Goursat (1858–1936), suppose that

$$\eta(R) := \int_{\partial^+ R} f(z)\, dz \neq 0, \tag{4.13}$$

and divide R in four equal rectangles $R_1^{(1)}, \dots R_1^{(4)}$. Since the integrals along the common segments of the boundaries of every two adjacent rectangles cancel, we find

$$\eta(R) = \sum_{i=1}^{4} \eta(R_1^{(i)}).$$

Therefore at least for one of them $R_1 := R_1^{(j)}$ we have

$$|\eta(R_1)| \geq \frac{1}{4} |\eta(R)|.$$

Dividing R_1 in four parts and proceeding in this way by induction, we find a decreasing sequence of rectangles $\{R_n\}$ such that $R_{n+1} \subset R_n$, $\mathrm{diag}\,(R_n) = 2^{-n}\mathrm{diag}\,(R)$, $\mathrm{perimeter}\,(R_n) = 2^{-n}\mathrm{perimeter}\,(R)$ and

$$|\eta(R_n)| \geq 4^{-n}|\eta(R)|. \tag{4.14}$$

Set $z^* = \cap_n \overline{R}_n$. Since f is $\mathbb{C}$-differentiable at z^*, for every $\epsilon > 0$ there exist $\delta > 0$ and $\overline{n}$ such that for every $n \geq \overline{n}$ we have $\overline{R}_n \subset B(z^*, \delta)$ and

$$|f(z) - f(z^*) - f'(z^*)(z - z^*)| \leq \epsilon |z - z^*| \qquad \forall z \in B(z^*, \delta).$$

Therefore, for n large we have

$$|\eta(R_n)| = \left| \int_{\partial^+ R_n} \Big(f(z) - f(z^*) - f'(z^*)(z - z^*) \Big)\, dz \right|$$

$$\leq \epsilon \int_{\partial^+ R_n} |z - z^*|\, dz \leq \epsilon\, \mathrm{diag}\,(R_n)\, \mathrm{perimeter}\,(R_n)$$

$$\leq c\, 4^{-n} \epsilon.$$

This together with (4.14) yields $|\eta(R)| \leq c\,\epsilon$, i.e., $\eta(R) = 0$. $\qquad \square$

b. Elementary domains and Goursat's theorem

An open and connected set in $\mathbb{C}$ is called a *domain* of $\mathbb{C}$. We recall that a domain A is said to be *regular* if its boundary is the union of the images of a finite number of simple curves of class C^1 that meet eventually at the extreme points. In this case, for all but finitely many points of ∂A, the exterior unit normal to ∂A is well defined, and we denote by $\partial^+ A$ a piecewise

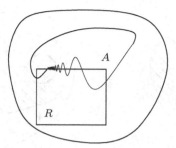

Figure 4.4. An elementary domain A of $\mathbb{C}$ for which $A \cap R$ is not an elementary domain for R.

smooth and simple curve that travels the boundary ∂A of A anticlockwise, i.e., leaving on the right the exterior unit normal to ∂A. Though $\partial^+ A$ is not uniquely defined, we call it the *counterclockwise oriented boundary of* A as, for any continuous function $f : \Omega \to \mathbb{C}$, the integral

$$\int_{\partial^+ A} f(z)\, dz$$

does not depend on the parameterization of ∂A (provided it is counterclockwise).

Let Ω be an open set. If $A \subset\subset \Omega$ is a regular domain and $R \subset\subset \Omega$ is a rectangle, in general $A \cap R$ is not a regular domain of $\mathbb{C}$, see Figure 4.4. We say that a domain A is an *elementary domain* of Ω if $A \subset\subset \Omega$, and we can square off $\mathbb{C}$ with squares with sides parallel to the axes in such a way that for each of its open squares R with $R \cap A \neq \emptyset$ we have

(i) $R \subset \Omega$,
(ii) $R \cap A$ is a regular domain of $\mathbb{C}$.

We do not dwell any further on this definition, since it is only a technical means and *a posteriori* it will be superfluous. We only notice that the rectangles and the balls inside Ω are elementary for Ω. We use in the sequel the following.

4.16 Proposition. *Let Ω be an open set of $\mathbb{C}$, let A, B be elementary domains for Ω with $A \subset\subset B \subset\subset \Omega$, and let $z_0 \in A$. Then $A \setminus B$ is an elementary domain of $\Omega \setminus \{z_0\}$.*

4.17 Proposition. *Let $f : \Omega \to \mathbb{C}$ be a continuous function in an open set of $\Omega \subset \mathbb{C}$. The following claims are equivalent.*

(i) *For every rectangle $R \subset\subset \Omega$ with sides parallel to the coordinate axes we have $\int_{\partial^+ R} f(z)\, dz = 0$.*
(ii) *For every elementary domain A of Ω we have*

$$\int_{\partial^+ A} f(z)\, dz = 0.$$

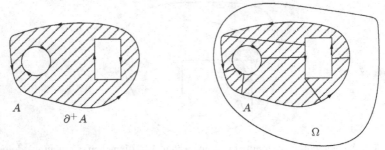

Figure 4.5. A domain $A \subset \mathbb{C}$ that is an elementary domain of Ω.

Proof. It suffices to prove that (i) implies (ii). Let R be a rectangle as in (i); from (i) and Theorem 4.12 we infer that f has a holomorphic primitive F_R in R, hence

$$\int_\gamma f(z)dz = \int_0^1 f(\gamma(t))\gamma'(t)\, dt = \int_0^1 \frac{d}{dt} F_R(\gamma(t))\, dt = F_R(\gamma(1)) - F_R(\gamma(0))$$

for every curve γ in R that is piecewise regular. In particular,

$$\int_{\partial^+ A} f(z)\, dz = 0$$

for every elementary domain A of R, since $\partial^+ A$ is a closed curve. Now, split A as $A = \cup_{i=1}^N A_i$ where the A_i's are domains with disjoint interiors each contained in a rectangle $R_i \subset \Omega$, the segments common to more than one A_i are traveled exactly twice with opposite orientation when traveling along the boundaries of the A_i's; consequently the integrals over these segments cancel to get

$$\int_{\partial^+ A} f(z)\, dz = \sum_{i=1}^N \int_{\partial^+ A_i} f(z)\, dz = 0.$$

$\square$

From Goursat's lemma and Proposition 4.17 we then infer the following.

4.18 Theorem (Goursat). *Let Ω be a domain of $\mathbb{C}$ and $f \in \mathcal{H}(\Omega)$. Then*

$$\int_{\partial^+ A} f(z)\, dz = 0$$

for every elementary domain A of Ω.

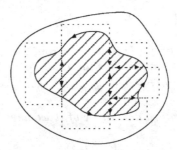

Figure 4.6. Illustration of the proof of Proposition 4.17.

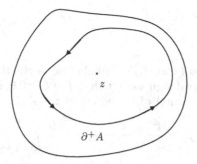

Figure 4.7. $\partial^+ A$.

The converse of Theorem 4.18 is also true as we shall prove later on.

4.19 ¶. Let Ω be a domain of $\mathbb{C}$, let A, B be elementary domains of Ω such that $A \subset\subset B$ and let $z_0 \in A$. Show that

$$\int_{\partial^+ A} f(z)\, dz = \int_{\partial^+ B} f(z)\, dz \qquad \forall f \in \mathcal{H}(\Omega \setminus \{z_0\}).$$

c. Cauchy formula and power series development

4.20 Theorem (Cauchy formula, I). *If f is holomorphic in a domain Ω, then for every elementary domain A of Ω and every $z \in A$ we have*

$$f(z) = \frac{1}{2\pi i} \int_{\partial^+ A} \frac{f(\zeta)}{\zeta - z}\, d\zeta.$$

Proof. Let δ_0 be such that $B(z, \delta_0) \subset A$. For every δ, $0 < \delta \leq \delta_0$, the set $A \setminus \overline{B(z,\delta)}$ is an elementary domain for $\Omega \setminus \{z\}$ and the function $\zeta \to f(\zeta)/(\zeta - z)$ is holomorphic in $\Omega \setminus \{z\}$, hence

$$\int_{\partial^+ (A \setminus \overline{B(z,\delta)})} \frac{f(\zeta)}{\zeta - z}\, d\zeta = 0$$

on account of Goursat's theorem, Theorem 4.18. Therefore

$$\int_{\partial^+ A} \frac{f(\zeta)}{\zeta - z}\, d\zeta = \int_{\partial^+ B(z,\delta)} \frac{f(\zeta)}{\zeta - z}\, d\zeta$$

$$= f(z) \int_{\partial^+ B(z,\delta)} \frac{1}{\zeta - z}\, d\zeta + \int_{\partial^+ B(z,\delta)} \frac{f(\zeta) - f(z)}{\zeta - z}\, d\zeta$$

$$= 2\pi i f(z) + \int_{\partial^+ B(z,\delta)} \frac{f(\zeta) - f(z)}{\zeta - z}\, d\zeta \qquad \forall \delta \leq \delta_0.$$

In particular, the function

$$\delta \to \int_{\partial^+ B(z,\delta)} \frac{f(\zeta) - f(z)}{\zeta - z}\, d\zeta$$

is constant and the claim follows if

$$\lim_{\delta \to 0} \int_{\partial^+ B(z,\delta)} \frac{f(\zeta) - f(z)}{\zeta - z}\, d\zeta = 0.$$

The latter claim is in fact true: On account of the continuity of f at z we have

$$\left| \int_{\partial^+ B(z,\delta)} \frac{f(\zeta) - f(z)}{\zeta - z} \, d\zeta \right| \leq \frac{1}{\delta} (2\pi \, \delta) \sup_{\zeta \in B(z,\delta)} |f(\zeta) - f(z)| = o(1) \qquad \text{as } \delta \to 0.$$

□

As a consequence of Cauchy's formula and of the theorem of differentiation under the integral sign we get that $f \in C^\infty(\Omega)$ if $f \in \mathcal{H}(\Omega)$ and for every elementary domain A of Ω and every $z \in A$

$$f^{(k)}(z) = \frac{1}{2\pi i} \int_{\partial^+ A} f(\zeta) \frac{d^k}{dz^k} \frac{1}{(\zeta - z)} \, d\zeta = \frac{k!}{2\pi i} \int_{\partial^+ A} \frac{f(\zeta)}{(\zeta - z)^{k+1}} \, d\zeta. \quad (4.15)$$

Actually, we have more.

4.21 Theorem. *Let $\Omega \subset \mathbb{C}$ be an open set, let $f \in \mathcal{H}(\Omega)$, let $z_0 \in \Omega$ and $\rho := \operatorname{dist}(z_0, \partial\Omega)$. Then*

$$f(z) = \sum_{k=0}^{\infty} a_k (z - z_0)^k, \qquad \forall z \in B(z_0, \rho),$$

where for every $k \in \mathbb{N}$

$$a_k := \frac{1}{2\pi i} \int_{\partial^+ A} \frac{f(\zeta)}{(\zeta - z_0)^{k+1}} \, d\zeta, \qquad (4.16)$$

A being an elementary domain of Ω containing z_0.

4.22 Lemma. *Let $f : \overline{B(z_0, r)} \subset \mathbb{C} \to \mathbb{C}$ be a continuous function such that*

$$f(z) = \frac{1}{2\pi i} \int_{\partial^+ B(z_0, r)} \frac{f(\zeta)}{\zeta - z} \, d\zeta \qquad \forall z \in B(z_0, r).$$

Then

$$f(z) = \sum_{k=0}^{\infty} a_k (z - z_0)^k, \qquad \forall z \in B(z_0, r)$$

where for every $k \in \mathbb{N}$

$$a_k := \frac{1}{2\pi i} \int_{\partial^+ B(z_0, r)} \frac{f(\zeta)}{(\zeta - z_0)^{k+1}} \, d\zeta.$$

Proof. We write for ζ with $|\zeta - z_0| = r$

$$\frac{1}{\zeta - z} = \frac{1}{\zeta - z_0} \frac{1}{1 - \frac{z - z_0}{\zeta - z_0}} = \frac{1}{\zeta - z_0} \sum_{k=0}^{\infty} \left(\frac{z - z_0}{\zeta - z_0} \right)^k$$

with uniform convergence (when ζ varies) since $\left| \frac{z - z_0}{\zeta - z_0} \right| = \frac{|z - z_0|}{r} < 1$. By integrating term by term, we then find

$$f(z) = \frac{1}{2\pi i} \int_{\partial^+ B(z_0, r)} \frac{f(\zeta)}{\zeta - z} \, d\zeta = \sum_{k=0}^{\infty} \left(\frac{1}{2\pi i} \int_{\partial^+ B(z_0, r)} \frac{f(\zeta)}{(\zeta - z_0)^{k+1}} \, d\zeta \right) (z - z_0)^k.$$

□

Proof of Theorem 4.21. If $z \in B(z_0, \rho)$ and r is such that $|z - z_0| < r < \rho$, the Cauchy formula,

$$f(w) = \frac{1}{2\pi i} \int_{\partial^+ B(z_0, r)} \frac{f(\zeta)}{\zeta - w} \, d\zeta \qquad \forall w \in B(z_0, r),$$

then yields $f(z) = \sum_{k=0}^{\infty} a_k (z - z_0)^k$ with

$$a_k := \frac{1}{2\pi i} \int_{\partial^+ B(z_0, r)} \frac{f(\zeta)}{(\zeta - z_0)^{k+1}} \, d\zeta.$$

Choose now $\epsilon > 0$ such that $B(z_0, \epsilon) \subset A$. Since $g(\zeta) := f(\zeta)/(\zeta - z_0)^{k+1}$ is holomorphic in $\Omega \setminus \{z_0\}$, we may apply Goursat's theorem to g on the elementary domains $B(z_0, r) \setminus B(z_0, \epsilon)$ and $A \setminus B(z_0, \epsilon)$ of $\Omega \setminus \{z_0\}$ to get

$$\int_{\partial^+ B(z_0, r)} g(\zeta) \, d\zeta = \int_{\partial^+ B(z_0, \epsilon)} g(\zeta) \, d\zeta = \int_{\partial^+ A} g(\zeta) \, d\zeta.$$

$\square$

4.23 Remark. We have in fact proved that any function for which the Cauchy formula holds true has locally a power series development.

4.24 Theorem. *If $S(z) = \sum_{k=0}^{\infty} a_k (z - z_0)^k$, $z \in B(z_0, r)$, $r > 0$, is the sum of a power series, then S is of class $C^\infty(B(z_0, r))$ and has complex derivatives $S^{(k)}(z)$ of any order in $B(z_0, r)$; in particular, $S(z)$ is holomorphic. Moreover, for every $k \in \mathbb{N}$ we have*

$$S^{(k)}(z) = \sum_{n=k}^{\infty} n(n-1) \ldots (n - k + 1) \, a_n (z - z_0)^{n-k} \qquad \forall z \in B(z_0, \rho),$$

hence $S^{(k)}(z)$ is holomorphic and

$$S^{(k)}(z_0) = k! \, a_k \qquad \forall k > 0. \tag{4.17}$$

This follows by applying inductively the following.

4.25 Proposition. *If $S(z) = \sum_{k=0}^{\infty} a_k (z - z_0)^k$, $z \in B(z_0, r)$, $r > 0$, then $S \in C^1(B(z_0, r)) \cap \mathcal{H}(B(z_0, r))$ and*

$$S'(z) = \sum_{k=1}^{\infty} k \, a_k (z - z_0)^{k-1} \qquad \forall z \in B(z_0, r).$$

Proof. Since the radius of convergence of $\sum_{k=1}^{\infty} k \, a_k \, (z - z_0)^{k-1}$ is the same as the radius of $\sum_{k=0}^{\infty} a_k (z - z_0)^k$, it is at least r. In $B(z_0, r)$ set then $T(z) := \sum_{k=1}^{\infty} k \, a_k \, (z - z_0)^{k-1}$. For every $z, w \in B(z_0, r)$ and every curve $\gamma : [0, 1] \to B(z_0, r)$ piecewise of class C^1 with $\gamma(0) = w$ and $\gamma(1) = z$, we have

$$\sum_{k=1}^{p} a_k (z - z_0)^k - \sum_{k=1}^{p} a_k (w - z_0)^k = \int_{\gamma} \sum_{k=1}^{p} k \, a_k (\zeta - z_0)^{k-1} \, d\zeta,$$

as $D((z - z_0)^k) = k(z - z_0)^{k-1}$. Since the sequence of the partial sums of a power series converges uniformly in compact sets of the domain of convergence, when $p \to \infty$ we get

Figure 4.8. Joseph Liouville (1809–1882) and Giacinto Morera (1856–1909).

$$\sum_{k=1}^{p} a_k (z - z_0)^k \;\to\; S(z) - a_0,$$

$$\sum_{k=1}^{p} a_k (w - z_0)^k \;\to\; S(w) - a_0,$$

$$\int_\gamma \sum_{k=1}^{p} k\, a_k (\zeta - z_0)^{k-1}\, d\zeta \;\to\; \int_\gamma T(\zeta)\, d\zeta,$$

hence

$$S(z) - S(w) = \int_\gamma T(\zeta)\, d\zeta.$$

Since $z, w \in B(z_0, r)$ and γ are arbitrary, Theorem 4.9, says that S is holomorphic in $B(z_0, r)$ with $S'(z) = T(z) \; \forall z \in B(z_0, r)$. $\qquad\qquad\square$

4.26 Corollary. *Let Ω be an open set of $\mathbb{C}$. If $f \in \mathcal{H}(\Omega)$, then all the derivatives of f exist and are holomorphic functions in Ω.*

Finally, we can state the following.

4.27 Corollary. *Let $f : \Omega \subset \mathbb{C} \to \mathbb{C}$ be a function. The following claims are equivalent.*

(i) $f \in \mathcal{H}(\Omega)$,
(ii) f *admits locally holomorphic pimitives.*
(iii) $\int_{\partial^+ A} f(z)\, dz = 0$ *for every elementary domain A of Ω,*
(iv) *Cauchy's formula holds: for every elementary domain A of Ω we have*

$$f(z) = \frac{1}{2\pi i} \int_{\partial^+ A} \frac{f(\zeta)}{\zeta - z}\, d\zeta \qquad \forall z \in A,$$

(v) f *is locally the sum of a power series.*

Proof. (i) $\Rightarrow$ (iii) is Theorem 4.18, and (ii) and (iii) are equivalent by Proposition 4.17. The implications (i) $\Rightarrow$ (iv), (iv) $\Rightarrow$ (v), (v) $\Rightarrow$ (i) are, respectively, Theorems 4.20, 4.21 and 4.24.

It remains to prove that (ii) $\Rightarrow$ (i). If (ii) holds, f is locally the derivative of a holomorphic function, thus in turn it is a holomorphic function because of Corollary 4.26.
□

4.28 Remark. The implication f holomorphic $\Rightarrow f \in C^1(\Omega)$ is known in the literature as *Goursat's lemma*, and the equivalence (i) $\Leftrightarrow$ (iii) as *Morera's theorem*.

4.3.2 Liouville's theorem

From (4.17) and (4.16) or from (4.15) we infer the following.

4.29 Proposition (Cauchy's estimates). *If* $f \in \mathcal{H}(\Omega)$, *then*

$$|f^{(k)}(z_0)| \le \frac{k!}{r^k} \max_{\partial B(z_0, r)} |f(z)| \qquad (4.18)$$

for every $z_0 \in \Omega$, $k \in \mathbb{N}$ *and* $r < \text{dist}(z_0, \partial\Omega)$.

As a corollary we get the following.

4.30 Theorem (Liouville). *The only bounded and holomorphic functions in the whole complex plane are the constants.*

Proof. If $|f(z)| \le M \ \forall z \in \mathbb{C}$, we have for all $z \in \mathbb{C}$ and $r > 0$

$$|f'(z)| \le \frac{M}{r}$$

on account of Cauchy's estimate for f'. Letting $r \to \infty$, we infer $f'(z) = 0$ for all $z \in \mathbb{C}$, i.e., f is constant.
□

As an application of Liouville's theorem we find another proof of the *fundamental theorem of algebra*.

4.31 Theorem (fundamental of algebra). *A complex polynomial of degree* n, $n \ge 1$, *has* n *roots.*

Proof. It suffices to prove that a nonconstant polynomial has at least one root. Suppose that $P(z)$ is a nonconstant polynomial such that $P(z) \ne 0 \ \forall z$. Then $1/P(z)$ is holomorphic in $\mathbb{C}$ and bounded by the Weierstrass theorem since $\lim_{|z|\to\infty} |P(z)| = +\infty$. It follows that $1/P(z) = \text{const}$, a contradiction.
□

Actually, we have proved more.

4.32 Theorem. *Let* $f : \mathbb{C} \to \mathbb{C}$ *be holomorphic in* $\mathbb{C}$ *and assume that* $\liminf_{z\to\infty} |f(z)| > 0$. *Then, either* $f(z)$ *is constant or* f *has a zero.*

4.3.3 The unique continuation principle

If f is holomorphic in the open set Ω with all derivatives vanishing at a point $z_0 \in \Omega$, by (4.17) f vanishes in a neighborhood of z_0. In other words, the set

$$X(f) := \left\{ z \in \Omega \mid f^{(k)}(z) = 0 \ \forall k \right\}$$

is open. On the other hand $X(f)$ is closed since all derivatives of f are continuous. Therefore $X(f)$ is the connected component of Ω that contains z_0. This is known as the *unique continuation* or *identity principle*.

4.33 Theorem (The identity principle). *Let f and g be two holomorphic functions in a domain Ω. Suppose that at $z_0 \in \Omega$ we have $f^{(k)}(z_0) = g^{(k)}(z_0)$ for all $k = 0, 1, \ldots$, then $f = g$ in Ω.*

Denote the set of zeros of a function f in Ω by

$$Z(f) := \left\{ z \in \Omega \,\Big|\, f(z) = 0 \right\}.$$

4.34 Theorem. *Let Ω be a domain of $\mathbb{C}$ and $f \in \mathcal{H}(\Omega)$. If f is not identically zero, then $Z(f)$ is discrete and without accumulation points in Ω.*

Proof. Since $Z(f)$ is closed in Ω, it suffices to prove that $Z(f)$ is discrete. Let $z_0 \in Z(f)$ and let k be the first nonnegative integer such that $f^{(k)}(z_0) \neq 0$. In a neighborhood of z_0 we have

$$f(z) = \sum_{j=k}^{\infty} \frac{f^{(k)}(z_0)}{k!}(z - z_0)^j = (z - z_0)^k g(z)$$

with $g(z_0) \neq 0$. Since $g(z) \neq 0$ in a neighborhood U of z_0, clearly U has no zero other than z_0. $\qquad\square$

In conclusion we can state the identity principle as follows.

4.35 Theorem (The identity principle). *Let f and g be two holomorphic functions in the domain Ω of $\mathbb{C}$. The following claims are equivalent*

(i) $f = g$ in Ω
(ii) *There exists $z_0 \in \Omega$ such that $f^{(k)}(z_0) = g^{(k)}(z_0)$ for all k.*
(iii) *The set $\{ z \in \Omega \mid f(z) = g(z) \}$ has at least an accumulation point in Ω.*

4.3.4 Holomorphic differentials

Let Ω be an open set of $\mathbb{C}$ and let $f := u + iv : \Omega \to \mathbb{C}$ be a function of class C^1. It is easy to check that the two differential 1-forms

$$\omega_1 := u\,dx - v\,dy, \qquad \omega_2 := v\,dx + u\,dy \tag{4.19}$$

are closed in Ω if and only if the Cauchy–Riemann equations for f

$$\begin{cases} v_x = u_y, \\ v_y = -u_x \end{cases}$$

hold in Ω. Since holomorphic functions are of class C^1, see Corollary 4.26, on account of Theorem 4.9 we therefore conclude the following.

4.36 Proposition. *Let* $f = u + iv : \Omega \subset \mathbb{C} \to \mathbb{C}$ *be of class* C^1*. Then* $f \in \mathcal{H}(\Omega)$ *if and only if the two differential 1-forms*

$$\omega_1 := u\,dx - v\,dy, \qquad \omega_2 = v\,dx + u\,dy$$

are closed C^1 *forms.*

Suppose now that $\alpha, \beta : \Omega \to \mathbb{R}$ of class C^1 are potentials, respectively, of ω_1 and ω_2, that is,

$$\begin{cases} \alpha_x = u, \\ \alpha_y = -v, \end{cases} \qquad \begin{cases} \beta_x = v, \\ \beta_y = u. \end{cases}$$

Setting $F := \alpha + i\beta$, we have

$$\begin{cases} F_x = \alpha_x + i\beta_x = u + iv = f, \\ F_y = \alpha_y + i\beta_y = -v + iu = if, \end{cases}$$

that is $F \in \mathcal{H}(\Omega)$, and $F' = f$ in Ω. We therefore conclude the following.

4.37 Proposition. *The function* $f = u + iv \in \mathcal{H}(\Omega)$ *has a holomorphic primitive in* Ω *if and only if the forms* ω_1 *and* ω_2 *in (4.19) are exact in* Ω*. Moreover,* $F \in \mathcal{H}(\Omega)$ *and* $F' = f$ *in* Ω *if and only if* $\alpha := \Re(F(z))$ *and* $\beta := \Im(F(z))$ *are the potentials of respectively,* ω_1 *and* ω_2*.*

The theory of differential forms then applies to holomorphic functions. In particular, the following holds. Since a holomorphic function has locally holomorphic primitives on account of Goursat's lemma, we then have the following.

4.38 Theorem (Homotopy invariance). *Let* Ω *be a domain of* $\mathbb{C}$ *and* $f \in \mathcal{H}(\Omega)$*. If* $\gamma, \delta : [0,1] \to \Omega$ *are two homotopic curves in* Ω *of class* C^1*, then*

$$\int_\gamma f(z)\,dz = \int_\delta f(z)\,dz.$$

4.39 Corollary. *Let* Ω *be a simply connected domain. Then every* $f \in \mathcal{H}(\Omega)$ *has a holomorphic primitive in* Ω*.*

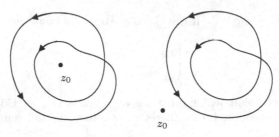

Figure 4.9. From the left: (a) $I(\gamma, z_0) = 2$ and (b) $I(\gamma, z_0) = 0$.

a. Winding number

Let $\gamma : [0,1] \to \Omega$ be a closed curve that is piecewise of class C^1 and let $z \notin \gamma([0,1])$. The *winding number* of γ *around* z or the *index* of γ with respect to z is

$$I(\gamma, z) := \frac{1}{2\pi i} \int_\gamma \frac{d\zeta}{\zeta - z}.$$

For example, if $\gamma(t) := z + e^{ikt}$, $t \in [0, 2\pi]$ and $k \in \mathbb{Z}$, then

$$I(\gamma, z) = \frac{1}{2\pi i} \int_0^{2\pi} \frac{ike^{ikt}}{e^{ikt}} \, dt = k.$$

If $\gamma : [0,1] \to \mathbb{C}$ is a smooth curve with $z \notin \gamma([0,1])$, then γ is homotopic to

$$\delta(t) := z + \frac{\gamma(t) - z}{|\gamma(t) - z|}$$

in $\mathbb{C} \setminus \{z\}$, a homotopy h being given by

$$h(t, s) = (1 - s)\gamma(t) + s\delta(t), \qquad t, s \in [0, 1].$$

Since $\zeta \to \frac{d\zeta}{\zeta - z}$ is a holomorphic differential on $\mathbb{C} \setminus \{z\}$, Theorem 4.38 yields that *homotopic curves in* $\mathbb{C} \setminus \{z\}$ *have the same index*,

$$I(\gamma, z) = I(\delta, z).$$

We therefore see that the index of γ with respect to z is the *topological degree* of the map $t \to \frac{\gamma(t)-z}{|\gamma(t)-z|}$ from $[0, 2\pi]$ into $S^1 = \partial B(0, 1)$ and the following holds, see [GM3].

4.40 Proposition. *Let* $z, z_0 \in \mathbb{C}$, $z \neq z_0$ *and let* $\pi_1(\mathbb{C} \setminus \{z\}, z_0)$ *be the first homotopy group of* $\mathbb{C} \setminus \{z\}$ *with base point* z_0. *The winding number is surjective and injective as a map from* $\pi_1(\mathbb{C} \setminus \{z\}, z_0)$ *into* $\mathbb{Z}$. *In particular, we have:*

(i) *Homotopic curves have the same index.*
(ii) *The index is an integer.*
(iii) *For every* $k \in \mathbb{Z}$ *there is a curve through* z_0 *avoiding* z *and with index with respect to* z *that equals* k.

(iv) *Two curves are homotopic if and only if they have the same index.*

(v) *Let $\gamma : [0,1] \to \mathbb{C}$ be a closed curve. The winding number map $z \to I(\gamma, z)$, $z \in \mathbb{C} \setminus \gamma([0,1])$, is continuous, hence constant on each connected component of $\mathbb{C} \setminus \gamma([0,1])$.*

4.41 ¶. Prove that

(i) $I(\gamma, z) = 0$ for every z in the unbounded connected component of $\mathbb{C} \setminus \gamma([0,1])$.

(ii) $I(\gamma, z)$ is locally constant in $\mathbb{C} \setminus \mathrm{Supp}\,(\gamma)$, hence on each connected component of $\mathbb{C} \setminus \mathrm{spt}\,\gamma$.

(iii) $I(\partial^+ B(0,1), z) = 0$ if $z \notin \overline{B(0,1)}$, and $I(\partial^+ B(0,1), z) = 1$ if $z \in B(0,1)$.

4.42 Theorem (Cauchy formula, II). *Let $\Omega \subset \mathbb{C}$ be open and let $f \in \mathcal{H}(\Omega)$. For every closed curve $\gamma : [0,1] \to \Omega$ that is piecewise of class C^1 and for all $z \notin \gamma([0,1])$, we have*

$$I(\gamma, z) f(z) = \frac{1}{2\pi i} \int_\gamma \frac{f(\zeta)}{\zeta - z}\, d\zeta.$$

Proof. Let $r > 0$ be so that $B(z, r) \subset \Omega$ and set $k := I(\gamma, z)$. The curve γ is homotopic in $\mathbb{C} \setminus \{z\}$ to $\delta(t) := z + r e^{ikt}$, $t \in [0, 2\pi]$ since $I(\gamma, z) = k = I(\delta, z)$. Using the periodicity of $t \to e^{it}$ and Theorem 4.18, we then compute

$$\int_\gamma \frac{f(\zeta)}{\zeta - z}\, d\zeta = \int_\delta \frac{f(\zeta)}{\zeta - z}\, d\zeta = \int_0^{2\pi} \frac{f(e^{ikt})}{e^{ikt}} i k e^{ikt}\, dt = ik \int_0^{2\pi} f(e^{ikt})\, dt$$

$$= k \int_{\partial^+ B(z,r)} \frac{f(\zeta)}{\zeta - z}\, d\zeta = 2\pi\, k\, i\, f(z).$$

$\square$

b. Stokes's formula and Cauchy's and Morera's theorems

Let Ω be an open set of $\mathbb{C}$ and let $f \in \mathcal{H}(\Omega)$. Since f is of class C^1, see Corollary 4.26, we may apply Stokes's formulas in the plane, see Proposition 3.38, to the closed differential forms $u\, dx - v\, dy$ and $v\, dx + u\, dy$ to get for every regular domain $A \subset\subset \Omega$

$$\int_{\partial^+ A} (u\, dx - v\, dy) = \iint_A \left(-\frac{\partial v}{\partial x} - \frac{\partial u}{\partial y} \right) dx\, dy = 0$$

$$\int_{\partial^+ A} (v\, dx + u\, dy) = \iint_A \left(\frac{\partial u}{\partial x} - \frac{\partial v}{\partial y} \right) dx\, dy = 0,$$

hence the following extensions of Cauchy's and Morera's theorems hold.

4.43 Theorem (Cauchy formula, III). *Let $A \subset\subset \Omega$ be a regular domain of $\mathbb{C}$ and let $f \in \mathcal{H}(\Omega)$. We have*

$$\int_{\partial^+ A} f(z)\, dz = 0$$

and, for all $z \in A$,

$$f(z) = \int_{\partial^+ A} \frac{f(\zeta)}{\zeta - z}\, d\zeta.$$

4.44 Remark. Actually, if we switch to Lebesgue's integral, Stokes's theorem holds for every admissible domain, see Section 3.4. Consequently, switching to Lebesgue's integral, we infer that Theorem 4.43 holds for every admissible domain $A \subset\subset \Omega$. Furthermore, if we also assume that Ω is admissible and $f \in \mathcal{H}(\Omega) \cap C^0(\overline{\Omega})$ with $|Df| \in \mathcal{L}^1(\Omega)$, then Theorem 4.43 also holds with $A = \Omega$.

4.45 Remark. Theorem 4.43 in particular says that Goursat's lemma for functions in $\mathcal{H}(\Omega) \cap C^1(\Omega)$ is a trivial consequence of Stokes's formula in the plane. Since *a priori* it is not evident that holomorphic functions are of class C^1, we need a proof that applies to solely holomorphic functions: this was done in the proof of Theorem 4.18 or can be done by an approximation procedure, see [GM5].

4.4 Examples of Holomorphic Functions

4.4.1 Some simple functions

Here we present basic examples of holomorphic functions.

4.46 $f(z) = z^2$. It is a holomorphic function from $\mathbb{C}$ to $\mathbb{C}$ with $f'(z) = 2z$. In real Cartesian coordinates we have

$$z^2 = (x^2 - y^2) + i2xy \qquad \text{if } z = x + iy,$$

and in polar coordinates

$$z^2 = r^2 e^{2i\theta} \qquad \text{if } z = re^{i\theta}.$$

It is easily seen that the transformation $z \to z^2$

   (i)   maps lines through the origin into half-lines from the origin,
  (ii)   maps circles around the origin into circles around the origin,
 (iii) maps the hyperbolas $x^2 - y^2 = k$ into vertical lines,
 (iv) maps the hyperbolas $2xy = k$ into horizontal lines.

4.47 The exponential function. The complex exponential is defined as

$$e^z = e^x(\cos y + i \sin y), \qquad z = x + iy \in \mathbb{C}.$$

It is a holomorphic function in $\mathbb{C}$ with $De^z = e^z$ since, for instance,

$$\frac{\partial e^z}{\partial x} = e^z, \qquad \frac{\partial e^z}{\partial y} = ie^z.$$

It is easily seen that the transformation $z \to e^z$

   (i)   maps horizontal lines into half-lines from the origin,
  (ii)   maps vertical lines into circles around the origin,
 (iii) satisfies $|e^z| = e^x$, in particular e^z is bounded on the half-planes $\{z = x + iy \mid x < x_0\}$, $x_0 \in \mathbb{R}$,

(iv) satisfies $e^z \neq 0 \ \forall z \in \mathbb{C}$,
(v) satisfies $e^{z+w} = e^z e^w \ \zeta, w \in \mathbb{C}$,
(vi) is not injective, in fact, $e^z = e^w$ if and only if $e^{z-w} = 1$, i.e., if and only if z and w have the same real part and imaginary part differing for a multiple of 2π

$$e^z = e^w \quad \text{if and only if} \quad z - w = i2\pi k, \ k \in \mathbb{Z}:$$

we say that e^z is periodic of period $2\pi i$,
(vii) is the sum of the power series

$$e^z = \sum_{n=0}^{\infty} \frac{z^n}{n!}, \quad z \in \mathbb{C}$$

that converges uniformly on the compact sets of $\mathbb{C}$.

4.48 Sinus and cosinus and hyperbolic sinus and cosinus. The functions $\sin z$ and $\cos z$, $z \in \mathbb{C}$, and the hyperbolic functions $\sinh z$, $\cosh z$, $z \in \mathbb{C}$ are defined by means of Euler's formulas

$$\cos z := \frac{e^{iz} + e^{-iz}}{2}, \qquad \sin z = \frac{e^{iz} - e^{-iz}}{2i},$$

$$\cosh z := \frac{e^z + e^{-z}}{2}, \qquad \sinh z = \frac{e^z - e^{-z}}{2}.$$

They are holomorphic in $\mathbb{C}$ with

$$D \sin z = \cos z, \qquad D \cos z = -\sin z,$$

$$D \sinh z = \cosh z, \qquad D \cosh z = \sinh z.$$

The functions $\cos z$ and $\sin z$ vanish only at points $z = \pi/2 + k\pi$, $k \in \mathbb{Z}$ and $z = k\pi$, $k \in \mathbb{Z}$, respectively, of the real axis; moreover they are unbounded in $\mathbb{C}$ since we have

$$\frac{e^{|y|} - e^{-|y|}}{2} \leq |\cos z| = \frac{|e^{ix}e^{-y} + e^{-ix}e^y|}{2} \leq \frac{e^y + e^{-y}}{2} = \cosh y$$

and, similarly,

$$\frac{e^{|y|} - e^{-|y|}}{2} \leq |\sin z| = \frac{|e^{ix}e^{-y} - e^{-ix}e^y|}{2} \leq \frac{e^y + e^{-y}}{2} = \cosh y.$$

The hyperbolic functions $\cosh z$ and $\sinh z$ are related to the trigonometric functions by

$$\cosh z = \cos(iz), \qquad \sinh z = -i \sin(iz).$$

Therefore they vanish respectively at the points $z = i(\pi/2 + k\pi)$, $k \in \mathbb{Z}$, and $z = ik\pi$, $k \in \mathbb{Z}$, on the imaginary axis.

Finally, trigonometric and hyperbolic functions are all sums in $\mathbb{C}$ of their power series, see [GM2],

$$\cos z = \sum_{k=0}^{\infty} (-1)^k \frac{z^{2k}}{(2k)!}, \qquad \sin z = \sum_{k=0}^{\infty} (-1)^k \frac{z^{2k+1}}{(2k+1)!}.$$

$$\cosh z = \sum_{k=0}^{\infty} \frac{z^{2k}}{(2k)!}, \qquad \sinh z = \sum_{k=0}^{\infty} \frac{z^{2k+1}}{(2k+1)!}.$$

4.49 Tangent and hyperbolic tangent. The map $\tan z := \frac{\sin z}{\cos z}$ is well defined and holomorphic in $\mathbb{C} \setminus \{z = \pi/2 + k\pi \mid k \in \mathbb{Z}\}$. Notice that $\tan z$ is bounded and away from zero as far as z stays away from the real axis; in fact, for $z = x + iy$ we have

$$|\tan z| = \frac{|e^{ix}e^{-y} - e^{-ix}e^y|}{|e^{ix}e^{-y} + e^{-ix}e^y|} \leq \frac{e^{-y} + e^y}{e^y - e^{-y}} = \coth y, \tag{4.20}$$

hence $|\tan z| \leq \coth y_0$ in $A := \{z \mid |\text{Im}(z)| \geq y_0\}$. Similarly,

$$|\cot z| = \frac{1}{|\tan z|} = \frac{|e^{ix}e^{-y} + e^{-ix}e^y|}{|e^{ix}e^{-y} - e^{-ix}e^y|} \leq \frac{e^{-y} + e^y}{e^y - e^{-y}} = \coth y. \tag{4.21}$$

4.4.2 Inverses of holomorphic functions

Let $\Omega \subset \mathbb{C}$. We recall that a function $f : \Omega \to \mathbb{C}$ is *locally invertible* if for every $x_0 \in \Omega$ there exists a neighborhood U_{x_0} of x_0 such that $f_{|U_{x_0}}$ is invertible. We say that $h : \Delta \to \mathbb{C}$ is a *local inverse* of f defined on Δ if $f(h(w)) = w$ for all $w \in \Delta$. We have

4.50 Theorem (Local inverse of a holomorphic function). *Let $\Omega \subset \mathbb{C}$ be an open set and let $f \in \mathcal{H}(\Omega)$ with $f'(z) \neq 0 \; \forall z \in \Omega$. Then*

(i) *f is open and locally invertible with continuous inverses.*
(ii) *If $\Delta \subset \mathbb{C}$ is open and $g : \Delta \to \mathbb{C}$ is a continuous inverse of f, then $g \in \mathcal{H}(\Delta)$ and*

$$g'(w) = \frac{1}{f'(g(w))} \qquad \forall w \in \Delta.$$

Proof. (i) By identifying $\mathbb{R}^2$ with $\mathbb{C}$, we denote by f also the real map $f : \Omega \subset \mathbb{R}^2 \to \mathbb{R}^2$. Since f is holomorphic,

$$\mathbf{D}f(z_0) = \begin{pmatrix} a & -b \\ b & a \end{pmatrix}$$

where $f_x := a + ib$. Consequently,

$$0 \neq |f'(z_0)|^2 = |f_x(z_0)|^2 = a^2 + b^2 = \det \mathbf{D}f(z_0)$$

for any $z_0 \in \Omega$. The (real) local invertibility theorem then yields a neighborhood U_{z_0} of z_0 such that $f_{|U_{z_0}}$ is open and invertible. It then follows that f is an open map and that $g := f_{|U_{z_0}}^{-1}$ is a continuous local inverse of f.

(ii) For $v, w \in \Delta$ we have

$$\frac{g(v) - g(w)}{v - w} = \frac{g(v) - g(w)}{f(g(v)) - f(g(w))} \to \frac{1}{f'(g(w))} \qquad \text{as } v \to w$$

since g is continuous. $\qquad \square$

As already noticed in Chapter 1, the condition $\det \mathbf{D}f(z) \neq 0$ at every point $z \in \Omega$ does not suffice to give the *global invertibility* of f, the exponential function being an example.

Theorem 4.50 (ii) reduces the existence of a holomorphic inverse of f to the existence of a continuous inverse of f. Therefore, covering maps and in particular Theorem 8.47 of [GM3] is a useful tool in discussing the existence of holomorphic inverses. We have in fact the following.

4.51 Theorem. *Let Ω be an open set in $\mathbb{C}$ and let $f : \Omega \to f(\Omega)$ be a covering of $f(\Omega)$ and a holomorphic function with $f'(z) \neq 0 \; \forall z \in \Omega$. Then f has a local inverse $h_\Delta \in \mathcal{H}(\Delta)$ for every connected and simply connected open set $\Delta \subset f(\Omega)$. Moreover, the number of distinct inverse maps of f on Δ agrees with the number of the connected components of $f^{-1}(\Delta)$.*

Proof. First observe that Δ is path-connected and locally path-connected since it is open and connected by assumption. Choose now a point $x_0 \in \Omega$ such that $y_0 := f(x_0) \in \Delta$ and let X_0 be the connected component of $f^{-1}(\Delta)$ that contains x_0. Clearly X_0 is open in $\mathbb{C}$, hence path-connected and locally path-connected.

We now claim that $f_{|X_0} : X_0 \to \Delta$ is onto, hence a covering of Δ. In fact, by Proposition 8.45 of [GM3], starting from a continuous curve α joining y_0 to $y \in \Delta$, there exists a continuous curve β on $f^{-1}(\Delta)$ with $\beta(0) = x_0$ and $f(b(t)) = \alpha(t) \, \forall t$. In particular we have $\beta(1) \in X_0$ and $f(\beta(1)) = \alpha(1) = y$.

Therefore, Theorem 8.47 of [GM3] yields that $f_{|X_0}$ is a homeomorphism from X_0 onto Δ. Thus $h := (f_{|X_0})^{-1}$ is a continuous local inverse of f, and h is holomorphic by Theorem 4.50 (ii).

By construction the number of continuous inverses defined on Δ is greater than or equal to the number of the connected components of $f^{-1}(\Delta)$. On the other hand, if $h : \Delta \to \mathbb{C}$ is a continuous inverse of f, then $h(\Delta)$ is connected, hence coincides with a connected component $\widehat{X}$ of $f^{-1}(\Delta)$, thus concluding $h = (f_{|\widehat{X}})^{-1}$. □

4.52 Remark. If f is locally invertible but not globally invertible, the equation $w = f(z)$ may have several solutions for a given w. In other words, the graph of $f(z)$

$$\left\{ (z, w) \in \mathbb{C} \times \mathbb{C} \,\middle|\, w = f(z) \right\} \tag{4.22}$$

is not the graph of a function $h(w)$ of the second variable w. However, the classic literature insists on seeing (4.22) as the graph of a *multifunction* f^{-1}, and refers to a local inverse $h : \Delta \to \mathbb{C}$ of f as to a *leaf on* Δ of the multifunction f^{-1}.

a. Complex logarithm

The previous considerations apply to the *complex logarithm*.

For $z \in \mathbb{C}$, $z \neq 0$, every $w \in \mathbb{C}$ such that $e^w = z$ is called a *complex logarithm* of z. Since $z \to e^z$ is $2\pi i$-periodic, there are infinitely many w such that $e^w = z$ differing by $2k\pi i$, $k \in \mathbb{Z}$. In other words, e^z is not globally invertible even, as we know, if it is locally invertible.

Observe that $f(z) := e^z$ *is a covering map* $f : \mathbb{C} \to \mathbb{C} \setminus \{0\}$ *of* $\mathbb{C} \setminus \{0\}$. Therefore, see Theorem 4.51, for any connected and simply connected open set $\Delta \subset \mathbb{C} \setminus \{0\}$, there exists at least a local inverse $\log_\Delta \in \mathcal{H}(\Delta)$ of $z \to e^z$, called also a *leaf on* Δ of the *complex logarithm* . By definition we have

$$\exp(\log_\Delta(w)) = w \qquad \forall w \in \Delta$$

and by Theorem 4.50 (ii),

$$D \log_\Delta(w) = \frac{1}{w} \qquad \forall w \in \Delta.$$

The complex logarithm has infinitely many leaves on Δ. In fact, if $h : \Delta \to \mathbb{C}$ is any leaf of the the complex logarithm, then

$$\exp(h(w)) = w = \exp(\log_\Delta(w)) \qquad \forall w \in \Delta$$

i.e.,

$$h(w) - \log_\Delta(w) = 2\pi i k(w) \qquad \forall w \in \Delta$$

for some integer valued function $k(w)$, actually an integer valued constant, since the left-hand side of the previous equation is continuous and Δ is connected. We therefore conclude the following.

4.53 Proposition. Let $\Delta \subset \mathbb{C} \setminus \{0\}$ be open, connected, and simply connected. Then there exist infinitely many local holomorphic inverses of $z \to e^z$ on Δ. Equivalently, there exist infinitely many leaves on Δ of the complex logarithm. Moreover, if $\varphi : \Delta \to \mathbb{C}$ is one of these inverses, then the functions $\varphi(z) + 2\pi k i$, $k \in \mathbb{Z}$ are distinct leaves on Δ of the complex logarithm, and any leaf on Δ has such a form.

As a special case, let R be the negative real axis,

$$R := \Big\{ z = x + iy \in \mathbb{C} \,\Big|\, y = 0, x \leq 0 \Big\},$$

and let

$$\Delta := \mathbb{C} \setminus R$$

which is open, connected and simply connected. The connected components of its inverse image

$$f^{-1}(\Delta) = \Big\{ z \in \mathbb{C} \,\Big|\, z \neq \pi + 2k\pi, \; k \in \mathbb{Z} \Big\}$$

are the sets

$$S_k := \Big\{ z = x + iy, \,\Big|\, (2k-1)\pi < y < (2k+1)\pi \Big\} \qquad k \in \mathbb{Z}.$$

Using Theorem 4.51 or directly, we infer that the map $z \to e^z$ when restricted to S_k has a holomorphic inverse defined on $\mathbb{C} \setminus R$ with values on S_k, that we call the *kth leaf on* $\mathbb{C} \setminus R$ of the logarithm; we denote it by $\log^{(k)}$.

The 0th leaf on $\mathbb{C} \setminus R$ is denoted simply by $z = \log w$; we also call it the *principal determination* of the logarithm, or the *principal logarithm*. By definition, $e^{\log^{(k)} w} = w \;\forall w \in \mathbb{C} \setminus R$ and

$$z = \log^{(k)} w \qquad \text{if and only if} \qquad \begin{cases} z \in S_k, \\ e^z = w. \end{cases}$$

In particular, $\log 1 = 0$ since $e^0 = 1$ and $0 \in S_0$. All the leaves of the logarithm on Δ agree up to an integer multiple of $2\pi i$. In particular,

$$\log^{(k)}(z) = \log z + i2k\pi \qquad \forall z \in \Delta, \; \forall k \in \mathbb{Z}.$$

Moreover,

$$\log(1+z) = \log^{(0)}(1+z) = \sum_{n=0}^{\infty} (-1)^n \frac{z^{n+1}}{n+1}, \qquad |z| < 1.$$

Notice that the difficulty in inverting the complex exponential is the same we encounter in inverting the uniform motion map, $t \to e^{it}$, $t \in \mathbb{R}$. In fact, for z with $|z| = 1$, the *argument of z* is the real number t defined modulus 2π such that $e^{it} = z$. For $k \in \mathbb{Z}$ and $z \in \mathbb{C}$, $z \neq -1$, define the *kth leaf on $\mathbb{C} \setminus R$ of the argument of z*, as the unique $t \in]-\pi + 2k\pi, \pi + 2k\pi[$ such that $e^{it} = z$ and denote it by $\arg^{(k)}(z)$. If $z = x + iy$ with $-\pi + 2k\pi < y < \pi + 2k\pi$ and $w \in \mathbb{C} \setminus R$, then

$$\begin{cases} w = e^z = e^x e^{iy}, \\ (2k-1)\pi < y < (2k+1)\pi \end{cases} \quad \text{iff} \quad \begin{cases} e^x = |w|, \\ e^{iy} = \frac{w}{|w|}, \\ (2k-1)\pi < y < (2k+1)\pi \end{cases}$$

(4.23)

which yields the *polar formula for the logarithm on $\mathbb{C} \setminus R$*

$$\log^{(k)} w := x + iy = \log|w| + i\arg^{(k)}\left(\frac{w}{|w|}\right) \qquad (4.24)$$

$\forall k \in \mathbb{Z}$ and $\forall w \in \mathbb{C} \setminus \{0\}$.

From (4.23) it easily follows that $\log z$ has a constant jump of $2\pi i$ through R. In fact, if $z_0 \in R$, $z_0 = x_0 + i0$, $x_0 \neq 0$, then

$$\lim_{\substack{z \to z_0 \\ \Im z > 0}} \log z = \log|z_0| + i\pi,$$

$$\lim_{\substack{z \to z_0 \\ \Im z < 0}} \log z = \log|z_0| - i\pi.$$

Finally, since $\log z$ takes its values on S_0, a special care is needed in computing with it: for instance, from the polar formula for the logarithm, we have

$$\log(zw) = \log z + \log w + \begin{cases} \pi i & \text{if } -2\pi < \arg(z) + \arg(w) \leq -\pi, \\ 0 & \text{if } -\pi < \arg(z) + \arg(w) < \pi, \\ -\pi i & \text{if } \pi \leq \arg(z) + \arg(w) < 2\pi. \end{cases}$$

b. Real powers

Let Δ be a connected and simply connected set of $\mathbb{C} \setminus \{0\}$ and let $\log_\Delta : \Delta \to \mathbb{C}$ be a leaf on Δ of the logarithm. We define the leaves of $z^\alpha : \Delta \to \mathbb{C}$, $\alpha \in \mathbb{R}$, by means of the leaves of the logarithm by

$$z^\alpha := e^{\alpha \log_\Delta z}, \qquad z \in \Delta.$$

Of course, each leaf of z^α is holomorphic on Δ with

$$D(z^\alpha) = e^{\alpha \log_\Delta z} \frac{\alpha}{z} = \alpha z^{\alpha-1}, \qquad z \in \Delta.$$

In general z^α has at most infinitely many leaves as the complex logarithm. Let us compute the number of distinct leaves of z^α on Δ. Let h_1 and h_2 be two leaves of the logarithm that, we know, differ by $2\pi ki$ for some $k \in \mathbb{Z}$. The corresponding leaves of z^α then agree if and only if $\alpha(h_2(w) - h_1(w)) = \alpha 2\pi ki$ is an integer multiple of $2\pi i$, i.e., if and only if αk is an integer. Therefore, we distinguish three cases:

(i) $\alpha \in \mathbb{Z}$. In this case, αk is always an integer; hence, all the leaves of z^α are the same, and

$$z^\alpha = \begin{cases} \underbrace{z \cdot z \ldots \cdot z}_{|\alpha| \text{ times}} & \text{if } \alpha \geq 0, \\ \underbrace{\frac{1}{z} \cdot \frac{1}{z} \cdots \cdots \frac{1}{z}}_{|\alpha| \text{ times}} & \text{if } \alpha < 0. \end{cases}$$

(ii) $\alpha \in \mathbb{Q}$, $\alpha = p/q$ with p, q coprime. In this case, αk is an integer if and only if k is a multiple of q. Hence, z^α has q distinct leaves. If $p = 1$, then $z^{1/q}$ denotes the local inverses defined on Δ of $z \to z^q$, since $z^{1/q} = \exp\left(\frac{1}{q} \log_\Delta z\right)$, and

$$(z^{1/q})^q = \overbrace{z^{1/q} \cdots \cdots z^{1/q}}^{q \text{ times}} = \sum_{i=1}^{q} \exp\left(\frac{1}{q} \log_\Delta(z)\right)$$

$$= \exp\left(\log_\Delta(z)\right) = z \qquad \forall z \in \Delta.$$

(iii) α is irrational. In this case there are infinitely many distinct leaves since αk is not integer for any k.

Finally, notice that in a fixed leaf on Δ, in general

$$(zw)^\alpha \neq z^\alpha w^\alpha.$$

In fact,

$$\frac{(zw)^\alpha}{z^\alpha w^\alpha} = \exp\left(\alpha(\log(zw) - \log z - \log w)\right)$$

$$= \begin{cases} \exp(\pi i \alpha) & \text{if } -2\pi < \arg(z) + \arg(w) \leq \pi, \\ 1 & \text{if } -\pi < \arg(z) + \arg(w) < \pi, \\ \exp(-\pi i \alpha) & \text{if } \pi \leq \arg(z) + \arg(w) < 2\pi. \end{cases}$$

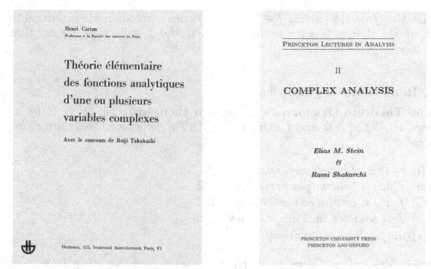

Figure 4.10. Two textbooks on holomorphic functions.

4.5 Singularities

Let us begin stating a remark that will be useful for the sequel on the zeros of a holomorphic function. Let $\Omega \subset \mathbb{C}$ be an open set, $f \in \mathcal{H}(\Omega)$ and $z_0 \in \Omega$. As we have seen, f agrees with its power series expansion in a neighborhood of z_0. We say that f has a *zero of order* m at z_0 if

$$f(z) = \sum_{k=m}^{\infty} a_k(z - z_0)^k = (z - z_0)^m \sum_{k=0}^{\infty} a_{k+m}(z - z_0)^k$$

with $a_m \neq 0$.

4.54 Proposition. *Let $\Omega \subset \mathbb{C}$ be an open set, $f \in \mathcal{H}(\Omega)$, and $z_0 \in \Omega$. The following claims are equivalent.*

  (i) *f has a zero of order m at z_0.*
  (ii) *$f(z_0) = f'(z_0) = f''(z_0) = \cdots = f^{(m-1)}(z_0) = 0$ and $f^{(m)}(z_0) \neq 0$.*
  (iii) *There exists $g \in \mathcal{H}(\Omega)$ such that $f(z) = (z - z_0)^m g(z)$ with $g(z_0) \neq 0$.*
  (iv) *m is the largest integer k such that $f(z)/(z - z_0)^k$ extends to a holomorphic function on Ω.*

4.55 ¶. Prove Proposition 4.54.

Let $\Omega \subset \mathbb{C}$ be an open set and let $z_0 \in \Omega$. If $f \in \mathcal{H}(\Omega \setminus \{z_0\})$, we say that z_0 is a *singularity* for f.

We say that $f \in \mathcal{H}(\Omega \setminus \{z_0\})$ has a continuous (resp. holomorphic) extension to z_0 if there is a map $F \in C^0(\Omega)$ (respectively $F \in \mathcal{H}(\Omega)$) such that $F = f$ on $\Omega \setminus \{z_0\}$. If $f \in \mathcal{H}(\Omega \setminus \{z_0\})$ has a holomorphic extension to Ω, we say that z_0 is a *removable singularity* for f, otherwise, we say that z_0 is a *singular point* for f.

a. Removable singularities

4.56 Theorem (Riemann's extension theorem). *Let $\Omega \subset \mathbb{C}$ be an open set, let $z_0 \in \Omega$ and $f \in \mathcal{H}(\Omega \setminus \{z_0\})$. The following claims are equivalent*

(i) *z_0 is a removable singularity for f.*
(ii) *f has a holomorphic extension to Ω.*
(iii) *f has a continuous extension to Ω.*
(iv) *f is bounded in a neighborhood of z_0.*
(v) *$\lim_{z \to z_0} (z - z_0) f(z) = 0$.*

Proof. Trivially (i) $\Rightarrow$ (ii) $\Rightarrow$ (iii) $\Rightarrow$ (iv) $\Rightarrow$ (v). Let us prove that (v) $\Rightarrow$ (i). Set

$$g(z) := \begin{cases} (z - z_0)f(z) & \text{if } z \in \mathbb{C} \setminus \{z_0\} \\ 0 & \text{if } z = z_0 \end{cases}, \quad \text{and} \quad h(z) := (z - z_0)g(z).$$

The claim (v) is equivalent to the continuity of $g(z)$ at z_0, hence

$$h(z) - h(z_0) = h(z) = (z - z_0)g(z_0) + (z - z_0)o(1) \quad \text{as } z \to z_0.$$

In other words, $h(z)$ is $\mathbb{C}$-differentiable at z_0 with $h(z_0) = 0$ and $h'(z_0) = g(z_0) = 0$. It follows that $h \in \mathcal{H}(\Omega)$, and, by Proposition 4.54,

$$h(z) = (z - z_0)^2 k(z)$$

for some $k \in \mathcal{H}(\Omega)$. Therefore

$$(z - z_0)^2 f(z) = h(z) = (z - z_0)^2 k(z),$$

and $k(z)$ is a holomorphic extension of f to Ω. $\qquad\square$

4.57 Corollary. *Let $\Omega \subset \mathbb{C}$ be an open set, $z_0 \in \Omega$ and $f \in \mathcal{H}(\Omega \setminus \{z_0\})$. Then*

(i) *z_0 is a removable singularity for f if and only if*

$$\limsup_{z \to z_0} |f(z)| < +\infty.$$

(ii) *z_0 is a singular point for f if and only if*

$$\limsup_{z \to z_0} |f(z)| = +\infty.$$

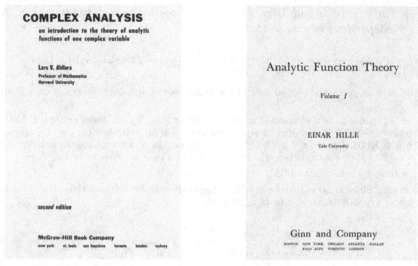

Figure 4.11. Two more textbooks on holomorphic functions.

b. Poles

4.58 Definition. *Let $\Omega \subset \mathbb{C}$ be an open set, $z_0 \in \Omega$, $f \in \mathcal{H}(\Omega \setminus \{z_0\})$ and let m be a positive integer. We say that z_0 is a* pole of order m *for f if z_0 is a removable singularity of $(z - z_0)^m f(z)$ but not of $(z - z_0)^{m-1} f(z)$.*

As a consequence of Riemann's extension theorem we get the following.

4.59 Proposition. *Let $f \in \mathcal{H}(\Omega \setminus \{z_0\})$. z_0 is a pole of order m for f if and only if m is the smallest integer k for which $|(z - z_0)^k f(z)|$ is bounded in a neighborhood of z_0.*

Pole singularities are well-characterized.

4.60 Theorem. *Let $\Omega \subset \mathbb{C}$ be an open set, $z_0 \in \Omega$ and $f \in \mathcal{H}(\Omega \setminus \{z_0\})$. Then f has a pole at z_0 if and only if $|f(z)| \to +\infty$ as $z \to z_0$. Moreover, for any integer $m \geq 1$, the following claims are equivalent.*

(i) *f has a pole of order m at z_0.*
(ii) *There exists $g \in \mathcal{H}(\Omega)$ with $g(z_0) \neq 0$ such that*

$$f(z) = \frac{g(z)}{(z - z_0)^m} \qquad \forall z \in \Omega \setminus \{z_0\}.$$

(iii) *There exists $r > 0$ such that $f(z) = \sum_{k=-m}^{\infty} a_k (z - z_0)^k \ \forall B(z_0, r) \setminus \{z_0\}$.*
(iv) *There exist a ball $B(z_0, r) \subset \Omega$ and $h \in \mathcal{H}(B(z_0, r))$, $h \neq 0$, such that $f(z) = \frac{1}{(z-z_0)^m h(z)} \ \forall z \in B(z_0, r) \setminus \{z_0\}$.*

(v) *There exist a ball $B(z_0, r) \subset \Omega$ and positive constants $0 < \lambda < \Lambda$ independent of r such that*

$$\lambda \frac{1}{|z - z_0|^m} \le |f(z)| \le \Lambda \frac{1}{|z - z_0|^m} \qquad \forall z \in B(z_0, r) \setminus \{z_0\}.$$

Proof. Let us prove the second part of the claim, as the first part follows at once from (v).

(i) $\Rightarrow$ (ii). Since z_0 is a removable singularity for $(z - z_0)^m f(z)$, there exists $g \in \mathcal{H}(\Omega)$ such that $(z - z_0)^m f(z) = g(z) \ \forall z \ne z_0$. Moreover, if $g(z_0) = 0$, then $g(z) = (z - z_0)\widehat{g}(z)$ with $\widehat{g} \in \mathcal{H}(\Omega)$, hence $(z - z_0)^{m-1} f(z) = \widehat{g}(z)$, i.e., z_0 is a removable singularity for $(z - z_0)^{m-1} f(z)$, contradicting the fact that f has a pole of order m at z_0.

(ii) is trivially equivalent to (iii).

(ii) $\Rightarrow$ (iv). Since $g(z_0) \ne 0$ and $g \in \mathcal{H}(\Omega)$, then there exists $r > 0$ such that $B(z_0, r) \subset \Omega$ and $h := 1/g$ is holomorphic in $B(z_0, r) \subset \Omega$.

(iv) $\Rightarrow$ (v) Set

$$\lambda := \inf_{z \in B(z_0, r/2)} \frac{1}{|h(z)|}, \qquad \Lambda := \sup_{z \in B(z_0, r/2)} \frac{1}{|h(z)|}.$$

Then $0 < \lambda \le \Lambda < \infty$ and, since $f(z) = \frac{1}{(z-z_0)^m h(z)}$ in $B(z_0, \rho)$, we infer

$$\lambda \frac{1}{|z - z_0|^m} \le |f(z)| \le \Lambda \frac{1}{|z - z_0|^m} \qquad \forall z \subset B(z_0, r/2) \setminus \{z_0\}.$$

(v) $\Rightarrow$ (i) The estimate $|(z-z_0)^m f(z)| \le \Lambda$ implies by Riemann's extension theorem that z_0 is a removable singularity for $(z - z_0)^m f(z)$, and the estimate $|(z - z_0)^{m-1} f(z)| \ge \lambda |z - z_0|^{-1}$ implies that $(z - z_0)^{m-1} f(z)$ is unbounded around z_0. Again by Riemann's extension theorem z_0 is not a removable singularity for $(z - z_0)^{m-1} f(z)$. $\square$

4.61 ¶. Let P, Q be two polynomials. Suppose that z_0 is a zero of order m for Q and $P(z_0) \ne 0$. Show that z_0 is a pole of order m for $f(z) := \frac{P(z)}{Q(z)}$.

c. Essential singularities

Let $\Omega \subset \mathbb{C}$ be an open set, $z_0 \in \Omega$, and $f \in \mathcal{H}(\Omega \setminus \{z_0\})$. If z_0 is neither a removable singularity nor a pole for f, we say that z_0 is an *essential singularity*. From Corollary 4.57 and Theorem 4.60 z_0 is an essential singularity if and only if

$$\liminf_{z \to z_0} |f(z)| < +\infty, \qquad \text{and} \qquad \limsup_{z \to z_0} |f(z)| = +\infty.$$

Actually, the following holds.

4.62 Proposition. *Let $\Omega \subset \mathbb{C}$ be an open set, $z_0 \in \Omega$, and $f \in \mathcal{H}(\Omega \setminus \{z_0\})$. z_0 is an essential singularity for f if and only if*

$$\liminf_{z \to z_0} |f(z)| = 0, \qquad \limsup_{z \to z_0} |f(z)| = +\infty.$$

Proof. In fact, if $\liminf_{z \to z_0} |f(z)| > 0$, then $1/|f(z)|$ is bounded in a neighborhood of z_0, hence z_0 is a removable singularity for $1/f$. Consequently $|f(z)| \to L$ ($L = \infty$ or $L \in \mathbb{C}$), and z_0 needs to be a removable singularity or a pole for f, a contradiction. $\square$

In other words, f has an essential singularity at z_0 if, roughly, $|f(z)|$ oscillates between zero and infinity in *every* neighborhood of z_0. The following theorem is even stronger.

4.63 Theorem (Casorati–Weierstrass). *If $f \in \mathcal{H}(\Omega \setminus \{z_0\})$ has an essential singularity at z_0, then for all $\delta > 0$ the set $f(B(z_0, \delta) \setminus \{z_0\})$ of values of $f_{|B(z_0,\delta)\setminus\{z_0\}}$ is dense in $\mathbb{C}$.*

Proof. Suppose that for a $c \in \mathbb{C}$ and an $\epsilon_0 > 0$ we have $|f(z) - c| \geq \epsilon_0$ for all $z \in B(z_0, \delta) \setminus \{z_0\}$. Then

$$\varphi(z) := \frac{f(z) - c}{z - z_0}$$

has a pole at z_0 since $|\varphi(z)| = |z - z_0|^{-1}|f(z) - c| \to \infty$ as $z \to z_0$. Consequently, there exists an integer $m \geq 1$ such that $|z - z_0|^m|f(z)| \to 0$, i.e., z_0 is a removable singularity for $(z - z_0)^m f(z)$, a contradiction. $\quad\square$

We also state without proof the following celebrated result about essential singularities.

4.64 Theorem (Picard). *If $f \in \mathcal{H}(\Omega \setminus \{z_0\})$ has an essential singularity at z_0, then for all $\delta > 0$ the set $f(B(z_0, \delta) \setminus \{z_0\})$ of values of $f_{|B(z_0,\delta)\setminus\{z_0\}}$ leaves out at most one point.*

4.65 ¶. Show that $e^{1/z}$ has an essential singularity at 0.

4.66 ¶. Show that $1/\sin(z)$ has poles at the points $z = k\pi$, $k \in \mathbb{Z}$.

4.67 ¶. Show that $\frac{z}{e^z-1}$ has a removable singularity at 0 and poles of order 1 at the points $z_k := 2k\pi i$, $k \in \mathbb{Z} \setminus \{0\}$. Consequently show that

$$\frac{z}{e^z - 1} = \sum_{k=0}^{\infty} \frac{B_k}{k!} z^k, \qquad \forall z, \ |z| < 2\pi.$$

The numbers $\{B_k\}$ are called *Bernoulli's numbers*; they are characterized by the recursive formulas

$$\begin{cases} B_0 := 1, \\ \sum_{j=0}^{n} \binom{n+1}{j} B_j = 0 \quad \forall n \geq 1, \end{cases} \tag{4.25}$$

see [GM2].

d. Singularities at infinity

4.68 Definition. *We say that $f : \{|z| > R\} \to \mathbb{C}$ has a removable singularity, a pole, or an essential singularity at infinity if $f(1/z)$ has respectively a removable singularity, a pole, or an essential singularity at 0.*

4.69 Example. For example
  (i) z^n has a pole of order n at infinity,
  (ii) e^z has an essential singularity at infinity,
  (iii) $\frac{z}{1+z}$ has a removable singularity at infinity.

4.70 ¶. Show that a nonconstant function $f \in \mathcal{H}(\mathbb{C} \setminus \{z_1, z_2, \ldots, z_n\})$ has at least a singular point in the plane or at infinity.

e. Singular points at boundary and radius of convergence

The notion of a singular point extends also to boundary points. For a holomorphic function $f \in \mathcal{H}(\Omega)$ we say that $z_0 \in \partial\Omega$ is a *singular point* for f at $\partial\Omega$, if there is no neighborhood $B(z_0, \delta)$ of z_0 and no holomorphic function $\widehat{f} \in \mathcal{H}(B(z_0, \delta))$ such that $\widehat{f} = f$ on $B(z_0, \delta) \cap \Omega$.

4.71 Theorem. *Let $f : B(z_0, \rho) \to \mathbb{C}$ be the sum of a power series,*

$$f(z) = \sum_{k=0}^{\infty} a_k (z - z_0)^k,$$

with convergence radius $\rho > 0$. Then there exists at least a point $\zeta \in \partial B(z_0, r)$ that is singular for f.

Proof. Let $\Omega \supset B(z_0, \rho)$ be the largest open set in which f can be holomorphically extended. Since the extension of f has a power series development around z_0 with radius of convergence $r := \operatorname{dist}(z_0, \partial\Omega)$ and $\partial\Omega$ is closed, we find $\zeta \in \partial\Omega$ such that $|\zeta - z_0| = r$. By construction ζ is a singular point of f. $\square$

4.72 ¶. Show that $f(z) := \sum_{n=1}^{\infty} \frac{z^n}{n}$ has a singularity at $z = 1$.

4.73 ¶. Show that $\sum_{n=1}^{\infty} \frac{z^n}{n^2}$ has a singularity at $z = 1$.

f. Laurent series development

A *Laurent series* around z_0 is the sum of a power series in the variable $z - z_0$ with radius of convergence ρ_2 and of a power series in the variable $\frac{1}{z - z_0}$ of radius $1/\rho_1$ with $\rho_1 < \rho_2$,

$$\sum_{k=-\infty}^{\infty} a_k (z - z_0)^k := \sum_{k=0}^{\infty} a_k (z - z_0)^k + \sum_{k=1}^{\infty} a_{-k} \frac{1}{(z - z_0)^k}. \qquad (4.26)$$

We call the series $\sum_{k=1}^{\infty} a_{-k} \frac{1}{(z-z_0)^k}$ the *singular part* of the Laurent series (4.26).

From the theorems about power series we find:

(i) The Laurent series (4.26) converges absolutely in the open annulus

$$A(z_0, \rho_1, \rho_2) := \Big\{ z \,\Big|\, \rho_1 < |z - z_0| < \rho_2 \Big\},$$

and uniformly on compact sets $K \subset A(z_0, \rho_1, \rho_2)$.

(ii) The sum of the Laurent series (4.26) is holomorphic in $A(z_0, \rho_1, \rho_2)$.

From Cauchy's formula, we immediately get that every function $f \in \mathcal{H}(B(z_0, r) \setminus \{z_0\})$ with a pole of order m at z_0 has a Laurent series development on the annulus $B(z_0, r) \setminus \{z_0\}$,

$$f(z) = \frac{a_{-m}}{(z - z_0)^m} + \frac{a_{-m+1}}{(z - z_0)^{m-1}} + \cdots + \frac{a_{-1}}{z - z_0} + a_0 + a_1(z - z_0) + \cdots.$$

Actually, we have the following.

4.74 Theorem. *Let* $0 \leq \rho_1 < \rho_2 \leq \infty$ *and let* $f \in \mathcal{H}(A(z_0, \rho_1, \rho_2))$. *Then*

$$f(z) = \sum_{k=-\infty}^{\infty} a_k(z - z_0)^k \qquad \forall z \in A(z_0, \rho_1, \rho_2)$$

where $\forall k \in \mathbb{Z}$

$$a_k = \frac{1}{2\pi i} \int_{\partial^+ B(z_0, r)} \frac{f(\zeta)}{(\zeta - z_0)^{k+1}} \, d\zeta, \tag{4.27}$$

r being arbitrary in $]\rho_1, \rho_2[$.

Proof. The uniqueness of the Laurent series development follows from the identity principle, and the calculus of the development follows from Cauchy's formula. For $z \in A(z_0, \rho_1, \rho_2)$, choose $r_1 < r_2$ such that $\rho_1 < r_1 < |z - z_0| < r_2 < \rho_2$. From Cauchy's formula

$$\begin{aligned} f(z) &= \frac{1}{2\pi i} \int_{\partial^+ A(z_0, r_1, r_2)} \frac{f(\zeta)}{\zeta - z} \, d\zeta \\ &= \frac{1}{2\pi i} \int_{\partial^+ B(z_0, r_2)} \frac{f(\zeta)}{\zeta - z} \, d\zeta - \frac{1}{2\pi i} \int_{\partial^+ B(z_0, r_1)} \frac{f(\zeta)}{\zeta - z} \, d\zeta. \end{aligned} \tag{4.28}$$

If $\zeta \in \partial B(z_0, r_2)$ we have

$$\frac{1}{\zeta - z} = \frac{1}{\zeta - z} \sum_{k=0}^{\infty} \left(\frac{z - z_0}{\zeta - z_0} \right)^k = \sum_{k=0}^{\infty} \frac{(z - z_0)^k}{(\zeta - z_0)^{k+1}}$$

where the series converges uniformly on $\partial B(z_0, r_2)$, and, similarly, for $z \in \partial B(z_0, r_1)$

$$\begin{aligned} \frac{1}{\zeta - z} &= \frac{-1}{z - z_0} \sum_{k=0}^{\infty} \left(\frac{\zeta - z_0}{z - z_0} \right)^k = \frac{-1}{z - z_0} \sum_{k=0}^{\infty} \left(\frac{\zeta - z_0}{z - z_0} \right)^k \\ &= - \sum_{k=-\infty}^{-1} \frac{(z - z_0)^k}{(\zeta - z_0)^{k+1}} \end{aligned}$$

uniformly in $\partial B(z_0, r_1)$. Therefore, by interchanging the series and the integral signs, from (4.28) we infer

$$f(z) = \sum_{k=-\infty}^{\infty} a_k(z - z_0)^k$$

in $A(z_0, r_1, r_2)$ with

$$a_k = \begin{cases} \dfrac{1}{2\pi i} \displaystyle\int_{\partial^+ B(z_0, r_2)} \dfrac{f(\zeta)}{(\zeta - z_0)^{k+1}} \, d\zeta & \text{if } k \geq 0, \\[3mm] \dfrac{1}{2\pi i} \displaystyle\int_{\partial^+ B(z_0, r_1)} \dfrac{f(\zeta)}{(\zeta - z_0)^{k+1}} \, d\zeta & \text{if } k < 0. \end{cases}$$

Since

$$\int_{\partial^+ B(z_0, r)} \frac{f(\zeta)}{(\zeta - z_0)^{k+1}} \, d\zeta$$

does not depend on r for $\rho_1 < r < \rho_2$ ($\zeta \to \frac{f(\zeta)}{(\zeta - z_0)^k}$ is holomorphic in $A(z_0, \rho_1, \rho_2)$), the claim in the theorem follows. $\qquad \square$

4.75 Laurent and Fourier series. Let f be holomorphic on $A(0; 1 - \epsilon, 1 + \epsilon)$, where $\epsilon > 0$, and let

$$\sum_{n=-\infty}^{+\infty} c_n z^n = f(z)$$

be its Laurent series development. As we know,

$$c_n = \frac{1}{2\pi i} \int_{|\zeta|=1} \frac{f(\zeta)}{\zeta^{n+1}} \, d\zeta = \frac{1}{2\pi} \int_0^{2\pi} f(e^{i\theta}) e^{-in\theta} \, d\theta.$$

If we set $\varphi(t) := f(e^{it}) = \sum_{n=-\infty}^{+\infty} c_n e^{int}$, then we see that the Laurent series of f at e^{it} is the Fourier series of $\varphi(t)$ at $t \in \mathbb{R}$.

Conversely, every trigonometric series in the complex variable z

$$\frac{a_0}{2} + \sum_{k=1}^{\infty} (a_k \cos kz + b_k \sin kz)$$

can be written, by the change of variable $e^{iz} := \zeta$, as the Laurent series

$$\sum_{-\infty}^{+\infty} c_n \zeta^n$$

with

$$c_0 := \frac{a_0}{2}, \qquad c_n := \begin{cases} \dfrac{a_n - i b_n}{2} & \text{if } n > 0 \\[2mm] \dfrac{a_{-n} + i b_{-n}}{2} & \text{if } n < 0. \end{cases}$$

If the last series converges in the annulus $\{z \mid r < |\zeta| < R\}$, $r < 1 < R$, then $\sum_{-\infty}^{+\infty} c_n \zeta^n$ is a Laurent series with sum a holomorphic function. Consequently the trigonometric series converges in the strip $\log r < -y < \log R$ parallel to the real axis and has a holomorphic function as sum. In the limit case, $r = R = 1$, the Fourier series may or may not converge, see [GM3].

4.76 ¶. Write the Fourier series of

$$\varphi(t) = \frac{a \sin t}{1 - 2a \cos t + a^2}, \qquad |a| < 1.$$

[*Hint:* Notice that $\varphi(t) = f(e^{it})$ where

$$f(z) := \frac{1 - z^2}{2i\left[z^2 - (a + \frac{1}{a})z + 1\right]},$$

then compute the Laurent series of $f(z)$ to find $\varphi(t) = \sum_{k=1}^{\infty} a^n \sin nt$.]

4.6 Residues

Let Ω be open, $z_0 \in \Omega$, and $f \in \mathcal{H}(\Omega \setminus \{z_0\})$. Goursat's lemma tells us that the number

$$\int_{\partial^+ B(z_0,r)} f(z)\, dz$$

is independent on r as far as $B(z_0,r) \subset \Omega$. The number

$$\operatorname{Res}(f, z_0) := \frac{1}{2\pi i} \int_{\partial^+ B(z_0,r)} f(z)\, dz$$

is called the *residue* of f at z_0. Of course, by Goursat's lemma

$$\operatorname{Res}(f, z) = 0$$

if f is holomorphic in a neighborhood of z.

Similarly, if A is bounded and $f \in \mathcal{H}(\mathbb{C} \setminus \overline{A})$, the *residue of f at infinity* is the number

$$\operatorname{Res}(f, \infty) := -\frac{1}{2\pi i} \int_{\partial^+ B(0,r)} f(z)\, dz$$

where r is such that $\overline{A} \subset B(0,r)$. If we change variable, see Exercise 4.145, we find

$$\operatorname{Res}(f, \infty) = -\frac{1}{2\pi i} \int_{\partial^+ B(0,2r)} f(z)\, dz$$

$$= -\frac{1}{2\pi i} \int_{\partial^+ B(0,1/(2r))} f\left(\frac{1}{\zeta}\right) \frac{1}{\zeta^2}\, d\zeta \qquad (4.29)$$

$$= -\operatorname{Res}\left(f\left(\frac{1}{z}\right) \frac{1}{z^2}, 0\right).$$

As a consequence of Goursat's lemma, we then get the following at once.

4.77 Theorem (Residue theorem, I). *Let $\Omega \subset \mathbb{C}$ be open, $z_1, \ldots, z_n \in \Omega$, $f \in \mathcal{H}(\Omega \setminus \{z_1, z_2, \ldots, z_n\})$ and let $A \subset\subset \Omega$ be a regular domain such that $\{z_1, z_2, \ldots, z_n\} \subset A$. Then*

$$\int_{\partial^+ A} f(z)\, dz = 2\pi i \sum_{j=1}^{n} \operatorname{Res}(f, z_j).$$

4.78 Theorem (Residue theorem, II). *Let $K \subset \mathbb{C}$ be a compact set, let $\Omega := \mathbb{C} \setminus K$, let $A \subset \Omega$ be a bounded regular domain, and let $f \in \mathcal{H}(\Omega \setminus \{z_1, z_2, \ldots, z_n\})$ where $z_1, \ldots, z_n \in A$. Then*

$$\int_{\partial^+ A} f(z)\, dz = -2\pi i \left(\operatorname{Res}(f, \infty) + \sum_{j=1}^{n} \operatorname{Res}(f, z_j)\right).$$

4.79 Corollary. *If $f \in \mathcal{H}(\mathbb{C} \setminus \{z_1, z_2, \ldots, z_n\})$, then*

$$\operatorname{Res}(f, \infty) + \sum_{i=1}^{n} \operatorname{Res}(f, z_i) = 0.$$

a. Calculus of residues

On account of Theorem 4.74, we have

4.80 Corollary. *Let f be the sum of a Laurent series,*

$$f(z) = \sum_{k=-\infty}^{\infty} a_k(z - z_0)^k$$

on $B(z_0, r) \setminus \{z_0\}$, $r > 0$. *Then* $\operatorname{Res}(f, z_0) = a_{-1}$.

Let us discuss a few cases.

(i) Trivially, we have

$$\operatorname{Res}\left(\frac{1}{(z - z_0)^m}, z_0\right) = \begin{cases} 1 & \text{if } m = 1, \\ 0 & \text{otherwise.} \end{cases}$$

(ii) If f has a removable singularity at z_0, then $\operatorname{Res}(f, z_0) = 0$.

(iii) Suppose that f has a pole of order one at z_0,

$$f(z) = \frac{a_{-1}}{z - z_0} + a_0 + a_1(z - z_0) + \ldots.$$

Multiplying by $z - z_0$, we find

$$(z - z_0)f(z) = a_{-1} + O(1) \qquad \text{as } z \to z_0$$

hence

$$\operatorname{Res}(f, z_0) = \lim_{z \to z_0} (z - z_0)f(z).$$

In the special case $f(z) = g(z)/h(z)$ where g, h are holomorphic and $h(z)$ has a simple zero at z_0 we have $h'(z_0) \neq 0$ and

$$(z - z_0)\frac{g(z)}{h(z)} = \frac{z - z_0}{h(z) - h(z_0)}g(z) \to \frac{g(z_0)}{h'(z_0)} \qquad \text{as } z \to z_0,$$

thus concluding

$$\operatorname{Res}\left(\frac{g(z)}{h(z)}, z_0\right) = \frac{g(z_0)}{h'(z_0)}.$$

(iv) If f is holomorphic in $B(z_0, \delta) \setminus \{z_0\}$ and has a pole of order $m > 1$ at z_0, we have

$$f(z) = \frac{g(z)}{(z - z_0)^m},$$

where $g \in \mathcal{H}(B(z_0, \delta))$ with $g(z_0) \neq 0$. It follows that the coefficient a_{-1} of the Laurent series of f is the coefficient of $(z - z_0)^{m-1}$ of the power series development of g. Consequently

$$\operatorname{Res}(f, z_0) = a_{-1} = \frac{D^{m-1}(g)(z_0)}{(m - 1)!}$$

$$= \frac{1}{(m - 1)!} \lim_{z \to z_0} D^{m-1}\Big((z - z_0)^m f(z)\Big).$$

(v) If f is holomorphic in $B(z_0,\delta)\setminus\{z_0\}$ and has a pole of order $m>1$ at z_0, we can also proceed by computing inductively the singular part of the development of f. In fact, if

$$f(z) = \frac{a_{-m}}{(z-z_0)^m} + \cdots + \frac{a_{-1}}{z-z_0} + h(z)$$

with $h \in \mathcal{H}(B(z_0,\delta))$, then

$$\begin{cases} a_{-m} = \lim_{z\to z_0}(z-z_0)^m f(z), \\ a_{-m+1} = \lim_{z\to z_0}(z-z_0)^{m-1}\left(f(z) - \frac{a_{-m}}{(z-z_0)^m}\right), \\ \cdots \\ a_{-1} = \lim_{z\to z_0}(z-z_0)\left(f(z) - \sum_{k=-m}^{-2} a_k(z-z_0)^k\right), \end{cases}$$

and we may proceed as follows. For $f(z) = g(z)/(z-z_0)^m$ we set $h_m(z) := g(z)$, and inductively when $j = m, m-1, \ldots, 1$

$$\begin{cases} \lambda_j := g_j(z_0), \\ g_{j-1}(z) := \frac{g(z)-g(z_0)}{z-z_0}. \end{cases}$$

Then

$$f(z) = \frac{g(z)}{(z-z_0)^m} = \frac{\lambda_m}{(z-z_0)^m} + \cdots + \frac{\lambda_1}{z-z_0} + h_0(z).$$

(vi) If f is the quotient of two polynomials, one can also use Hermite's algorithm to compute the singular part of the Laurent development, see [GM2].

b. Definite integrals by the residue method

A number of integrals can be computed by means of the residue theorem. In fact, if the domain of integration is a nonclosed curve $\gamma : [0,1] \to \mathbb{C}$ as for instance, an interval, we may think this trajectory as part of the oriented boundary of a domain A. If f extends as a holomorphic function with possibly singularities on a domain $\Omega \supset A$, and we are able to compute the integral of f on $\partial^+ A \setminus \gamma([0,1])$, then the method of residues applies for computing the integral over γ. In trying to do that, of course, there is no general rule. Here we collect some significant cases.

4.81 Trigonometric integrands. Consider a definite integral of the type

$$\int_0^{2\pi} R(\cos t, \sin t)\, dt,$$

where R is a rational function. We may interpret it as an integral on $\partial B(0,1)$. In fact, since

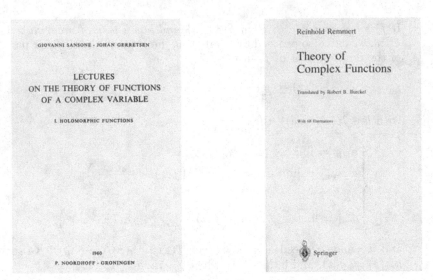

Figure 4.12. Frontispieces of two treatises on holomorphic functions.

$$\frac{1}{2}\left(z+\frac{1}{z}\right) = \cos\theta, \qquad \frac{1}{2i}\left(z-\frac{1}{z}\right) = \sin\theta, \qquad \text{if } z := e^{i\theta},$$

by setting

$$f(z) := \frac{1}{iz} R\left(\frac{1}{2}\left(z+\frac{1}{z}\right), \frac{1}{2i}\left(z-\frac{1}{z}\right)\right)$$

we get

$$\int_0^{2\pi} R(\cos t, \sin t)\, dt = \int_{\partial^+ B(0,1)} f(z)\, dz.$$

If f has no singular point on $\partial B(0,1)$, equivalently, if $t \to R(\cos t, \sin t)$ is continuous on $[0, 2\pi]$, Theorem 4.77 yields

$$\int_0^{2\pi} R(\cos t, \sin t)\, dt = \int_{\partial^+ B(0,1)} f(z)\, dz = 2\pi i \sum_{z \in B(0,1)} \text{Res}\,(f, z).$$

4.82 ¶. If $p_1, p_2, \ldots, p_k$ are the poles of f on $\partial B(0,1)$, compute the integral along the oriented boundary of the domain $B(0,1) \setminus \cup_i B(p_i, \epsilon)$, $\epsilon << 1$. Infer, as in the proof of the residue theorem, that when $\epsilon \to 0$ one has

$$\int_0^{2\pi} R(\cos t, \sin t)\, dt = \int_{\partial^+ B(0,1)} f(z)\, dz$$
$$= 2\pi i \sum_{z \in B(0,1)} \text{Res}\,(f, z) + \pi i \sum_{z \in \partial B(0,1)} \text{Res}\,(f, z).$$

4.83 Example. Let us show that for $a > |b|$ we have

$$\int_0^{2\pi} \frac{1}{a + b\sin\theta}\, d\theta = \frac{2\pi}{\sqrt{a^2 - b^2}}.$$

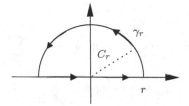

Figure 4.13. Path integrations for improper and Fourier type integrals.

Writing $\sin y = \frac{e^{iy} - e^{-iy}}{2i}$ and rewriting the integral as a line integral on the boundary of the unit ball, we get

$$\int_0^{2\pi} \frac{1}{a + b\sin\theta}\, d\theta = \int_{\partial + B(0,1)} \frac{dz}{iz(a + b(z - z^{-1})/2i)} = \int_{\partial + B(0,1)} \frac{2\, dz}{bz^2 + 2iaz - b}$$

The function $bz^2 + 2iaz - b$ has exactly two zeros

$$z_1 := \frac{-a + \sqrt{a^2 - b^2}}{b} \cdot i, \qquad z_2 := \frac{-a - \sqrt{a^2 - b^2}}{b} i$$

and only z_1 belongs to the disk. It is a pole of order one for $f(z) = \frac{2}{bz^2 + 2iaz - b}$ hence

$$\mathrm{Res}\,(f, z_1) = \frac{2}{2bz_1 + 2ia} = \cdots = \frac{1}{i\sqrt{a^2 - b^2}}.$$

Therefore, from the residue theorem we get

$$\int_0^{2\pi} \frac{d\theta}{a + b\sin\theta} = \frac{2\pi i}{i\sqrt{a^2 - b^2}} = \frac{2\pi}{\sqrt{a^2 - b^2}}.$$

4.84 Improper integrals. Consider an integral of the type

$$\int_{-\infty}^{+\infty} f(t)\, dt := \lim_{r \to +\infty} \int_{-r}^{r} f(t)\, dt$$

where f is continuous on $\mathbb{R}$. Suppose moreover that f extends as a function $f(z)$ that is holomorphic except for at most finitely many points on a neighborhood of the upper half-plane $\overline{A} := \{z \mid \Im z \geq 0\}$ and such that $|zf(z)| \to 0$ as $|z| \to \infty$, $z \in \overline{A}$. Since f is continuous on the real line, singularities of $f(z)$ do not lie on the real line by assumption, and, moreover, the singularies of f with positive imaginary part are contained in a ball $B(0, R)$ for a suitably large R since $zf(z) \to 0$ as $z \to \infty$, $z \in \overline{A}$. Then for $r > R$ we have

$$\int_{-r}^{r} f(x)\, dx + \int_{\gamma_r} f(z)\, dz = 2\pi i \sum_{\Im z > 0} \mathrm{Res}\,(f, z)$$

where γ_r is the counterclockwise oriented boundary of the half-disk C_r in Figure 4.13. From the assumption,

$$\left| \int_{\gamma_r} f(z)\, dz \right| \leq M(r) \cdot \pi r \to 0 \qquad \text{per } r \to \infty$$

where $M(r) := \sup_{z \in \gamma_r} |f(z)|$, hence

$$\int_{-\infty}^{\infty} f(x)\, dx = \lim_{r \to \infty} \int_{-r}^{r} f(x)\, dx = 2\pi i \sum_{\Im z > 0} \operatorname{Res}(f, z).$$

4.85 Example. The above applies to compute

$$\int_0^{\infty} \frac{dx}{1 + x^6}.$$

In fact, if C_r, $r \gg 1$, is as above, of the six distinct roots of $z^6 + 1 = 0$, $z_k := e^{\frac{i(2k+1)\pi}{6}}$, $k = 0, \ldots, 5$, that are the (simple) poles of $f(z) := 1/(1 + z^6)$, only z_0, z_1, z_2 belong to C_r. For $k = 0, 1, 2$ we have

$$\operatorname{Res}(f, z_k) = \frac{1}{6z_k^5} = \frac{1}{6} e^{-i \frac{5(2k+1)\pi}{6}}.$$

Therefore, we have

$$\int_{-r}^{r} \frac{dx}{1 + x^6} + \int_{\gamma_r} \frac{dz}{1 + z^6} = \frac{2\pi i}{6} \sum_{k=0}^{2} e^{-i \frac{5(2k+1)\pi}{6}} = \frac{2\pi}{3},$$

where $\gamma_r(t) := re^{it}$, $t \in [0, \pi]$. Since $\int_{\gamma_r} \frac{dz}{1+z^6} \to 0$ when $r \to \infty$, we conclude

$$\int_0^{\infty} \frac{dx}{1 + x^6} = \frac{\pi}{3}.$$

4.86 Proposition (Fourier type integrals). *Let* $\overline{A} := \{z = x + iy \,|\, y \geq 0\}$ *and let* $f(z)$ *be a holomorphic function on a neighborhood of* $\overline{A}$ *except for a finite number of singularities none of which is real, such that* $|f(z)| \to 0$ *per* $|z| \to \infty$, $z \in \overline{A}$. *Then, if* $\omega > 0$, *we have*

$$\int_{-\infty}^{\infty} f(x)\, e^{i\omega x}\, dx = 2\pi i \sum_{z \in \Im z > 0} \operatorname{Res}(f(z) e^{i\omega z}, z). \qquad (4.30)$$

4.87 Lemma. *Let* $\overline{A} = \{z = x + iy \,|\, y \geq 0\}$ *and let* $f : \overline{A} \cap B(0, R)^c \to \mathbb{C}$ *be a continuous function such that* $|f(z)| \to 0$ *as* $z \to \infty$, $z \in \overline{A}$. *Then, if* $\omega > 0$,

$$\int_{\gamma_r} f(z)\, e^{i\omega x}\, dz \to 0 \qquad \text{as } r \to \infty$$

where γ_r, $r > R$, *denotes the counterclockwise oriented upper bound of the half-disk* C_r *in Figure 4.13.*

Proof. For $r > R$ we have

$$\int_{\gamma_r} f(z) e^{i\omega z}\, dz = \int_0^{\pi} f(re^{i\theta}) e^{i\omega r \cos \theta} e^{-\omega r \sin \theta} i r e^{i\theta}\, d\theta,$$

hence

$$\left| \int_{\gamma_r} f(z) e^{i\omega z}\, dz \right| \leq M(r) \int_0^{\pi} e^{-\omega r \sin \theta} r\, d\theta$$

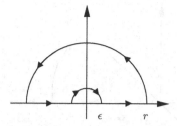

Figure 4.14. Path integrations for integrating e^{it}/t and for Euler's integral in Exercise 4.91.

where $M(r) := \sup_{z \in \gamma_r} |f(z)|$. Since from Jordan's inequality $\frac{2}{\pi} \le \frac{\sin \theta}{\theta} \le 1$ for $0 \le \theta \le \frac{\pi}{2}$, we have

$$\int_0^\pi e^{-\omega r \sin \theta} r \, d\theta = 2 \int_0^{\pi/2} e^{-\omega r \sin \theta} r \, d\theta \le \frac{\pi}{2\omega} (1 - e^{-r\omega}) \le \frac{\pi}{2\omega},$$

the result follows.   □

Proof of Proposition 4.86. Choosing R large enough so that the poles of f with positive imaginary part lie in C_R, for $r > R$ we infer from the residue theorem

$$\int_{-r}^r f(x) e^{i\omega x} \, dx + \int_{\gamma_r} f(z) e^{i\omega z} \, dz = 2\pi i \sum_{\Im z > 0} \text{Res} \left(f(z) e^{i\omega z}, z \right)$$

and, when $r \to \infty$, the claim on account of Lemma 4.87.   □

Applying Proposition 4.86 to $f(-z)$ we also have the following.

4.88 Proposition. *Let $\overline{B} := \{z = x + iy \,|\, y \le 0\}$ and let $f(z)$ be a holomorphic function on a neighborhood of $\overline{B}$ except for a finite number of singularities none of which is real, such that $|f(z)| \to 0$ as $|z| \to \infty$, $z \in \overline{B}$. Then, if $\omega > 0$, we have*

$$\int_{-\infty}^\infty f(x) \, e^{-i\omega x} \, dx = -2\pi i \sum_{z \in \Im z < 0} \text{Res} \left(f(z) e^{-i\omega z}, z \right). \tag{4.31}$$

4.89 Example. For $k > 0$ we have

$$\int_0^\infty \frac{\cos kx}{1 + x^2} \, dx = \frac{\pi}{2} e^{-k}.$$

The only poles of $f(z) := e^{ikz}/(1+z^2)$ are simple and at $z = \pm i$. Integrating along the curves in Figure 4.13, we find

$$\int_{-r}^r \frac{e^{ikx} \, dx}{1 + x^2} + \int_{\gamma_r} \frac{e^{ikz}}{1 + z^2} \, dz = \text{Res} \left(f, i \right) = 2\pi i \frac{e^{-k}}{2i} = \pi e^{-k}.$$

Similarly, we find

$$\int_{-\infty}^\infty \frac{e^{-ikx}}{1 + x^2} \, dx = \pi e^{-k},$$

and, in conclusion,

$$2 \int_0^\infty \frac{\cos kx}{1 + x^2} \, dx = \int_{-\infty}^\infty \frac{\cos kx}{1 + x^2} \, dx = \pi e^{-k}.$$

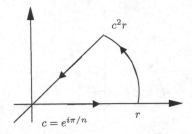

Figure 4.15. The integration curve to integrate Fresnel integrals.

4.90 ¶ Laplace's formulas. Prove that for $\alpha, \beta > 0$

$$\int_0^{+\infty} \frac{\beta \cos(\alpha x)}{x^2 + \beta^2}\, dx = \int_0^{+\infty} \frac{\beta \sin(\alpha x)}{x^2 + \beta^2}\, dx = \frac{\pi}{2} e^{-\alpha\beta}.$$

4.91 ¶ Euler's integral. For $\alpha = 1$ and $\beta = 0$, Laplace's formulas suggest that

$$\int_0^\infty \frac{\sin x}{x}\, dx = \lim_{r \to \infty} \int_0^r \frac{\sin x}{x}\, dx = \frac{\pi}{2}.$$

[*Hint:* Write $\sin z = \frac{e^{iz} - e^{-iz}}{2i}$ and show integrating along the line $\gamma_{\epsilon,r}$ in Figure 4.14 that for $\epsilon \to 0$ and $r \to \infty$

$$\int_{-\infty}^\infty \frac{e^{iz}}{z}\, dz = \lim_{r \to \infty} \lim_{\epsilon \to 0} \int_{\epsilon < |x| < r} \frac{e^{ix}}{x}\, dx = \pi i.]$$

4.92 Example (Fresnel integrals). Let us prove that

$$\int_0^\infty \sin x^2\, dx = \int_0^\infty \cos x^2\, dx = \frac{1}{2}\sqrt{\frac{\pi}{2}}.$$

We first compute

$$\int_0^\infty e^{ix^2}\, dx = \frac{1}{2}\sqrt{\frac{\pi}{2}}(1 + i).$$

We integrate the function $f(z) := e^{iz^2}$ holomorphic on $\mathbb{C}$ along the curve in Figure 4.15, then we split such a curve as the sum of the three curves

$$\gamma_1(t) = t,\ t \in [0, r], \qquad \gamma_2(t) = \frac{t(1 + i)}{\sqrt{2}},\ t \in [0, r]$$

and $\gamma_3(t) := re^{it}$, $t \in [0, \pi/4]$. Goursat's theorem yields

$$\int_{\gamma_1} f(z)\, dz - \int_{\gamma_2} f(z)\, dz + \int_{\gamma_3} f(z)\, dz = 0 \qquad (4.32)$$

where

$$\int_{\gamma_1} f(z)\, dz = \int_0^r e^{ix^2}\, dx,$$

$$\int_{\gamma_2} f(z)\, dz = \frac{1 + i}{\sqrt{2}} \int_0^r e^{-t^2}\, dt \to \frac{1 + i}{\sqrt{2}} \int_0^\infty e^{-t^2}\, dt = \frac{1}{2}\sqrt{\frac{\pi}{2}}(1 + i),$$

$$\int_{\gamma_3} f(z)\, dz = \int_0^{\pi/4} ire^{ir^2\theta^2} e^{i\theta}\, d\theta.$$

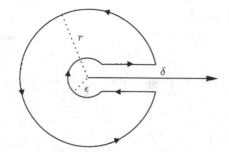

Figure 4.16. Path integrations to integrate Mellin integrals.

Since $i(\cos\theta + i\sin\theta)^2 = i\cos(2\theta) - \sin 2\theta$, we can estimate the modulus of the third integrals by

$$\int_0^{\pi/4} re^{-r^2\sin(2\theta)}\, d\theta \leq \int_0^{\pi/4} re^{-r^2\frac{4}{\pi}\theta}\, d\theta = \frac{\pi}{4r}(1 - e^{-r^2}),$$

thus it converges to zero as $r \to \infty$. From (4.32) we conclude when $r \to \infty$ that the improper integral of e^{ix^2} exists on $(0, +\infty)$ and

$$\int_0^\infty e^{ix^2}\, dx = \lim_{r\to\infty}\int_0^r e^{ix^2}\, dx = \frac{1}{2}\sqrt{\frac{\pi}{2}}(1 + i).$$

Similarly, we have

$$\int_0^\infty e^{-ix^2}\, dx = \frac{1}{2}\sqrt{\frac{\pi}{2}}(1 - i),$$

thus the claim, using Euler's formulas for $\sin x$ and $\cos x$. Finally, we notice that the change of variables $x = \sqrt{t}$, $t > 0$, yields also

$$\int_0^\infty \frac{\sin t}{\sqrt{t}}\, dt = \sqrt{\frac{\pi}{2}}, \qquad \int_0^\infty \frac{\cos t}{\sqrt{t}}\, dt = \sqrt{\frac{\pi}{2}}.$$

4.93 Proposition (Mellin integrals). *Let f be a holomorphic function in $\mathbb{C}$ minus finitely many points leaving outside $R_+ := \{z = x + iy\,|\, y = 0,\ x \geq 0\}$ and let α be a real number with $0 < \alpha < 1$. Suppose that $f(x) \to 0$ as $x \to \infty$, $x \in \mathbb{R}$. Then*

$$(1 - e^{2\pi i\alpha})\int_0^\infty \frac{f(x)}{x^\alpha}\, dx = 2\pi i\sum_{z\neq 0}\mathrm{Res}\left(\frac{f(z)}{z^\alpha}, z\right)$$

where z^α denotes the leaf of z^α on $\mathbb{C}\setminus R_+$ such that $(-1)^\alpha = e^{\alpha\log(-1)} = e^{i\pi\alpha}$.

Proof. Set $g(z) := \frac{f(z)}{z^\alpha}$, $z \in \mathbb{C}\setminus\{0\}$. Denote by $\gamma_{r,\epsilon,\delta}$ the oriented boundary of $D_{r,\epsilon,\delta}$ in Figure 4.16 where $r \gg 1$, $\epsilon \ll 1$ and $\delta \ll \epsilon$ in such a way that all singularities of f but zero are contained in $D_{r,\epsilon,\delta}$. The residue theorem yields

$$\int_{\gamma_{r,\epsilon,\delta}} g(z)\, dz = 2\pi i\sum_{z\neq 0}\mathrm{Res}\left(g(z), z\right)$$

for all $r \gg 1$, $\epsilon \ll 1$ and $\delta \ll \epsilon$. Denoting with γ_+ and γ_- the two horizontal parts of $\gamma_{r,\epsilon,\delta}$ and noticing that for $z = x + iy$, $x > 0$ and $y \to 0^+$, we have

$$z^\alpha = e^{\alpha \log z} \to e^{\alpha \log x} = x^\alpha \qquad \text{as } y \to 0^+$$
$$z^\alpha = e^{\alpha \log z} \to e^{\alpha \log x + i2\pi\alpha} = e^{2\pi i\alpha} x^\alpha \qquad \text{as } y \to 0^-,$$

we deduce for $\delta \to 0$

$$\int_{\gamma_{r,\epsilon,\delta}} g(z)\, dz \to (1 - e^{2\pi i\alpha}) \int_\epsilon^r g(x)\, dx + \int_{\partial^+ B(0,r)} g(z)\, dz - \int_{\partial^+ B(0,\epsilon)} g(z)\, dz.$$

Consequently, we have

$$(1 - e^{2\pi i\alpha}) \int_\epsilon^r \frac{f(x)}{x^\alpha}\, dx + \int_{\partial^+ B(0,r)} \frac{f(z)}{z^\alpha}\, dz - \int_{\partial^+ B(0,\epsilon)} \frac{f(z)}{z^\alpha}\, dz = 2\pi i \sum_{z \neq 0} \mathrm{Res}\left(\frac{f(z)}{z^\alpha}, z\right).$$

$$(4.33)$$

On the other hand, by Lemma 4.87 we have $\int_{\partial^+ B(0,r)} \frac{f(z)}{z^\alpha}\, dz \to 0$ as $r \to \infty$, and

$$\left| \int_{\partial^+ B(0,\epsilon)} \frac{f(z)}{z^\alpha}\, dz \right| \leq M(\epsilon) \epsilon^{-\alpha} 2\pi\epsilon \to 0 \qquad \text{as } \epsilon \to 0.$$

Letting $\epsilon \to 0$ and $r \to \infty$ in (4.33) we get the result.   □

4.94 ¶. Show that $\int_0^\infty \frac{dx}{\sqrt{x}(1+x)} = \pi$.

c. Sums of series by the residue method

4.95 Gauss's sums. For $n \geq 1$, *Gauss's sums* are defined as

$$S_n := \sum_{k=0}^{n-1} e^{\frac{2\pi i k^2}{n}}.$$

For instance, $S_2 = 2$, $S_3 = 1 + i\sqrt{3}$, $S_4 = 2(1 + i)$. For large n, consider the function

$$f(z) := 2 \frac{\exp(2\pi i z^2/n)}{e^{2\pi i z} - 1}$$

which has poles at $0, \pm 1, \pm 2, \ldots$. All poles are simple with residues $\frac{1}{\pi i} e^{2\pi i k^2/n}$ respectively. Using the periodicity of $t \to e^{it}$, integrating along the path in Figure 4.17, and letting $\omega \to \infty$, a long computation[1] yields

$$S_n = 2i(1 + i^{3n})\sqrt{n} \int_0^\infty e^{-2\pi i t^2}\, dt,$$

in particular,

$$2(1 + i) = S_4 = 8i \int_0^\infty e^{-2\pi i t^2}\, dt$$

hence

$$\sum_{k=0}^{n-1} e^{\frac{2\pi i k^2}{n}} = \frac{1}{2}(1 + i)(1 + i^{3n})\sqrt{n} = \frac{1 + (-i)^n}{1 - i}\sqrt{n}.$$

4.96 ¶. Compute the asymptotic development of $\int_x^\infty e^{-t} t\, dt$.

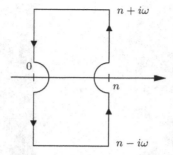

Figure 4.17. Path integration for Gauss's sums.

4.97 Theorem (Sums of series). *Let $f(z)$ be a holomorphic function in $\mathbb{C}$ except on a finite number of isolated singularities at points different from $\pm 1, \pm 2, \ldots$. Moreover, suppose that for some $M, \alpha > 1$ we have $|f(z)| \leq M/|z|^\alpha$ for all z with $|z| \gg 1$. Then the series $\displaystyle\sum_{\substack{n=-\infty \\ n \neq 0}}^{+\infty} |f(n)|$ converges and we have*

$$\sum_{\substack{n=-\infty \\ n \neq 0}}^{+\infty} f(n) = - \sum_{\substack{z \text{ singularity of } f \\ \text{or } z=0}} \operatorname{Res}\left(\frac{\pi f(z)}{\tan(\pi z)}, z\right),$$

$$\sum_{\substack{n=-\infty \\ n \neq 0}}^{+\infty} (-1)^n f(n) = - \sum_{\substack{z \text{ singularity of } f \\ \text{or } z=0}} \operatorname{Res}\left(\frac{\pi f(z)}{\sin(\pi z)}, z\right).$$

To prove the previous theorem, we observe the following.

4.98 Proposition. *The functions $\cot(\pi z)$ and $1/\sin(\pi z)$ are bounded on the boundary of the square*

$$Q_n := \left\{ z = x + iy \,\middle|\, |x|, |y| \leq n + \frac{1}{2} \right\}$$

independently on n.

Proof. Let $z = x + iy \in \partial^+ Q_n$. We distinguish two cases. If $|y| \geq 1/2$ then, see (4.21),

$$|\cot(\pi z)| \leq \coth(\pi|y|) = \frac{1 + e^{-2\pi|y|}}{1 - e^{-2\pi|y|}} \leq \frac{1 + e^{-\pi}}{1 - e^{-\pi}} =: C_1,$$

whereas, if $|y| < 1/2$, then necessarily $|x| = n + 1/2$, hence $\cot(\pi(x + iy)) = \cot(\pi/2 + i\pi y) = \tanh(\pi y)$, from which

$$|\cot(\pi z)| \leq \tanh \pi|y| \leq 1.$$

Therefore $|\cot(\pi z)| \leq C := \max(C_1, 1)$ on ∂Q_n. Similarly, one proceeds to prove that $1/\sin(\pi z)$ is bounded on ∂Q_n independently on n. $\qquad\square$

---

[1] see, e.g., G. Sansone, J. Gerretsen, *Lectures on the Theory of Functions of a Complex Variable*, P. Noordhoff, Gröningen, 1960, vol. 1, p. 139–141.

Figure 4.18. Leonhard Euler (1707–1783) and Gösta Mittag-Leffler (1846–1927).

Proof of Theorem 4.97. The decay at infinity of $|f|$ clearly implies the convergence of the two series $\sum_{n=1}^{\infty} |f(n)|$ and $\sum_{n=1}^{\infty} |f(-n)|$. Let us prove the first equality; one can similarly prove the second using instead the boundedness of $1/\sin(\pi z)$. Set $g(z) := f(z)\pi \cot(\pi z)$. Since $|\cot z| \leq C$ on ∂Q_n with C independent on n, we have

$$\left| \int_{\partial^+ Q_n} f(z) \cot(\pi z)\, dz \right| \leq \frac{8C\,M}{\left(n + \frac{1}{2}\right)^{\alpha}} \left(n + \frac{1}{2}\right) \to 0 \qquad (4.34)$$

as $n \to \infty$. On the other hand, f has no singularites outside Q_n for n large and g has poles only at $0, \pm 1, \pm 2, \ldots$ or inside Q_n for n large. From the residue theorem and (4.34) we infer that

$$\sum_{z \in Q_n} \mathrm{Res}\left(\pi f(z) \cot(\pi z), z\right) \to 0 \qquad \text{as } n \to \infty.$$

Since the singular points of $\cot(\pi z)$ different from zero are simple poles and f is holomorphic in a neighborhood of those points, we find for $k \in \mathbb{Z}$, $k \neq 0$

$$\mathrm{Res}\,(g(z), k) = \mathrm{Res}\left(f(z)\pi \frac{\cos(\pi z)}{\sin(\pi z)}, k\right) = f(k) \frac{\pi \cos(\pi k)}{\pi \cos(\pi k)} = f(k).$$

hence

$$\sum_{\substack{k=-n \\ k \neq 0}}^{n} f(k) + \sum_{\substack{z \in Q_n \\ z \text{ singularity of } f \text{ or } z=0}} \mathrm{Res}\,(g(z), z) \to 0 \qquad \text{as } n \to \infty,$$

i.e., the result. $\square$

4.99 Theorem (Mittag-Leffler). *Let f be a holomorphic function in $\mathbb{C}$ minus a sequence of points $\{a_n\}$, all simple poles for f and without accumulation points. Set $b_n := \mathrm{Res}\,(f(z), a_n)$. Suppose there is a sequence of radii $\{r_n\}$ with $r_n \to \infty$ such that the restriction of f to $\partial B(0, r_n)$ is continuous and for some $M > 0$ we have $|f(z)| \leq M \ \forall z \in \partial B(0, r_n) \ \forall n$. Then for all z and $\zeta \in \mathbb{C} \setminus \{a_n\}$, the series*

$$\sum_{n=1}^{\infty} b_n \left(\frac{1}{a_n - z} - \frac{1}{a_n - \zeta} \right)$$

converges, and

$$f(z) - f(\zeta) = -\sum_{n=1}^{\infty} b_n \left(\frac{1}{a_n - z} - \frac{1}{a_n - \zeta} \right) \qquad \forall z, \zeta \in \mathbb{C} \setminus \{a_n\}. \quad (4.35)$$

Moreover, equality holds in the sense of uniform convergence on compact subsets of $\mathbb{C} \times \mathbb{C}$.

Proof. If $w \ne a_n$ $\forall n$, the function $g(t) := f(t)/(t - w)$

(i) is holomorphic in $\mathbb{C} \setminus \{a_1, a_2, \ldots, w\}$,
(ii) has a simple pole at each a_n with residue given by

$$\operatorname{Res} \left(\frac{f(z)}{z - w}, a_n \right) = \lim_{z \to a_n} \frac{(z - a_n) f(z)}{z - w} = \frac{b_n}{a_n - w},$$

(iii) has a simple pole at w with residue given by

$$\operatorname{Res} \left(\frac{f(z)}{z - w}, w \right) = \lim_{z \to w} \frac{(z - w) f(z)}{z - w} = f(w).$$

From the residue theorem we can deduce

$$f(w) + \sum_{a_k \in B(0, r_n)} \frac{b_k}{a_k - w} = \frac{1}{2\pi i} \int_{\partial + B(0, r_n)} \frac{f(\eta)}{\eta - w} \, d\eta.$$

Evaluating with $w = z$ and ζ and subtracting we get

$$f(z) - f(\zeta) + \sum_{a_k \in B(0, r_n)} b_k \left(\frac{1}{a_k - z} - \frac{1}{a_k - \zeta} \right) \qquad (4.36)$$

$$= \frac{1}{2\pi i} \int_{\partial + B(0, r_n)} f(t) \left(\frac{1}{t - z} - \frac{1}{t - \zeta} \right) dt$$

$$= \frac{1}{2\pi i} \int_{\partial + B(0, r_n)} f(t) \frac{\zeta - z}{(t - z)(t - \zeta)} \, dt.$$

If now $r_n \ge \max(|z|, |\zeta|)$ and $|t| = r_n$, we have $|(t - z)(t - \zeta)| \ge (r_n - |z|)(|r_n| - |\zeta|)$
hence

$$\left| \int_{\partial + B(0, r_n)} f(t) \left(\frac{1}{t - z} - \frac{1}{t - \zeta} \right) dt \right| \le \frac{M|\zeta - z| 2\pi r_n}{(r_n - |z|)(r_n - \zeta)} \to 0 \qquad \text{as } n \to \infty$$

uniformly on $\mathbb{C} \times \mathbb{C}$. It follows from (4.36) that for all z and ζ in $\mathbb{C} \setminus \{a_n\}$,

$$f(z) - f(\zeta) + \sum_{a_k \in B(0, r_n)} b_k \left(\frac{1}{a_k - z} - \frac{1}{a_k - \zeta} \right) \to 0$$

as $n \to \infty$ uniformly on bounded sets of $\mathbb{C} \times \mathbb{C}$, i.e., the convergence of $\sum_{n=1}^{\infty} b_n \left(\frac{1}{a_n - z} - \frac{1}{a_n - \zeta} \right)$ and the (4.35). $\qquad \square$

4.100 Example (The Euler formula for $\cot z$). The function

$$f(z) = \cot z - \frac{1}{z}$$

has singular points at $z = k\pi$, $k \in \mathbb{Z}$. Since

$$\cot z - \frac{1}{z} = \frac{z \cos z - \sin z}{z \sin z},$$

f has a removable singularity at zero, and we may assume that $f(0) = 0$, has simple poles at the points $z_k = k\pi$, $k = \pm1, \pm2, \dots$ and, by Proposition 4.98, is bounded independently on n on ∂Q_n, $Q_n := \{x + iy \,|\, |x|, |y| \leq n + 1/2\}$; finally,

$$\text{Res}\,(f(z), k\pi) = \frac{k\pi \cos(k\pi) - \sin(k\pi)}{\sin(k\pi) + k\pi\cos(k\pi)} = 1.$$

Therefore, by Mittag-Leffler's theorem we have at any point $z \neq k\pi$, $k \in \mathbb{Z}$, $k \neq 0$,

$$\cot z - \frac{1}{z} = f(z) - f(0) = -\sum_{\substack{n=-\infty,\infty \\ n \neq 0}} \left(\frac{1}{n\pi - z} - \frac{1}{n\pi} \right)$$

$$= \sum_{\substack{n=-\infty,\infty \\ n \neq 0}} \left(\frac{1}{z - n\pi} + \frac{1}{n\pi} \right).$$

Rearranging the sum by first summing the terms with indices ±1, $\pm2, \dots$, we find

$$\cot z - \frac{1}{z} = \sum_{n=1}^{\infty} \left(\frac{1}{z - n\pi} + \frac{1}{z + n\pi} \right) = 2z \sum_{n=1}^{\infty} \frac{1}{z^2 - n^2\pi^2}, \qquad (4.37)$$

i.e., the celebrated *Euler's formula for cotangent: the series* $\sum_{n=1}^{\infty} \frac{1}{z^2 - n^2\pi^2}$ *converges for every* $z \neq k\pi$, $k \in \mathbb{Z}$, $k \neq 0$, *uniformly on bounded sets and*

$$z \cot z - 1 = 2 \sum_{n=1}^{\infty} \frac{z^2}{z^2 - n^2\pi^2} \qquad \forall z \neq k\pi, \ k \in \mathbb{Z}, \ k \neq 0. \qquad (4.38)$$

Integrating (4.37), we get for every $z \neq k\pi$, $k \in \mathbb{Z}$, $k \neq 0$,

$$\log\left(\frac{\sin z}{z} \right) = \sum_{k=1}^{\infty} \log\left(1 - \frac{z^2}{k^2\pi^2} \right)$$

uniformly on bounded sets of $\mathbb{C}$. Here log denotes a leaf of the complex logarithm containing 1 and $\sin(z)/z$ with $\log 1 = 0$. Finally, by taking the exponential, we get the *Euler formula for* $\sin z$

$$\sin z = z \prod_{n=1}^{\infty} \left(1 - \frac{z^2}{n^2\pi^2} \right) \qquad \forall z \neq k\pi, \ k \in \mathbb{Z},$$

uniformly on bounded sets of $\mathbb{C}$.

d. $\mathcal{Z}$-transform

Let $a = \{a_n\}$ be a sequence that grows at most exponentially, i.e., there are C and R such that $|a_n| \leq CR^n$, so that

$$r := \limsup_{n \to \infty} \sqrt[n]{|a_n|} \leq R < +\infty.$$

To the sequence $\{a_n\}$ one can associate the power series $\sum_{n=0}^{\infty} a_n w^n$, which, as we know, converges in the disk $\{z \,|\, |z| < 1/r\}$ to a holomorphic function $S(w)$,

$$S(w) = \sum_{n=0}^{\infty} a_n w^n \qquad \forall w, \ |w| < 1/r.$$

The function $S(w)$, or its unique holomorphic maximal extension, which we denote again by S, is sometimes called in the applications the *generating function* of the sequence $\{a_n\}$. Trivially, it is also well possible to consider the negative power series $\sum_{n=0}^{\infty} a_n \frac{1}{z^n}$, which in turn converges in $\{z \,|\, |z| > r\}$ to a function $A(z)$

$$A(z) = \sum_{n=0}^{\infty} a_n \frac{1}{z^n}, \qquad \forall z, \ |z| > r \qquad (4.39)$$

which is holomorphic in $|z| > r$ since trivially

$$A(z) = S\left(\frac{1}{z}\right) \qquad \forall z, \ |z| > r.$$

The function $A(z)$, or its unique holomorphic maximal extension, which we denote again by A, is called the $\mathcal{Z}$-*transform* of the sequence $\{a_n\}$. The number r is called *the radius of convergence* of the $\mathcal{Z}$-transform.

The generating function and the $\mathcal{Z}$-transform are trivially equivalent tools that are useful in many circumstances as for instance in combinatorics, probability theory, or when studying sampling or digital filters. We state here a few facts as they follow easily from the theory of power series and holomorphic functions by changing variable $w \to z = \frac{1}{w}$.

4.101 Proposition. *Let $A(z)$ be the $\mathcal{Z}$-transform of $\{a_n\}$ and let $r_a := \limsup_{n \to \infty} \sqrt[n]{|a_n|}$. Then the series $\sum_{n=0}^{\infty} a_n \frac{1}{z^n}$*

(i) *converges absolutely if $|z| > r_a$,*
(ii) *does not converge if $|z| < r_a$,*
(iii) *converges to $A(z)$ uniformly on any closed set strictly contained in $\{z \,|\, |z| > r_a\}$.*

We have

$$A'(z) = -\sum_{n=0}^{\infty} a_n \frac{n}{z^{n+1}}, \qquad \forall z, |z| > r_a.$$

4.102 Proposition. *Let $\{a_n\}$ and $\{b_n\}$ be two sequences and let $A(z)$ and $B(z)$ be their $\mathcal{Z}$-transforms with radii of convergence respectively r_a and r_b.*

(i) (LINEARITY) *Let $\lambda, \mu \in \mathbb{C}$. The $\mathcal{Z}$-transform $C(z)$ of $\{\lambda a_n + \mu b_n\}$ is defined at least on $\{z \,|\, |z| > \max(r_a, r_b)|\}$ and*

$$C(z) = \lambda A(z) + \mu B(z).$$

(ii) (CONVOLUTION PRODUCT) *Let $\{(a * b)_n\}$, $(a * b)_n := \sum_{k=0}^{n} a_k b_{n-k}$, be the convolution product of the sequences $\{a_n\}$ and $\{b_n\}$. The $\mathcal{Z}$-transform $C(z)$ of $\{(a * b)_n\}$ is defined at least on $\{z \,|\, |z| > \max(r_a, r_b)\}$ and*

$$C(z) = A(z)\, B(z).$$

(iii) (PRODUCT) *The $\mathcal{Z}$-transform $C(z)$ of the sequence $\{a_n b_n\}$ is defined at least on $\{z\,|\,|z| > r_a r_b|\}$ and*

$$C(z) = \frac{1}{2\pi i} \int_{\partial + B(0,\rho)} A(w) B\left(\frac{z}{w}\right) dw$$

where $r_a < \rho < |z|/r_b$.

4.103 Example. Let us collect a few examples.

(i) Let δ_k be the sequence with 1 at place k and zero otherwise, often called the Kronecker sequence. Its $\mathcal{Z}$-transform is $A(z) := \frac{1}{z^k}$ with $r = 0$.

(ii) (CONSTANT SAMPLES) The $\mathcal{Z}$-transform of $\{a_n\}$ with $a_n := 1 \; \forall n$ is

$$A(z) = \sum_{n=0}^{\infty} \frac{1}{z^n} = \frac{1}{1 - \frac{1}{z}} = \frac{z}{z - 1}, \qquad |z| > 1.$$

(iii) (LINEAR SAMPLES) The $\mathcal{Z}$-transform of $\{a_n\}$, $a_n := n \; \forall n$ is

$$A(z) = \sum_{n=0}^{\infty} \frac{z}{z^n} = z \sum_{n=0}^{\infty} n z^{-n-1} = -z \sum_{n=0}^{\infty} D(z^{-n})$$

$$= -zD\frac{z}{z-1} = -\frac{z}{(z-1)^2}, \qquad |z| > 1.$$

(iv) (EXPONENTIAL SAMPLES) The $\mathcal{Z}$-transform of $\{q^n\}$ is

$$A(z) = \sum_{n=0}^{\infty} q^n \frac{1}{z^n} = \frac{z/q}{z/q - 1} = \frac{z}{z - q}, \qquad |z| > q.$$

For instance $A(z) = \frac{z}{z - e^{i\omega}}$ if $a_n = e^{in\omega}$.

4.104 Example. The following examples illustrate how the action on a sequence operates on the corresponding $\mathcal{Z}$-transform.

(i) (FORWARD SHIFT) Let $a_n := \{\underbrace{0, \dots, 0}_{k \text{ times}}, a_1, a_2, \dots\}$ be the *forward shift* of k places of the sequence $\{a_n\}$. We have $r_a = r_b$ and

$$B(z) = \sum_{n=k}^{\infty} a_n \frac{1}{z^{n+k}} = \frac{1}{z^k} A(z), \qquad |z| > r_a.$$

(ii) (BACKWARD SHIFT) If $b_n := a_{n+k}$ defines the *backward shifting* of k places, then again $r_b = r_a$ and

$$B(z) = \sum_{n=0}^{\infty} a_{n+k} \frac{1}{z^n} = z^k \left(A(z) - a_0 - \frac{a_1}{z} - \cdots - \frac{a_{k-1}}{z^{k-1}} \right), \qquad |z| > r_b.$$

(iii) (LINEAR SAMPLING) If $b_n = n a_n \; \forall n$, then $r_b = r_a$ and

$$B(z) = \sum_{n=0}^{\infty} n b_n \frac{1}{z^n} = -z A'(z), \qquad |z| > r_b.$$

(iv) (EXPONENTIAL SAMPLING) If $b_n = q^n a_n$, then $r_b = |q| r_a$ and

$$B(z) = A\left(\frac{z}{q}\right), \qquad |z| > r_b.$$

(v) (PERIODIC SAMPLING) If $\{b_n\}$ is periodic of period p, i.e., $b_{n+p} = b_n$ $\forall n$, and

$$a_n := \begin{cases} b_n & \text{if } 0 \le n < p, \\ 0 & \text{if } n \ge p, \end{cases}$$

then

$$B(z) = \sum_{n=0}^{\infty} b_n \frac{1}{z^n} = \sum_{k=0}^{\infty} \left(\sum_{n=kp}^{(k+1)p-1} b_n \frac{1}{z^n} \right) = \sum_{k=0}^{\infty} \sum_{n=0}^{p-1} b_n \frac{1}{z^{n+kp}}$$

$$= \sum_{k=0}^{\infty} \frac{1}{z^{kp}} \left(\sum_{n=0}^{p-1} b_n \frac{1}{z^n} \right) = \sum_{k=0}^{\infty} \frac{1}{z^{kp}} \left(\sum_{n=0}^{\infty} a_n \frac{1}{z^n} \right) = \frac{z^p}{z^p - 1} A(z), \qquad |z| > 1.$$

Let $A(z)$ be a holomorphic function on $\{z \,|\, |z| > r\}$ with a removable singularity at infinity, i.e.,

$$\lim_{w \to 0} w A(1/w) = \lim_{z \to \infty} \frac{A(z)}{z} = 0,$$

then $A(1/w)$ is the sum of a power series $A(1/w) = \sum_{n=0}^{\infty} a_n w^n$ in the ball $B(0, 1/r)$, and

$$A(z) = \sum_{n=0}^{\infty} a_n \frac{1}{z^n}, \qquad |z| > r.$$

According to Theorem 4.74 and (4.29)

$$a_n = \frac{1}{2\pi i} \int_{\partial + B(0,t)} \frac{S(s)}{s^{n+1}} ds = \text{Res} \left(\frac{S(z)}{z^{n+1}}, 0 \right) = -\text{Res} \left(A(z) z^{n-1}, \infty \right)$$

where $0 < t < 1/r$. Thus we can state the following.

4.105 Proposition. *If $A(z) = \sum_{n=0}^{\infty} a_n \frac{1}{z^n}$, $|z| > r$, then*

$$a_n = -\text{Res} \left(A(z) z^{n-1}, \infty \right) \qquad \forall n.$$

Finally, we notice that if A is holomorphic in $\mathbb{C}$ but a finite number of points $\{p_1, \ldots, p_k\}$, as in the case of the quotient of two polynomials, Corollary 4.79 yields

$$a_n = -\text{Res} \left(z^{n-1} A(z), \infty \right) = \sum_{j=1}^{k} \text{Res} \left(A(z) z^{n-1}, p_j \right).$$

Notice that for every j, p_j is a singularity of $A(z) z^{n-1}$ of order independent of n, so the computation of all residues $\text{Res} \left(A(z) z^{n-1}, p_j \right)$ $\forall n$ at p_j can be done in one single step.

The $\mathcal{Z}$-transform is particularly useful when dealing with *difference linear equations*.

4.106 Example (Fibonacci numbers). The sequence $\{f_n\}$ of *Fibonacci numbers* is defined by

$$\begin{cases} f_{n+2} = f_{n+1} + f_n, & n \geq 0, \\ f_0 = 0, \; f_1 = 1, \end{cases}$$

and one computes, see [GM2],

$$f_n := \frac{1}{\sqrt{5}}\left(\left(\frac{1+\sqrt{5}}{2}\right)^n - \left(\frac{1-\sqrt{5}}{2}\right)^n\right), \qquad n \geq 0. \tag{4.40}$$

We may get the same result by means of the $\mathcal{Z}$-transform. We notice that the $\mathcal{Z}$-transform of $\{f_n\}$,

$$F(z) := \sum_{n=0}^{\infty} f_n \frac{1}{z^n},$$

converges at least at each z with $|z| > 2$. Multiplying by $1/z^n$ each equation and summing on n, we find

$$z^2(F(z) - f_0 - f_1 1/z) = z(F(z) - f_0) - F(z) = 0,$$

i.e.,

$$F(z) = \frac{z}{z^2 - z - 1} \qquad \forall z, \; |z| > 2.$$

Therefore,

$$f_n = \frac{1}{2\pi i} \int_{\partial^+ B(0,r)} \frac{z^n}{z^2 - z - 1} dz$$

where $r > 2$, or

$$f_n = -\operatorname{Res}(g_n(z), \infty), \qquad g_n(z) := \frac{z^n}{z^2 - z - 1}.$$

Now the computation of f_n is only apparently iterative. The functions $g_n(z)$ are holomorphic on $\mathbb{C} \setminus \{a, b\}$ where

$$a = \frac{1+\sqrt{5}}{2}, \qquad b = \frac{1-\sqrt{5}}{2}$$

are the simple roots of the equation $z^2 - z - 1 = 0$. From Corollary 4.79

$$-\operatorname{Res}(g_n(z), \infty) = \operatorname{Res}(g_n(z), a) + \operatorname{Res}(g_n(z), b)$$

i.e.,

$$f_n = \operatorname{Res}(g_n(z), a) + \operatorname{Res}(g_n(z), b) = a^n \frac{1}{2a-1} + b^n \frac{1}{2b-1}$$

$$= \frac{1}{\sqrt{5}} a^n - \frac{1}{\sqrt{5}} b^n.$$

hence (4.40).

In general, consider the *linear difference equation*

$$a_n x_{n+k} + a_{k-1} x_{n+k-1} + \cdots + a_0 x_n = f_{n+1}, \qquad n \geq 0. \tag{4.41}$$

Suppose that the $\mathcal{Z}$-transform $X(z)$ and $F(z)$ of the sequences x_n and f_n (where $f_0 = 0$) have radius of convergence r_x and r_f. Then by the linearity and the formula for backward shifting we find

$$\sum_{n=0}^{\infty} f_{n+1} \frac{1}{z^n} = zF(z),$$

$$\sum_{n=0}^{\infty} \left(a_k x_{n+k} + a_{k-1} x_{n+k-1} + \cdots + a_0 x_n \right) \frac{1}{z^n}$$

$$= \sum_{n=0}^{\infty} \sum_{j=0}^{k} a_j x_{n+j} \frac{1}{z^n} = \sum_{j=0}^{k} \sum_{n=0}^{\infty} a_j x_{n+j} \frac{1}{z^n}$$

$$= \sum_{j=0}^{k} a_j z^j \left(X(z) - \sum_{i=0}^{j-1} \frac{x_i}{z^i} \right)$$

hence the equality

$$P(z)X(z) - \sum_{j=0}^{k} a_j z^j \left(\sum_{i=0}^{j-1} \frac{x_i}{z^i} \right) = z\,F(z),$$

where $P(z) := \sum_{j=0}^{k} a_j z^j$ is the *characteristic equation* of (4.41), which allows us to compute the Z-transform of the sequence $\{x_n\}$ in terms of $P(z)$, $F(z)$ and of the first k terms of the sequence.

4.107 Proposition. *Let $P(z)$ be the characteristic polynomial of (4.41) and $F(z)$ the Z-transform of $\{f_n\}$, $f_0 = 0$. If $1/P(z)$ and $F(z)$ are holomorphic respectively in $\{|z| > r_P\}$ and $\{z \,|\, |z| > r_f\}$, then the Z-transform $X(z)$ of a sequence $\{x_n\}$ satisfying (4.41) exists at least on $\{z \,|\, |z| > \max(r_F, r_P)\}$ and*

$$X(z) = \frac{1}{P(z)} \left(z\,F(z) + \sum_{j=0}^{k} a_j z^j \left(\sum_{i=0}^{j-1} \frac{x_i}{z^i} \right) \right).$$

e. Z-transform of a sequence of vectors

The method of Z-transform is not limited to scalar equations. We may extend it to sequences in a normed space, in particular to sequences in $\mathbb{C}^n$.

Consider the series

$$\sum_{k=0}^{\infty} f_k z^k := \sum_{k=0}^{\infty} z^k f_k, \qquad f_k \in \mathbb{C}^N. \tag{4.42}$$

It is easily seen that the series (4.42) converges absolutely for every z with $|z| < \rho$ where

$$\frac{1}{\rho} := \limsup_{n \to \infty} \sqrt[n]{|f_n|}.$$

Consequently the series (4.42) converges absolutely to a function $F(z)$ with values in $\mathbb{C}^N$

$$F(z) := \sum_{k=0}^{\infty} f_k z^k \in \mathbb{C}^N, \qquad |z| < \rho$$

or, in coordinates, if $f_k := (f_k^1, f_k^2, \ldots, f_k^N)$ and $F(z) := (F^1(z), F^2(z), \ldots, F^N(z))$,

$$F^i(z) = \sum_{k=0}^{\infty} f_k^i z^k, \qquad |z| < \rho$$

for all $i = 1, \ldots, N$.

Consequently, we may apply the theory of residues to conclude for instance that

$$f_n^i = \operatorname{Res}\left(\frac{F^i(z)}{z^{n+1}}, 0\right) \qquad \forall i = 1, \ldots, N, \ \forall n \geq 0,$$

or in vectorial notation

$$f_n = \operatorname{Res}\left(\frac{F(z)}{z^{n+1}}, 0\right) \qquad \forall n \geq 0.$$

Similarly, we may define the $\mathcal{Z}$-transform of a sequence $\{f_n\}$, $f_n \in \mathbb{C}^N$, with $|f_n| \leq CR^n$ for some C and $R > 0$, as the series

$$\sum_{n=0}^{\infty} f_n \frac{1}{z^n}.$$

It converges absolutely at every z with $|z| > r$ where

$$r := \limsup_{n \to \infty} \sqrt[n]{|f_n|},$$

to a function $S(z)$,

$$S(z) = \sum_{n=0}^{\infty} f_n \frac{1}{z^n}, \qquad |z| > r$$

with values on $\mathbb{C}^N$. Each component of $S(z)$ is holomorphic, and by the residue formula we have

$$f_n = -\operatorname{Res}\left(z^{n-1} S(z), \infty\right) \qquad \forall n \geq 0.$$

Moreover, if $S(z)$ is holomorphic on $\mathbb{C}^n$ except for a finite number of singularities at $p_1, p_2, \ldots, p_k$, then

$$f_n = -\operatorname{Res}\left(z^{n-1} S(z), \infty\right) = \sum_{j=1}^{k} \operatorname{Res}\left(z^{n-1} S(z), p_j\right).$$

4.108 Remark. A special interesting case is the case of series with matrix coefficients $\mathbf{F}_k \in M_{M \times N}(\mathbb{C})$,

$$\sum_{n=0}^{\infty} \mathbf{F}_n z^n. \tag{4.43}$$

The power series (4.43) converges absolutely for all z with $|z| < \rho$,

$$\frac{1}{\rho} := \limsup_{n \to \infty} \sqrt[n]{|\mathbf{F}_n|}$$

where $|\mathbf{F}|$ is the norm of the associated operator F, $x \to \mathbf{F}x$, i.e.,

$$|\mathbf{F}| := \sup_{\substack{x \in \mathbb{C}^N \\ x \neq 0}} \frac{|\mathbf{F}x|}{|x|}$$

to a function $\mathbf{F} : B(0, \rho) \subset \mathbb{C} \to M_{M,N}(\mathbb{C})$,

$$\mathbf{F}(z) := \sum_{n=0}^{\infty} \mathbf{F}_n z^n \in M_{M,N}(\mathbb{C}), \qquad |z| < \rho,$$

or, term by term,

$$\mathbf{F}_j^i(z) = \sum_{n=0}^{\infty} (\mathbf{F}_n)_j^i z^n, \qquad |z| < \rho.$$

f. Systems of recurrences and $\mathcal{Z}$-transform

Let $\mathbf{A} \in M_{k \times k}$ and $\{F_n\} \subset \mathbb{C}^k$ with $F_0 = 0$. Consider the system of recurrences

$$\begin{cases} X_{n+1} = \mathbf{A}X_n + F_{n+1} & \forall n \geq 0, \\ X_0 \text{ given.} \end{cases} \tag{4.44}$$

Here we want to find its solution, given by

$$X_n = \mathbf{A}^n X_0 + \sum_{j=1}^{n} \mathbf{A}^{n-j} F_j \qquad \forall n \geq 1, \tag{4.45}$$

by means of the $\mathcal{Z}$-transform.

Let $X(z) = \sum_{n=0}^{\infty} X_n \frac{1}{z^n}$ and $F(z) = \sum_{n=0}^{\infty} F_n \frac{1}{z^n}$ be the $\mathcal{Z}$-transforms of $\{X_n\}$ and $\{F_n\}$ ($F_0 := 0$) with radius of convergence r_X and r_F, respectively. Multiplying the equations in (4.44) by $\frac{1}{z^n}$ we find

$$z(X(z) - X_0) = \mathbf{A}X(z) + zF(z), \qquad |z| \gg 1$$

i.e.,

$$\left(z \operatorname{Id} - \mathbf{A} \right) X(z) = z(F(z) + X_0).$$

If z is not an eigenvalue of $\mathbf{A}$, in particular if $|z|$ is sufficiently large, then $z\,\mathrm{Id} - \mathbf{A}$ is invertible, hence

$$X(z) = z\Big(z\,\mathrm{Id} - \mathbf{A}\Big)^{-1}(F(z) + X_0), \qquad |z| \gg 1 \qquad (4.46)$$

or, by Cramer's formula,

$$X(z) = \frac{1}{\det(z\,\mathrm{Id} - \mathbf{A})} z\mathbf{cof}(z\,\mathrm{Id} - \mathbf{A})(F(z) + X_0)$$

from which

$$X_n = -\mathrm{Res}\,(z^{n-1}X(z), \infty) \qquad \forall n \geq 0.$$

Cramer's formula shows us that the singularities of $X(z)$ are the eigenvalues of $\mathbf{A}$ and the possible singularities of $F(z)$. In particular, if $F(z)$ is holomorphic on the whole of $\mathbb{C}$, we get

$$X_n = -\mathrm{Res}\,(z^{n-1}X(z), \infty) = \sum_{\lambda \text{ eigenvalue of } \mathbf{A}} \mathrm{Res}\,(z^{n-1}X(z), \lambda). \quad (4.47)$$

4.109 Example (Fibonacci numbers, II). Fibonacci's recurrence can be written as

$$\begin{cases} F_{n+1} = \mathbf{A}F_n \quad \forall n \geq 0, \\ F_0 = \begin{pmatrix} 0 \\ 1 \end{pmatrix} \end{cases} \qquad \mathbf{A} := \begin{pmatrix} 0 & 1 \\ 1 & 1 \end{pmatrix}.$$

If $F(z) := \sum_{n=0}^{\infty} F_n \frac{1}{z^n}$, multiplying by $\frac{1}{z^n}$ each equation we find

$$z(F(z) - F_0) = \mathbf{A}F(z),$$

i.e.,

$$F(z) = z(z\,\mathrm{Id} - \mathbf{A})^{-1}F_0$$

if z is not an eigenvalue for $\mathbf{A}$. Now

$$z\,\mathrm{Id} - \mathbf{A} = \begin{pmatrix} z & -1 \\ -1 & z-1 \end{pmatrix}$$

and by Cramer's rule

$$(z\,\mathrm{Id} - \mathbf{A})^{-1} = \frac{z}{z^2 - z - 1}\begin{pmatrix} z-1 & 1 \\ 1 & z \end{pmatrix}.$$

Thus

$$X(z) = \frac{z}{z^2 - z - 1}\begin{pmatrix} 1 \\ z \end{pmatrix}$$

and again, see Example 4.106,

$$f_n = X_n^1 = -\mathrm{Res}\left(\frac{z^n}{z^2 - z - 1}, \infty\right).$$

4.7 Further Consequences of Cauchy's Formula

a. The argument principle

4.110 Theorem (The argument principle). *Let $A \subset\subset \Omega$ be a regular domain of $\mathbb{C}$, $b_1, b_2, \ldots, b_k \in A$ and let $f \in \mathcal{H}(\Omega \setminus \{b_1, b_2, \ldots, b_k\})$ be continuous and nonzero on ∂A. Assume that $b_1, b_2, \ldots, b_k$ are poles of order respectively $q_1, q_2, \ldots, q_k$, and let $a_1, a_2, \ldots, a_h$ be the zeros of f in A with multiplicity respectively $p_1, p_2, \ldots, p_h$. Then*

$$\frac{1}{2\pi i} \int_{\partial^+ A} \frac{f'(\zeta)}{f(\zeta)} \, d\zeta = \sum_{j=1}^{h} p_j - \sum_{j=1}^{k} q_j$$

$$= \text{\# zeros - \# poles of } f \text{ according to multiplicity.}$$

Proof. The residue theorem yields

$$\frac{1}{2\pi i} \int_{\partial^+ A} \frac{f'(\zeta)}{f(\zeta)} \, d\zeta = \sum_{j=1}^{h} \text{Res}\left(\frac{f'}{f}, a_j\right) + \sum_{j=1}^{k} \text{Res}\left(\frac{f'}{f}, b_j\right).$$

In a neighborhood of a_j we have

$$f(z) = \varphi(z)(z - a_j)^{p_j}, \qquad \varphi \text{ holomorphic}, \ \varphi(a_j) \neq 0,$$

hence

$$\frac{f'(z)}{f(z)} = \frac{\varphi'(z)(z - a_j)^{p_j} + \varphi(z)\, p_j (z - a_j)^{p_j - 1}}{\varphi(z)(z - a_j)^{p_j}} = \frac{\varphi'(z)}{\varphi(z)} + \frac{p_j}{z - a_j},$$

therefore $\text{Res}\left(\frac{f'}{f}, a_j\right) = p_j$.

Similarly, in a neighborhood of b_j we have $f(z) = \psi(z)(z - b_j)^{-q_j}$ for a holomorphic function ψ with $\psi(b_j) \neq 0$. It follows

$$\frac{f'(z)}{f(z)} = \frac{\psi'(z)}{\psi(z)} - \frac{q_j}{z - b_j},$$

i.e., $\text{Res}\left(\frac{f'}{f}, b_j\right) = -q_j$.   □

4.111 Theorem. *As in Theorem 4.110, suppose moreover that g is a holomorphic function in Ω. Then we have*

$$\frac{1}{2\pi i} \int_{\partial^+ A} g(z) \frac{f'(\zeta)}{f(\zeta)} \, d\zeta = \sum_{j=1}^{h} p_j g(a_j) - \sum_{j=1}^{k} q_j g(b_j).$$

4.112 ¶. Prove Theorem 4.111.

4.113 ¶. Under the assumptions of Theorem 4.110, compute

$$\frac{1}{2\pi i} \int_{\partial^+ A} \frac{z f'(z)}{f(z)} \, dz, \qquad \frac{1}{2\pi i} \int_{\partial^+ A} \frac{z^2 f'(z)}{f(z)} \, dz.$$

4.114 ¶. By means of Theorem 4.111 prove the following.

Theorem (Jensen). *Let f be a holomorphic function with singularities in an open set Ω with finitely many poles $b_1, b_2, \ldots, b_h$ of order respectively $q_1, q_2, \ldots, q_h$ contained in a ball $B(0, R)$. Suppose moreover that f is continuous on $\partial B(0, R)$ and let $a_1, a_2, \ldots, a_k$ be the zeros of f of multiplicity respectively $p_1, p_2, \ldots, p_k$ on the ball $B(0, R)$. Suppose that $f(0)$ exists and $f(0) \neq 0$. Then*

$$\frac{1}{2\pi} \int_0^{2\pi} \log|f(re^{i\theta})| \, d\theta = \log|f(0)| + \sum_{i=1}^k p_i \log(R/|a_i|) - \sum_{i=1}^h q_i \log(R/|b_i|).$$

b. Rouché's theorem

4.115 Theorem (Rouché). *Let $\Omega \subset \mathbb{C}$ be an open set, let $f, g \in \mathcal{H}(\Omega)$, and let $A \subset\subset \Omega$ be a regular domain of $\mathbb{C}$. If*

$$|f(\zeta) - g(\zeta)| < |g(\zeta)| \qquad \forall \zeta \in \partial A,$$

then f and g have the same number of zeros in A counted according to their multiplicities.

Proof. For $t \in [0, 1]$ the function $h_t(z) := g(z) + t(f(z) - g(z))$ belongs to $\mathcal{H}(\Omega)$. From the assumption, we have

$$|h_t(\zeta)| = |g(\zeta) + t(f(\zeta) - g(\zeta))| \leq |\gamma(\zeta)| - t|f(\zeta) - g(\zeta)| > 0$$

for all $\zeta \in \partial A$. The argument principle yields

$$\int_{\partial + A} \frac{h_t'(\zeta)}{h_t(\zeta)} \, d\zeta = \text{\# zeros of } h_t \text{ in } A.$$

Since the quantity on the left is continuous in t, the number of zeros of h_t in A (counted according to their multiplicities) varies with continuity when t varies in $[0, 1]$. Since the number of zeros is an integer quantity, it is constant. Thus, we find

$$\text{\# zeros of } g = \text{\# zeros of } h_0 = \text{\# zeros of } h_1 = \text{\# zeros of } f \qquad \text{in } A.$$

$\square$

4.116 ¶. Let $f \in \mathcal{H}(\Omega)$ be nonconstant and let $z_0 \in \Omega$ be a root of multiplicty k of $f(z) = a$. Show that, for every sufficiently small neighborhood U of z_0, there is a neighborhood V of a such that for all $b \in V$ the equation $f(z) = b$ has exactly k distinct solutions in U. [*Hint:* Notice that there exists $\rho > 0$ such that $|f(z) - a| \geq \delta > 0$ on $\partial B(z_0, \rho)$ and that

$$2\pi i k = \int_{\partial + B(z_0, \rho)} \frac{f'(z)}{f(z) - a} \, dz = \int_\gamma \frac{1}{\zeta - a} \, d\zeta = I(\gamma, a)$$

where γ is the image of $\partial^+ B(z_0, \rho)$ by f and $I(\gamma, a)$ is the winding number of γ with respect to a. If b is close to a, we also have $|f(z) - b| \geq \delta/2 > 0 \;\forall z \in \partial B(z_0, \rho)$ and $I(\gamma, b) = I(\gamma, a)$, see Proposition 4.40. It follows that $f(z) = b$ has k roots in $B(z_0, \rho)$ when counted according to their multiplicities. They are simple since $f'(z) \neq 0$ in a sufficiently small neighborhood of z_0.]

c. Maximum principle

Let Ω be a bounded domain in $\mathbb{C}$ and $f \in \mathcal{H}(\Omega)$. If $B(z_0, r) \subset\subset \mathbb{C}$, Cauchy's formula

$$f(z_0) = \frac{1}{2\pi i} \int_{\partial^+ B(z_0, r)} \frac{f(\zeta)}{\zeta - z_0} \, d\zeta$$

rewrites as

$$f(z_0) = \frac{1}{2\pi} \int_0^{2\pi} f(z_0 + re^{i\theta}) \, d\theta, \tag{4.48}$$

i.e., $f(z_0)$ is the average of f on $\partial B(z_0, r)$ $\forall r$. This implies, or, better, is equivalent to saying that $f(z_0)$ is the average of f on $B(z_0, r)$, as one can easily prove. This is referred to as the *mean property of holomorphic functions*. As a consequence we have the following.

4.117 Theorem (Maximum principle). *Let $f \in \mathcal{H}(\Omega)$, Ω being a domain of $\mathbb{C}$. If $|f|$ has an interior local maximum point, then f is constant. Moreover, if $f \in \mathcal{H}(\Omega) \cap C^0(\overline{\Omega})$, then*

$$|f(z)| \leq \sup_{\partial\Omega} |f(z)| \qquad \forall z \in \overline{\Omega}$$

and, if f is not constant,

$$|f(z)| < \sup_{\partial\Omega} |f(z)| \qquad z \in \Omega.$$

Proof. Let us prove the first part of the claim, from which the second part follows at once.

If $f(z_0) = 0$ the claim is trivial. Otherwise, multiplying by $1/f(z_0)$, we can and do assume that $f(z_0) = 1$. In this case, we trivially have

$$\begin{aligned} \Re(1 - f(z)) &\geq \Re(1 - |f(z)|) \geq 0 \qquad \text{for } z \in B(z_0, r_0), \\ \Re(1 - f(z)) &- 0 \qquad \text{if and only if} \qquad f(z) = 1. \end{aligned} \tag{4.49}$$

We deduce from the mean property that

$$\int_0^{2\pi} \Re(1 - f(z_0 + re^{i\theta})) \, d\theta = 0;$$

while, from the first of (4.49), that $f(z) = 1$ on $\partial B(z_0, r)$. Since r is arbitrary, $f = 1$ in a ball $B(z_0, r_0) \subset \Omega$, hence in Ω, since Ω is connected. $\qquad \square$

4.118 Corollary. *Let Ω be a domain of $\mathbb{C}$, $B(z_0, r) \subset\subset \Omega$, and let $f \in \mathcal{H}(\Omega)$. If*

$$|f(z_0)| < \min\Big\{ |f(\zeta)| \,\Big|\, \zeta \in \partial B(z_0, r) \Big\},$$

then f has a zero in $B(z_0, r)$.

Proof. Suppose $f \neq 0$ in $B(z_0, r)$, then $g(z) := 1/f(z)$ is holomorphic in some open set Ω' with $B(x_0, r) \subset\subset \Omega'$. From the maximum principle

$$|g(z_0)| \leq \sup_{\zeta \in \partial B(z_0, r)} |g(\zeta)|$$

i.e.,

$$\min\Big\{ |f(\zeta)| \,\Big|\, \zeta \in \partial B(z_0, r) \Big\} \leq |f(z_0)|,$$

a contradiction. $\qquad \square$

d. On the convergence of holomorphic functions

From Cauchy's formula and the maximum principle, we infer at once the following theorems.

4.119 Theorem (Weierstrass). *Let $\{f_k\} \subset \mathcal{H}(\Omega)$. If $\{f_k\}$ converges uniformly to f in Ω, then $f \in \mathcal{H}(\Omega)$.*

4.120 Theorem (Morera). *Let $\{f_k\} \subset \mathcal{H}(\Omega)$. If $\{f_k\}$ converges uniformly to f on compact subsets of Ω, then $f \in \mathcal{H}(\Omega)$ and for all integers j, $D_j f_k \to D_j f$ uniformly on compact sets of Ω.*

Cauchy's estimates, which give uniform equiboundedness on compact sets of each sequence of derivatives of a uniformly equibounded sequence of holomorphic functions, together with the Ascoli–Arzelà theorem yields at once the following.

4.121 Theorem (Montel). *Let $\{f_k\}$ be a sequence of holomorphic functions in Ω that are uniformly equibounded on the compact sets of Ω. There exists a subsequence of $\{f_k\}$ that converges uniformly on the compact sets of Ω to a function $f \in \mathcal{H}(\Omega)$.*

4.122 Theorem (Vitali). *Let $\{f_k\} \subset \mathcal{H}(\Omega)$ be an equibounded sequence on the compact sets of Ω and let $\{z_n\}$ be a sequence converging to $z_0 \in \Omega$. If $\{f_k\}$ converges pointwise in $\{z_n\} \cup \{z_0\}$, then $\{f_k\}$ converges uniformly on compact sets of Ω to $f \in \mathcal{H}(\Omega)$.*

Another classical theorem concerning the convergence of holomorphic functions is the following.

4.123 Theorem (Hurwitz). *Let $\{f_k\} \subset \mathcal{H}(\Omega)$ be a sequence that converges uniformly on compact subsets of Ω to $f \in \mathcal{H}(\Omega)$.*

  (i) *If $B(z_0, r) \subset\subset \Omega$ and $f(z) \neq 0$ on $\partial B(z_0, r)$, then there exists $\overline{n}$ such that f_n and f have the same number of zeros in $B(z_0, r)$,*
  (ii) *If every f_n is injective and f is nonconstant, then f is injective.*

Proof. Let $\delta := \inf\{|f(z)| \,|\, |z - z_0| = r\} > 0$. Since $f_k \to f$ uniformly on the compact sets of Ω, there exists $\overline{n}$ such that for all $n \geq \overline{n}$

$$|f(\zeta)| \geq \delta > \frac{\delta}{2} \geq |f_n(\zeta) - f(\zeta)| \qquad \forall \zeta \in \partial B(z_0, r).$$

(i) then follows from Rouché's theorem. Let us prove (ii). Suppose that f is nonconstant and noninjective. Then there exist two distinct points z and w such that $f(z) = f(w)$. Set $F(\zeta) := f(\zeta) - f(w)$ and $F_n(\zeta) := f_n(\zeta) - f_n(w)$. Since $F(z) = 0$ and F is nonconstant, we infer from Theorem 4.35 that z is an isolated zero of F, i.e., there exists $r < \min(\text{dist}\,(z, \partial\Omega), \text{dist}\,(z, w))$ such that $F(\zeta) \neq 0$ for all $\zeta \in \partial B(z, r)$. Since $F_n \to F$ uniformly on compact sets of Ω, from (i) we infer that F_n and F have the same number of zeros on $B(z_0, r)$, a contradiction since F_n is injective and $F(z) = 0$. $\qquad\square$

e. Schwarz's lemma

4.124 Theorem (Schwarz's lemma). *Let $f \in \mathcal{H}(D)$ where D is the unit disk $D := \{z \mid |z| < 1\}$. If*

$$f(0) = 0 \quad and \quad |f(z)| < 1 \text{ for all } z, \, |z| < 1,$$

then

(i) *$|f'(0)| \leq 1$ and $|f(z)| \leq |z| \; \forall z \in D$,*
(ii) *if $|f'(0)| = 1$ or $|f(z_0)| = |z_0|$ at $z_0 \neq 0$, then $f(z) = \alpha z \; \forall z \in D$ for some $\alpha \in \mathbb{C}$ with $|\alpha| = 1$.*

Proof. (i) The function

$$g(z) := \begin{cases} f(z)/z & \text{if } z \neq 0, \\ f'(0) & \text{if } z = 0, \end{cases}$$

is holomorphic in D. Since $|f(z)| \leq 1 \; \forall z \in D$, we have $|g(z)| \leq r^{-1}$ on $\partial B(0,r) \; \forall r$, $0 < r < 1$, and the maximum principle yields $|g(z)| \leq r^{-1}$ for all $z \in B(0,r)$. When $r \to 1$, it follows that $|g(z)| \leq 1$ for all $z \in D$, i.e., $|f(z)| \leq |z|$ and $|f'(0)| = |g(0)| \leq 1$.

(ii) If $|f(z_0)| = |z_0|$ for some $z_0 \in D$, $z_0 \neq 0$, or, if $|f'(0)| = 1$, the function $|g|$ attains its maximum at an interior point of D. Thus, by the maximum principle, g is constant, $g(z) = \alpha$ with $|\alpha| = 1$, consequently $f(z) = \alpha z$. □

f. Open mapping and the inverse theorem

The maximum principle yields a self-contained proof of the local invertibility theorem for holomorphic functions. We have the following.

4.125 Theorem. *Every nonconstant holomorphic function $f \in \mathcal{H}(\Omega)$ is an open map.*

Proof. Let $z_0 \in \Omega$ and $w_0 = f(z_0)$. We need to prove that for every $r > 0$ there exists $\delta > 0$ such that $B(w_0, \delta) \subset f(B(z_0, r))$. Since f is not constant, $z \to f(z) - w_0$ has an isolated zero in Ω. Therefore, for r small enough, we have $f(z) \neq 0$ in $\partial B(z_0, r)$. Set

$$0 < 2\delta := \min_{\zeta \in \partial B(z_0, r)} |f(z) - w_0|.$$

For all $w \in B(w_0, \delta)$ and all $\zeta \in \partial B(z_0, r)$ we have

$$|f(\zeta) - w| \geq |f(\zeta) - w_0| - |w_0 - w| \geq \delta$$

while $|f(z_0) - w| < \delta$. Consequently, for the holomorphic function $F_w(z) := f(z) - w$ we have

$$|F_w(z_0)| < \min_{\zeta \in \partial B(z_0, r)} |F_w(\zeta)|.$$

This implies that F_w has a zero in $B(z_0, r)$, i.e., for every $w \in B(w_0, \epsilon)$ there exists $z \in B(z_0, r)$ such that $f(z) = w$. In other words, $B(w_0, \delta) \subset f(B(z_0, r))$. □

4.126 Theorem. *Let $f \in \mathcal{H}(\Omega)$ be one-to-one. Then f' never vanishes, $f(\Omega)$ is open, and $f^{-1} : f(\Omega) \to \Omega$ is holomorphic with*

$$(f^{-1})'(w) f'(f^{-1}(w))) = 1 \quad \forall w \in f(\Omega).$$

Proof. Since f is open by Theorem 4.125, $f(\Omega)$ is open and f is a homeomorphism from Ω into $f(\Omega)$. Suppose now that $f'(z_0) \neq 0$ at some $z_0 \in \Omega$ and let $g = f^{-1}$ and $w_0 = f(z_0)$. We have

$$\frac{g(w) - g(w_0)}{w - w_0} = \frac{g(w) - g(w_0)}{f(g(w)) - f(g(w_0))} \to \frac{1}{f'(g(w_0))}$$

since $w \to g(w)$ is continuous. This proves that $g = f^{-1}$ is holomorphic on the open set $S := \{z \in \Omega \,|\, f'(z) \neq 0\}$ and

$$(f^{-1})' f(z) f'(z) = 1 \qquad \forall z \in S. \tag{4.50}$$

Let us show finally that $f' \neq 0$ in Ω. Since f is nonconstant, the zeros of f' form a closed and discrete subset of Ω. Therefore, $f(S)$ is closed and discrete, too. Moreover, as we have seen, f^{-1} is holomorphic on $f(\Omega \setminus S) = f(\Omega) \setminus f(S)$, thus $f^{-1} : f(\Omega) \to \Omega$ is a holomorphic function with eventually a closed, discrete set of point singularities, which are, in fact, removable since f^{-1} is continuous on $f(\Omega)$. Finally, passing to the limit, we extend (4.50) to all points $z_0 \in f(\Omega)$. □

4.8 Biholomorphisms

Let Ω be an open set of $\mathbb{C}$. A function $f : \Omega \to f(\Omega) \subset \mathbb{C}$ is called a *biholomorphism* between Ω and $f(\Omega)$ if f is holomorphic, invertible with holomorphic inverse. Of course, a biholomorphism is also a homeomorphism and, as we stated in Theorem 4.126, f is a biholomorphism between Ω and $f(\Omega)$ iff f is holomorphic and injective. A biholomorphism with $\Omega = f(\Omega)$ is called an *automorphism* of Ω. We now discuss automorphisms of the unit disk $D = B(0, 1)$.

4.127 Definition. *Let $a \in \mathbb{C}$, $|a| < 1$. The map*

$$\varphi_a(z) := \frac{z - a}{1 - \overline{a}z}, \qquad z \neq \frac{1}{\overline{a}}$$

is called a Möbius transformation.

It is easy to show that

(i) φ_a is holomorphic in $\{z \neq 1/\overline{a}\}$, in particular, $\varphi_a \in \mathcal{H}(D)$,
(ii) φ_a maps D one-to-one into D, and $\varphi_a^{-1} = \varphi_{-a}$ since $\varphi_a(\varphi_{-a}(z)) = z = \varphi_{-a}(\varphi_a(z))$,
(iii) $|\varphi_a(e^{i\theta})| = \frac{|e^{i\theta} - a|}{|e^{-i\theta} - \overline{a}|} = 1$, i.e., $\varphi_a : \partial D \to \partial D$ is injective and surjective.
(iv) $\varphi'(0) = 1 - |a|^2$, $\varphi_a'(a) = (1 - |a|^2)^{-1}$.

Essentially, Möbius transformations are all and the sole automorphisms of D.

4.128 Theorem. *If $f : D \to D$ is an automorphism of D, then $f = \alpha \varphi_a$ for some $a \in D$ and $\alpha \in \mathbb{C}$ with $|\alpha| = 1$, i.e., every automorphism of D is the composition of a Möbius transformation with a rotation. In particular, f extends to a biholomorphism in a neighborhood Ω of D, which is in particular a homeomorphism from $\overline{D}$ into $\overline{D}$.*

Proof. Suppose that $f(0) = 0$. Schwarz's lemma applied to f and f^{-1} yields

$$|f(z)| \le |z| = |f^{-1}(f(z))| \le |f(z)| \qquad \forall z \in D,$$

hence $|f(z)| = |z|$. Now, again by Schwarz's lemma $f(z) = cz$. For the general case it suffices to consider $F := f \circ \varphi_{-a}$, $a = f^{-1}(0)$. $\qquad\square$

The following also holds, but we shall not prove it.

4.129 Theorem. *We have*

(i) *All and the sole automorphisms of $\mathbb{C}$ are the maps*

$$z \to az + b, \qquad a \in \mathbb{C} \setminus \{0\}, \ b \in \mathbb{C},$$

(ii) *All and the sole automorphisms of $\mathbb{C} \setminus \{0\}$ are the maps of the type $z \to az$ or $z \to b/z$ with $a, b \in \mathbb{C} \setminus \{0\}$.*

a. Riemann mapping theorem

A natural question to ask is whether or when two given domains Ω and Ω' are biholomorphic. Of course, they need to be homeomorphic, but this does not suffice. We have the following.

4.130 Proposition. *$\mathbb{C}$ and the unit disk $\{|z| < 1\}$ are not biholomorphic.*

Proof. In fact, if $f : \mathbb{C} \to D$ is holomorphic, f is constant by Liouville's theorem. $\qquad\square$

We could also prove the following, but we shall not do it.

4.131 Proposition. *The annuli $\{r_1 < |z| < R_1\}$ and $\{r_2 < |z| < R_2\}$ are biholomorphic if and only if $R_1/r_1 = R_2/r_2$; in this case, a family of biholomorphisms is given by $z \to e^{i\theta} \lambda z$, $\lambda := r_2/r_1$, $\omega \in \mathbb{R}$.*

We shall only discuss the case of simply-connected domains Ω and Ω'.

4.132 Theorem (Riemann). *Every simply connected domain $\Omega \ne \mathbb{C}$ is biholomorphic to the unit disk. More precisely, for every $z_0 \in \Omega$ there exists a unique $f \in \mathcal{H}(\Omega)$ with $f(z_0) = 0$, $f'(z_0)$ real with $f'(z_0) > 0$ such that f is a biholomorphism between Ω and the unit disk $\{|z| < 1\}$.*

Proof. Uniqueness: The uniqueness is proved by contradiction: if f_1 and f_2 are two biholomorphisms between Ω and the unit disk D, then $f_2 \circ f_1^{-1}$ is an automorphism of the unit disk, and by Schwarz's lemma, $f_2 \circ f_1^{-1}(z) = z \ \forall z \in D$, i.e., $f_1 = f_2$.

Existence: The existence is proved by successive steps. Following Koebe[2], one considers the family

$$\mathcal{F} := \Big\{ g \in \mathcal{H}(\Omega),\ g \text{ injective},\ |g(z)| \le 1,\ g(z_0) = 0 \text{ and } g'(z_0) \in \mathbb{R},\ g'(z_0) > 0 \Big\}$$

We then show that there exists $f \in \mathcal{F}$ that maximizes $|f'(z_0)|$ and that such a function has the requested properties.

Step 1 $\mathcal{F} \ne \emptyset$. Choose $a \notin \Omega$ and, Ω being simply connected, consider in Ω a leaf of $\sqrt{z - a}$, which we denote by $h(z)$. The image $h(\Omega)$ is open hence covers a disk $B(h(z_0), \rho)$ of sufficiently small radius. Moreover, $h(\Omega)$ is contained in one of the two connected components of $f^{-1}(\Omega)$, $f(z) = z^2 + a$. Thus $h(z_0)$ and $-h(z_0)$ belong to different connected components of $f^{-1}(\Omega)$, hence for a sufficiently small ρ we also have $B(-h(z_0), \rho)) \cap \Omega = \emptyset$, i.e., $|h(z) + h(z_0)| \ge \rho \ \forall z \in \Omega$, in particular $2|h(z_0)| \ge \rho$. We now claim that the function

$$g_0(z) := \frac{\rho}{4} \frac{|h'(z_0)|}{|h(z_0)|^2} \frac{h(z_0)}{h'(z_0)} \frac{h(z) - h(z_0)}{h(z) + h(z_0)}$$

belongs to $\mathcal{F}$. In fact, it is holomorphic, $g_0(z_0) = 0$, $g'(z_0) \in \mathbb{R}$ and is positive, and $|g_0(z)| < 1 \ \forall z \in \Omega$, since

$$\left| \frac{h(z) - h(z_0)}{h(z) + h(z_0)} \right| = |h(z_0)| \left| \frac{1}{h(z_0)} - \frac{2}{h(z) + h(z_0)} \right| \le 4 \frac{|h(z_0)|}{\rho}.$$

Step 2. Let $\{g_n\} \subset \mathcal{F}$ be a sequence such that $|g_n'(z_0)| \to \sup_{g \in \mathcal{F}} |g'(z_0)|$. Since $\{g_n\}$ is equibounded, it has a subsequence, which we still denote by $\{g_n\}$, which converges uniformly on compact sets of Ω to a holomorphic function f. Clearly, in Ω we have $f(z_0) = 0$ and $f'(z_0) = \gamma$, $|\gamma| = \sup_{g \in \mathcal{F}} |g'(z_0)| < +\infty$ and f is not constant. Moreover, f is injective. In fact, for $g \in \mathcal{F}$ with $g(z) - g(z_1) \ne 0$ in $\Omega \setminus \{z_1\}$, Hurwitz's theorem grants that for the limit f we also have $f(z) - f(z_1) \ne 0$ in $\Omega \setminus \{z_1\}$. This proves that $f \in \mathcal{F}$, f maximizes $|g'(z_0)|$ among the functions in $g \in \mathcal{F}$, and, by Theorem 4.126, f is a biholomorphism between Ω onto $f(\Omega)$.

Step 3. We claim that the function f constructed in Step 2 maps onto the disk. Suppose this is not the case and let w_0 with $|w_0| < 1$ be such that $f(z) \ne w_0 \ \forall z \in \Omega$. Then we can define a leaf in Ω of

$$F(z) := \sqrt{\frac{f(z) - w_0}{1 - \overline{w_0} f(z)}}$$

that is again injective with $|F| \le 1$. Now set

$$G(z) := \frac{|F'(z_0)|}{F'(z_0)} \frac{F(z) - F(z_0)}{1 - \overline{F(z_0)} F(z)}$$

that belongs to $\mathcal{F}$ since it vanishes at z_0 and has positive derivative at z_0. It turns out that

$$G'(z_0) = \frac{1 + |w_0|}{2\sqrt{|w_0|}} \gamma > \gamma,$$

a contradiction. □

---

[2] See L. V. Ahlfors, *Complex Analysis*, McGraw-Hill, New York, 1966.

We may ask what happens when $z \to z_0 \in \partial\Omega$. It can be shown that in this case $f(z_n)$ converges to the boundary of the disk and that, if $f : \Omega \to \Omega'$ is a biholomorphism and $\partial\Omega$ and $\partial\Omega'$ are Jordan curves, then f extends to a homeomorphism from $\overline{\Omega}$ to $\overline{\Omega'}$. In general, the study of the boundary values of holomorphic functions is quite complicated and we skip this topic.

b. Harmonic functions and Riemann's mapping theorem

Consider the problem of solving the Dirichlet problem in a simply connected domain Ω for the Laplace equation

$$\begin{cases} \Delta u = 0 & \text{in } \Omega, \\ u = g & \text{su } \partial\Omega. \end{cases}$$

If $f : \Omega \to \Omega'$ is a biholomorphism that extends continuously to $\overline{\Omega}$ and

$$U(z) := u(f(z)), \qquad z \in \Omega,$$

it is easy to show that u is harmonic in Ω' if and only if U is harmonic in Ω and $U(z) = g(f(z))$ on $\partial\Omega'$. Therefore Riemann's mapping theorem transforms in this case the problem of solving the Dirichlet problem in a simply connected domain Ω into the problem of solving on the unit disk D the corresponding Dirichlet problem

$$\begin{cases} \Delta v = 0 & \text{in } D, \\ v = g(f) & \text{on } \partial D \end{cases}$$

for which an explicit solution is available, see Poisson's and Schwarz's formulas.

According to Riemann there is a strong connection between solving in the simply connected domain Ω the Dirichlet problem

$$\begin{cases} \Delta u = 0 & \text{in } \Omega, \\ u = g & \text{su } \partial\Omega \end{cases}$$

for all g and constructing a biholomorphism of Ω onto the unit disk.

In fact, let $f : \Omega \to D$ be a biholomorphism with only a zero at z_0, i.e., $f(z_0) = 0$, and suppose that z_0 is a simple zero, $f'(z_0) \neq 0$. In a neighborhood of z_0 we then have

$$f(z) = c_1(z - z_0) + \dots, \qquad c_1 := f'(z_0) \neq 0.$$

Consequently

$$\frac{f(z)}{z - z_0} = c_1 + c_2(z - z_0) + \dots$$

is holomorphic near z_0, hence in Ω, and nonzero (since f vanishes only at z_0), therefore

$$F(z) := -\log \frac{f(z)}{z - z_0}$$

is holomorphic in Ω, and its real part

$$u(z) := -\log \frac{|f(z)|}{|z - z_0|}$$

is harmonic. Therefore we conclude that, if a biholomorphism from Ω to D has to exists, then $|f(z)| = 1$ on ∂D, and u has to solve

$$\begin{cases} \Delta u = 0 & \text{in } \Omega, \\ u(z) = \log \frac{1}{|z - z_0|} & \text{on } \partial\Omega. \end{cases}$$

Conversely, if we are able to solve this problem, defining the *conjugate harmonic* of u as a primitive v of the differential form

$$-u_y \, dx + u_x \, dy$$

according to Cauchy–Riemann equations, $u + iv$ is holomorphic and actually $f(z) := u(z) + iv(z)$ is the biholomorphism we were looking for.

c. Schwarz's and Poisson's formulas

Let $f \in C^0(B(0, R)) \cap \mathcal{H}(B(0, R))$, $f(z) = u(x, y) + iv(x, y)$, $z = x + iy$. As we know functions u and v are harmonic. Moreover, the function v, called the *harmonic conjugate of u*, is determined apart from an additive constant by the values of u on $B(0, R)$, and, actually, by the values of u on $\partial B(0, R)$.

4.133 Theorem. *Let $B := B(0, R)$ and let $f \in \mathcal{H}(B) \cap C^0(\overline{B})$, $f(z) = u(r, \theta) + iv(r, \theta)$, (r, θ) being the polar coordinates in B. Then*

$$u(r, \theta) = \frac{1}{2\pi} \int_0^{2\pi} u(R, \varphi) \left(\frac{\zeta}{\zeta - z} - \frac{\overline{z}}{\overline{\zeta} - \overline{z}} \right) d\varphi,$$

$$i\,v(r, \theta) = i\,v(0) + \frac{1}{2\pi} \int_0^{2\pi} u(R, \varphi) \left(\frac{\zeta}{\zeta - z} - \frac{\overline{z}}{\overline{\zeta} - \overline{z}} \right) d\varphi.$$

Consequently we get Schwarz's *formula*

$$f(z) = iv(0) + \frac{1}{2\pi} \int_0^{2\pi} u(R, \varphi) \frac{\zeta + z}{\zeta - z} \, d\zeta,$$

where $\zeta = Re^{i\varphi}$, $z = re^{i\theta}$, or, equivalently, Poisson's *formula*

$$u(r, \theta) = \frac{1}{2\pi} \int_0^{2\pi} u(R, \varphi) \frac{R^2 - r^2}{R^2 - 2Rr\cos(\theta - \varphi) + r^2} \, d\varphi,$$

$$v(r, \theta) = v(0) + \frac{1}{2\pi} \int_0^{2\pi} u(R, \varphi) \frac{2Rr\sin(\theta - \varphi)}{R^2 - 2Rr\cos(\theta - \varphi) + r^2} \, d\varphi.$$

Proof. For $\zeta := Re^{i\varphi}$ and $B := B(0, R)$, we have

$$f(z) = \frac{1}{2\pi i} \int_{\partial+B} \frac{f(\zeta)}{\zeta - z} d\zeta = \frac{1}{2\pi} \int_0^{2\pi} f(\zeta) \frac{\zeta}{\zeta - z} d\varphi \qquad (4.51)$$

if $z \in B$ and

$$\frac{1}{2\pi i} \int_{\partial+B} f(\zeta) \frac{\zeta}{\zeta - z} d\varphi = 0 \qquad (4.52)$$

if $z \notin \overline{B}$. Now for

$$\overline{z} := \frac{r^2}{z} = \frac{\zeta\overline{\zeta}}{z}$$

(4.52) gives

$$0 = \frac{1}{2\pi} \int_0^{2\pi} f(\zeta) \frac{\overline{z}}{\overline{z} - \overline{\zeta}} d\varphi$$

and subtracting from (4.51)

$$f(z) = \frac{1}{2\pi} \int_0^{2\pi} f(\zeta) \left(\frac{\zeta}{\zeta - z} - \frac{\overline{z}}{\overline{z} - \overline{\zeta}} \right) d\varphi, \qquad (4.53)$$

while summing to (4.51)

$$f(z) = \frac{1}{2\pi} \int_0^{2\pi} f(\zeta) \left(\frac{\zeta}{\zeta - z} + \frac{\overline{z}}{\overline{z} - \overline{\zeta}} \right) d\varphi$$

$$= f(0) + \frac{1}{2\pi} \int_0^{2\pi} f(\zeta) \left(\frac{\zeta}{\zeta - z} - \frac{\overline{z}}{\overline{\zeta} - \overline{z}} \right) d\varphi, \qquad (4.54)$$

from which Schwarz's formula follows at once. $\qquad \square$

4.134 Remark. We notice that for every continuous function $u \in C^0(D)$ Schwarz's and Poisson's formulas yield a holomorphic function $f \in \mathcal{H}(D)$ with $\Re(f) = u$ on ∂D.

4.135 ¶. Develop $\frac{\zeta+z}{\zeta-z}$ as a geometric series to find

$$u(r, \theta) = \frac{1}{2}a_0 + \sum_{i=1}^{\infty} \left(\frac{r}{R}\right)^\nu (a_\nu \cos \nu\theta + b_\nu \sin \nu\theta),$$

$$v(r, \theta) = v(0) + \sum_{i=1}^{\infty} \left(\frac{r}{R}\right)^\nu (-b_\nu \cos \nu\theta + a_\nu \sin \nu\theta),$$

where

$$a_\nu := \frac{1}{\pi} \int_0^{2\pi} u(R, \varphi) \cos(n\varphi) \, d\varphi, \qquad b_\nu := \frac{1}{\pi} \int_0^{2\pi} u(R, \varphi) \sin(n\varphi) \, d\varphi.$$

d. Hilbert's transform

4.136 Theorem. *Let* $H := \{z \,|\, \Im z > 0\}$, $f \in \mathcal{H}(H) \cap C^0(\overline{H})$ *and let* u *and* v *denote respectively the real and the imaginary part of* f *on the real axis,* $f(x + i\,0) =: u(x) + iv(x)$. *If*

(i) $\lim_{|z| \to \infty} \dfrac{f(z)}{z} = 0$,

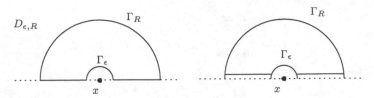

Figure 4.19. The domain $D_{\epsilon,R}$.

(ii) $f \in C^{0,\alpha}$ *locally in* $\overline{H}$, *i.e.*,

$$|f(z) - f(x + i0)| \leq C\,|z - x|^\alpha \qquad \text{for all } x + i0,\, z \in \overline{H},\ |x - z| \leq 1$$

for some constant $C > 0$,

then

$$f(x + i0) := \frac{1}{i\pi} \int_{-\infty}^{\infty} f(t)\frac{1}{t - x}\,dt,$$

or, equivalently,

$$\begin{cases} u(x) = -\dfrac{1}{\pi} \displaystyle\int_{-\infty}^{+\infty} \dfrac{v(t)}{x - t}\,dx, \\[2mm] v(x) = \dfrac{1}{\pi} \displaystyle\int_{-\infty}^{+\infty} \dfrac{u(t)}{x - t}\,dx, \end{cases}$$

where the integrals have to be interpreted as Cauchy's principal integrals[3]

$$\int_{-\infty}^{+\infty} \frac{u(t)}{t - x}\,dx = \lim_{\epsilon \to 0} \int_{\{|t-x|>\epsilon\}} \frac{u(t)}{t - x}\,dx.$$

Proof. Set $g(z) := \frac{f(z)}{z-x}$, $z \in \overline{H} \setminus \{x\}$ and let $0 < \epsilon < R$. Since g is holomorphic in H and continuous in $\overline{H} \setminus \{x\}$, we have

$$\int_{\partial^+ D_{\epsilon,R}} g(z)\,dz = 0,$$

where $D_{\epsilon,R}$ is the domain in Figure 4.19. Therefore,

$$\int_{\partial^+ \Gamma_R} \frac{f(z)}{z - x}\,dz - \int_{\partial^+ \Gamma_\epsilon} \frac{f(z)}{z - x}\,dz + \int_{-R}^{\epsilon} \frac{f(t)}{t - x}\,dt + \int_{\epsilon}^{R} \frac{f(t)}{t - x}\,dt = 0. \qquad (4.55)$$

From the growth of f at infinity, we infer

$$\int_{\partial^+ \Gamma_R} \frac{f(z)}{z - x}\,dz \to 0 \qquad \text{as } R \to \infty,$$

and from the Hölder-continuity of f at x, we infer

$$\int_{\partial^+ \Gamma_\epsilon} \frac{f(z)}{z - x}\,dz = f(x) \int_{\partial^+ \Gamma_\epsilon} \frac{1}{z - x}\,dz + \int_{\partial^+ \Gamma_\epsilon} \frac{f(z) - f(x)}{z - x}\,dz = i\pi f(x) + O(\epsilon^\alpha))$$

as $\epsilon \to 0$, and the claim follows from (4.55) when $R \to \infty$ and $\epsilon \to 0$. $\qquad\square$

---

[3] In the sense, for instance,

$$\int_{-\infty}^{+\infty} \frac{1}{x^3}\,dx = \lim_{\epsilon \to 0+} \left(\int_{-\infty}^{-\epsilon} \frac{1}{x^3}\,dx + \int_{\epsilon}^{+\infty} \frac{1}{x^3}\,dx \right) = 0.$$

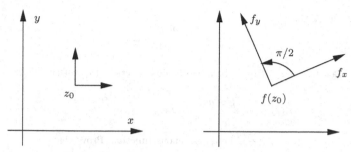

Figure 4.20. A $\mathbb{C}$-linear transformation.

4.9 Exercises

4.137 ¶. Let $f : \Omega \subset \mathbb{R}^2 \to \mathbb{C}$ be differentiable. Prove that
$$(f_x|f_y)_{\mathbb{R}^2} = -2\Im(f_z \overline{f_{\bar z}}), \qquad \det \mathbf{D}f = |f_z|^2 - |f_{\bar z}|^2.$$
[*Hint:* Infer from (4.1)
$$(f_x|f_y)_{\mathbb{R}^2} + i \det \mathbf{D}f = f_y \overline{f_x}. \,]$$

4.138 ¶. Let $f : \Omega \to \mathbb{C}$ be $\mathbb{C}$-differentiable. Prove that
$$|f'(z_0)|^2 = \det \mathbf{D}f(z_0).$$

4.139 ¶. Let $f \in \mathcal{H}(\Omega)$ where Ω is a connected open set. Prove that f is constant if $f'(z) = 0 \; \forall z \in \mathbb{C}$.

4.140 ¶. Let $f \in \mathcal{H}(\Omega)$. Then f is constant in Ω iff one of the following conditions holds:

    (i) $\Re f(z)$ is constant in Ω,
    (ii) $\Im f(z)$ is constant in Ω,
    (iii) $|f(z)|$ is constant in Ω.

4.141 ¶. Let $\ell : \mathbb{C} \simeq \mathbb{R}^2 \to \mathbb{C} \simeq \mathbb{R}^2$ be a $\mathbb{R}$-linear map with associated matrix $\mathbf{A}$. Prove that ℓ is $\mathbb{C}$-linear, $\ell(z) = az$, $a \in \mathbb{C}$, if and only if
$$\mathbf{AJ} = \mathbf{JA}$$
where $\mathbf{J}$ is the matrix $\begin{pmatrix} 0 & 1 \\ -1 & 0 \end{pmatrix}$ associated to the counterclockwise rotation of an angle $\pi/2$.

4.142 ¶. A matrix $\mathbf{A} \in M_{2,2}(\mathbb{R})$, $\mathbf{A} = \begin{pmatrix} a & c \\ b & d \end{pmatrix}$ is called *conform* if
$$a^2 + b^2 = c^2 + d^2, \qquad ac + bd = 0.$$
Prove that, if $\mathbf{A}$ is conform, then there exist $\lambda \in \mathbb{R}$ and a rotation matrix $\mathbf{R}$, $\mathbf{R}^T \mathbf{R} = \mathrm{Id}$, $\det \mathbf{R} = 1$, such that $\mathbf{A} = \lambda \mathbf{R}$. Moreover, prove in the case $\mathbf{A} \neq 0$ that $\mathbf{A}$ is conform iff $\mathbf{A}$ preserves the cosinus of the angle between two generic vectors:
$$\cos(\mathbf{A}u, \mathbf{A}v) = \frac{(\mathbf{A}u|\mathbf{A}v)}{|\mathbf{A}u|\,|\mathbf{A}v|} = \frac{(u|v)}{|u|\,|v|} = \cos(u, v).$$

4.143 ¶. Check that

$$4\frac{\partial^2 u}{\partial z \partial \bar{z}} = 4\frac{\partial^2 u}{\partial \bar{z} \partial z} = \Delta u.$$

4.144 ¶. Let $f : \partial B(z_0, R) \to \mathbb{C}$ be a continuous function. Prove that

$$\int_{\partial^+ B(z_0,R)} f(z)\, dz = -\int_{\partial^+ B(-z_0,R)} f(-z)\, dz.$$

4.145 ¶. Let $f : \partial B(0, R) \to \mathbb{C}$ be a continuous function. Prove that

$$\int_{\partial^+ B(0,R)} f(z)\, dz = \int_{\partial^+ B(0,1/R)} f\left(\frac{1}{w}\right)\frac{1}{w^2}\, dw.$$

4.146 ¶. Let $\Omega \subset \mathbb{C}$ be open and let A be an elementary domain for Ω. Prove

$$\frac{1}{2\pi i}\int_{\partial^+ A}\frac{f(\zeta)}{\zeta - z}\, d\zeta = \begin{cases} f(z) & \text{if } z \in A, \\ 0 & \text{if } z \notin \overline{A}. \end{cases}$$

4.147 ¶. Prove the following.

$$\cos(z_1 + z_2) = \cos z_1 \cos z_2 - \sin z_1 \sin z_2, \qquad \cos^2 z + \sin^2 z = 1,$$
$$\sin(z_1 + z_2) = \cos z_1 \sin z_2 + \cos z_2 \sin z_1, \qquad \cosh^2 z - \sinh^2 z = 1,$$
$$\cos(-z) = \cos z, \ \sin(-z) = -\sin z, \qquad \cos\left(z - \frac{\pi}{2}\right) = \sin z,$$
$$e^z = \cosh z + \sinh z, \qquad e^{iz} = \cos z + i\sin z,$$
$$\sin(x + iy) = \sin x \cosh y + i\cos x \sinh y, \qquad \cos(x + iy) = \cos x \cosh y - i\sin x \sinh y.$$

4.148 ¶. Compute the derivatives of the trigonometric and hyperbolic functions and try to find relationships among those functions.

4.149 ¶. Prove that the restriction of $\sin z$ to $\{z = x + iy \,|\, |x| < \pi/2\}$ is invertible with inverse given by $\sin^{-1} z = \frac{1}{i}\log(iz + \sqrt{1 - z^2})$.

4.150 ¶. Prove that the restriction of $\tanh z$ to $\{z = x + iy \,|\, |h| < \pi/2\}$ is invertible with inverse given by $\tanh^{-1} z = \frac{1}{2}\log\frac{1+z}{1-z}$.

4.151 ¶. Prove the following.

$$\int_0^{2\pi}\frac{\cos k\theta}{5 + 3\cos\theta}\, d\theta = \frac{1}{2}\frac{(-1)^k \pi}{3^k}, \qquad \int_{-\infty}^{\infty}\frac{e^{i\alpha x}}{a^2 + x^2}\, dx = \frac{\pi}{|a|}e^{-\alpha|a|}.$$

4.152 ¶. Compute the asymptotic development of $\int_x^{\infty} e^{-t}t\, dt$.

4.153 ¶. Compute the residues of

$$f(z) = \frac{z^2 - 2z}{(z + 1)^2(z^2 + 4)}, \qquad \text{and} \qquad f(z) = \frac{e^z}{\sin z}$$

both directly or by means of their Laurent development.

4.154 ¶. Compute

$$\frac{1}{2\pi i}\int_{\partial+B(0,3)}\frac{e^{zt}}{z^2(z^2+2z+2)}\,dz.$$

4.155 ¶. Prove that

$$\int_0^{2\pi}\frac{dt}{1-2p\cos t+p^2}=\begin{cases}\frac{2\pi}{1-p^2} & \text{if }|p|<1,\\[2mm]\frac{2\pi}{p^2-1} & \text{if }|p|>1,\end{cases}\qquad p\in\mathbb{C}\setminus\partial B(0,1),$$

$$\int_0^{2\pi}\frac{dt}{(p+\cos t)^2}=\frac{2\pi p}{(\sqrt{p^2-1})^3},\qquad p>1.$$

4.156 ¶. Prove that

$$\int_{-\infty}^{+\infty}\frac{x^2}{1+x^4}\,dx=\frac{\pi}{\sqrt{2}},$$

$$\int_{-\infty}^{+\infty}\frac{dx}{(x^4+a^4)^2}=\frac{3\sqrt{2}}{8}\frac{\pi}{a^7},\qquad a>0,$$

$$\int_{-\infty}^{\infty}\frac{x^2}{(x^2+1)^2(x^2+2x+2)}\,dx,$$

$$\int_0^{+\infty}\frac{x^{2p}}{1+x^{2q}}\,dx=\frac{1}{q}\frac{\pi}{\sin\left(\frac{2p+1}{2q}\pi\right)},$$

for all $p,q\in\mathbb{N}$, $0<p<q$.

4.157 ¶. Prove that, if $a>0$ and $b>0$, then

$$\int_{-\infty}^{+\infty}\frac{e^{iax}}{x-ib}\,dx=2\pi ie^{-ab},\qquad\int_{-\infty}^{+\infty}\frac{e^{iax}}{x+ib}\,dx=0.$$

Summing and subtracting, find again *Laplace's formulas*

$$\int_0^\infty\frac{\beta\cos\alpha x}{x^2+\beta^2}\,dx=\int_0^\infty\frac{x\sin\alpha x}{x^2+\beta^2}\,dx=\frac{1}{2}\pi e^{-\alpha\beta},\qquad\alpha,\beta>0.$$

4.158 ¶. Compute

$$\int_{-\infty}^{+\infty}\frac{x\sin x}{x^2-\sigma^2}\,dx.$$

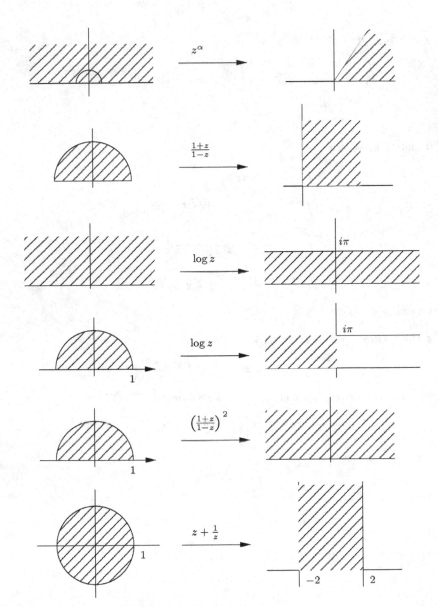

Figure 4.21. A few holomorphic transformations.

4.159 ¶. Prove that

$$\sum_{n=1}^{\infty} \frac{1}{n^2} = \frac{\pi^2}{6},$$

$$\sum_{n=1}^{\infty} \frac{1}{n^4} = \frac{\pi^4}{90},$$

$$\sum_{n=-\infty}^{\infty} \frac{1}{n^2 + a^2} = \frac{\pi}{a} \coth a, \ a > 0,$$

$$\sum_{n=1}^{\infty} \frac{(-1)^{n-1}}{n^2} = \frac{\pi^2}{12},$$

$$\sum_{n=-\infty}^{\infty} \frac{1}{n^4 + 4a^4} = \frac{\pi}{4a^3} \left(\frac{\sinh 2\pi a + \sin 2\pi a}{\cosh 2\pi a - \cos 2\pi a} \right), \ a > 0.$$

4.160 ¶. Prove that

$$\int_0^{\infty} \frac{\log(1 + x^2)}{1 + x^2} \, dx = \pi \log 2, \qquad \int_0^{\infty} \frac{(\log x)^2}{1 + x^2} \, dx = \frac{\pi^3}{8}.$$

4.161 ¶. Prove the following identities.

$$\frac{1}{\sin z} = \frac{1}{z} + 2z \sum_{k=1}^{\infty} \frac{(-1)^k}{z^2 - k^2 \pi^2},$$

$$\frac{1}{\cos z} = \pi \sum_{k=1}^{\infty} (-1)^k \frac{2k + 1}{\left((2k+1)\frac{\pi}{2} \right)^2 - z^2}.$$

4.162 ¶. Prove that, if g is holomorphic around 0 and γ_ϵ is the path given by the upper half-circle oriented anticlockwise with center 0 and radius $\epsilon > 0$, then

$$\lim_{\epsilon \to 0} \frac{1}{2\pi i} \int_{\gamma_\epsilon} \frac{g(z)}{z} \, dz = \frac{g(z_0)}{2} = \frac{1}{2} \operatorname{Res}\left(\frac{g(z)}{z}, 0 \right).$$

4.163 ¶. Show that

$$\int_{-\infty}^{\infty} \frac{\sin^2(\pi x)}{x^2} \, dx = \pi^2,$$

$$\int_{-\infty}^{\infty} \frac{\sin^2(\pi x)}{(a + x^2)(1 - x^2)} \, dx = \frac{\pi}{8}(1 - e^{-2\pi}) + \frac{\pi^2}{4},$$

$$\int_{-\infty}^{\infty} \frac{e^{\alpha x}}{e^{2x} - 1} \, dx = \frac{\pi}{2} \cot \frac{\pi \alpha}{2}.$$

4.164 ¶. Show that

$$\frac{\pi^2}{\sin^2(\pi z)} = \sum_{n=-\infty}^{\infty} \frac{1}{(z - n)^2}.$$

4.165 ¶. Compute

$$F(a) := \frac{1}{2\pi i} \int_{\partial B^+(0,1)} \frac{1}{\zeta(\zeta - 2))(\zeta - a)} \, d\zeta$$

for $a \in \mathbb{C}$, $|a| \neq 1$.

234   4. Holomorphic Functions

4.166 ¶. Show that the functions $F(z)$ below are holomorphic in the respective domains.

$$F(z) = \int_0^\infty \frac{e^{-tz}}{1+t^2}\, dt,\ \Re(z) > 0, \qquad F(z) = \int_0^\infty \frac{t^{z-1}}{t^2+1}\, dt,\ 0 < \Re(z) < 2,$$

$$F(z) = \int_0^1 \frac{\cot tz}{z+t}\, dt,\ \Re(z) \notin [-1,0], \qquad F(z) = \int_0^1 \frac{t\sin z}{t^2+z^2}\, dt,\ Re(z) > 0.$$

4.167 ¶. Show that 0 is a removable singularity for the function

$$\frac{\sin z}{z}, \qquad \frac{z}{\tan z}, \qquad \cot z - \frac{1}{z}, \qquad \frac{1}{e^z-1} - \frac{1}{\sin z}.$$

4.168 ¶. Show that $z = 0$ is a pole for the functions

$$\frac{z}{1-\cos z}, \qquad \frac{z}{(e^z-1)^2},$$

while $z = \infty$ for

$$\sin z, \qquad e^z, \qquad e^{-z^2}$$

and $z = 0$ for

$$z^2 \cos\frac{\pi}{z}, \qquad z(e^{1/z}-1)$$

are essential singularities.

4.169 ¶ Schwarz's lemma. Let $f : D \to D$, $D = B(0,1)$, be holomorphic. Prove that

$$\left|\frac{f(z)-f(z_0)}{1-f(z)\overline{f(z_0)}}\right| \le \frac{|z-z_0|}{|1-z\overline{z_0}|}, \qquad |f'(z_0)| \le \frac{1-|f(z_0)|^2}{|1-|z_0|^2}.$$

[*Hint:* Use a Möbius transformation both in the domain and the target disk.]

4.170 ¶. Let f be holomorphic in the strip $\Re(z) < \pi/4$, and such that $|f(z)| < 1$ and $f(0) = 0$. Prove that $|f(z)| \le |\tan z|$.

4.171 ¶ Schwarz's reflection principle. Let Ω be an open set, $\Omega \subset \{x+iy \,|\, y > 0\}$, let Ω^* be symmetric to Ω with respect to the real axis. Suppose that the intersection of $\overline{\Omega}$ with the real axis is an interval I. Prove that, if $f \in C^0(\Omega \cup I) \cap \mathcal{H}(\Omega)$, then

$$F(z) := \begin{cases} f(z) & \text{if } z \in \Omega \cup I, \\ \overline{f(\overline{z})} & \text{if } z \in \Omega^* \end{cases}$$

is holomorphic on $\Omega \cup I \cup \Omega^*$.

4.172 ¶. Let D be the unit disk and $f \in C^0(\overline{D}) \cap \mathcal{H}(D)$. Prove that f is constant if $|f(z)| = 1$ for every z with $|z| = 1$. [*Hint:* Extend f with $f(z) := \overline{f(1/\overline{z})}$, $z \in \mathbb{C} \setminus \overline{D}$ to the entire $\mathbb{C}$.]

4.173 ¶ A representation formula for the inverse. Let $f \in \mathcal{H}(B(0,R))$ be holomorphic in $|z| < r$ with $f(0) = 0$, $f'(0) \ne 0$ and $f(z) \ne 0$ in $0 < |z| < r$ and let $\rho < r$.

(i) Show that

$$g(w) := \frac{1}{2\pi i} \int_{\partial^+ B(0,\rho)} \frac{zf'(z)}{f(z)-w}\, dz$$

defines a holomorphic function on $\{w \,|\, |w| < \inf_{z \in \partial B(0,\rho)} |f(z)|\}$.

(ii) Prove that $f(g(w)) = w$ if $|w| < \inf_{\partial B(0,\rho)} |f|$.

[*Hint:* Notice that $|f(z)| > |w|$ if $|w| < \inf_{\partial B(0,\rho)} |f|$, consequently $f(z)$ and $f(z) - w$ have the same number of zeros in $B(0, \rho)$ by Rouché's theorem. From Theorem 4.111, infer that g is the inverse of f.]

4.174 ¶ Hadamard's three circles theorem. Let f be holomorphic in an open set $0 < R_1 < |z| < R_2$ that contains the annulus $R_1 \le |z| \le R_2$. Set

$$M(R) := \max\{|f(z)| \,\big|\, |z| = R\},$$

and prove that for $R_1 < R < R_2$,

$$M(R)^{\log \frac{R_2}{R_1}} \le M(R_1)^{\log \frac{R_2}{R}} M(R_2)^{\log \frac{R}{R_1}},$$

or in other words, prove that $\log M(R)$ is a convex function of $\log R$. [*Hint:* Notice that for all z, $R_1 < |z| < R_2$, we have

$$|z|^\alpha |f(z)| \le \max\Big\{ |z|^\alpha |f(z)| \,\big|\, |z| = R_1 \text{ or } |z| = R_2 \Big\}$$

and choose

$$\alpha := \frac{\log \frac{M(R_2)}{M(R_1)}}{\log \frac{R_1}{R_2}}.]$$

4.175 ¶. Let f be holomorphic in $H := \{z \,|\, \Im(z) > 0\}$ and continuous in $\{\Im(z) \ge 0\}$. Prove that f identically vanishes if it vanishes in $\{z = x + iy \,|\, y = 0, x \in [0, 1]\}$.

4.176 ¶. Let $f(z) = u(x, y) + iv(x, y)$ be holomorphic in a neighborhood of $[0, 1] \times [0, 1] \subset \mathbb{R}^2$. Suppose that $u(x, y) = 0$ in $[0, 1] \times \{0\}$ and $[0, 1] \times \{1\}$ and that $v(x, y) = 0$ in $\{0\} \times [0, 1]$ and $\{1\} \times [0, 1]$. Show that $f = 0$ in Ω. [*Hint:* Consider $f^2(z)$.]

5. Surfaces and Level Sets

In the first two sections of this chapter we discuss the notion of surface and, related to it, the inverse and the implicit function theorems. Applications as well as some aspects of the local theory of surfaces will be discussed in the last two sections.

5.1 Immersed and Embedded Surfaces

An important step for the development of analysis and geometry is the realization of the intuitive idea of a *regular surface* in $\mathbb{R}^n$. A sphere and a cylinder are regular surfaces in $\mathbb{R}^3$, the cone is not, at least near the vertex. The idea of a surface develops around the concept of a *diffeomorphism* and its analysis uses the *inverse function theorem*.

5.1.1 Diffeomorphisms

5.1 Definition. *Let $X \subset \mathbb{R}^r$ and $Y \subset \mathbb{R}^n$ be two sets. A map $\varphi : X \to Y$ is called a* diffeomorphism *if*

   (i) *φ is injective and surjective between X and Y,*
   (ii) *φ has an extension $\varphi : \Omega \to \mathbb{R}^n$ of class C^1 to an open set $\Omega \supset X$,*
   (iii) *φ^{-1} has an extension $\psi : \Delta \to \mathbb{R}^r$ of class C^1 to an open set $\Delta \supset Y$.*

Of course, φ is a diffeomorphism from X to Y if and only if φ^{-1} is a diffeomorphism from Y to X. Therefore, we say that X and Y are *diffeomorphic* if there exists a diffeomorphism between X and Y. Moreover, if a map $\varphi : X \to Y$ can be chosen in such a way that both φ and φ^{-1} are of class C^k, $k \geq 1$, respectively, in suitable open sets $\Omega \supset X$ and $\Delta \supset Y$, we say that X and Y are C^k-*diffeomorphic*.

In the literature, the definition is somewhat different, as in general one refers to *intrinsic differential structures* on X and Y. However, Definition 5.1 is more suitable when discussing *submanifolds* of $\mathbb{R}^n$.

M. Giaquinta and G. Modica, *Mathematical Analysis: An Introduction to Functions of Several Variables*, DOI: 10.1007/978-0-8176-4612-7_5,
© Birkhäuser Boston, a part of Springer Science + Business Media, LLC 2010

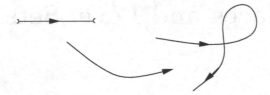

Figure 5.1. A regular noninjective map.

5.2 ¶. Let $\varphi : X \subset \mathbb{R}^r \to \mathbb{R}^n$ be a diffeomorphism between X and $Y = \varphi(X)$. Observe the following.

(i) φ and φ^{-1} extend to C^1 maps, in particular, φ is a *homeomorphism* between X and Y.
(ii) φ is also a diffeomorphism between A and $\varphi(A)$ for every $A \subset X$.

5.3 ¶. Show that being diffeomorphic is an equivalence relation, i.e.,

(i) X is diffeomorphic to X, $\forall X \subset \mathbb{R}^r$,
(ii) if $Y \subset \mathbb{R}^n$ is diffeomorphic to $X \subset \mathbb{R}^r$, then X is diffeomorphic to Y,
(iii) if $Z \subset \mathbb{R}^k$ is diffeomorphic to $Y \subset \mathbb{R}^n$ and Y is diffeomorphic to $X \subset \mathbb{R}^r$, then Z is diffeomorphic to X.

If $\Omega \subset \mathbb{R}^r$ is an open set, a map $\varphi : \Omega \to \mathbb{R}^r$ of class C^1 is a diffeomorphism onto $\varphi(\Omega)$ if there exists an open set $W \supset \varphi(\Omega)$ and a map $\psi : W \to \mathbb{R}^n$ of class $C^1(W)$ such that $\psi(\varphi(x)) = x \ \forall x \in \Omega$. In this case we say that $\varphi(\Omega)$ is an *embedded submanifold* of $\mathbb{R}^n$. The chain rule then yields

$$\mathbf{D}\psi(\varphi(x))\mathbf{D}\varphi(x) = \mathrm{Id} \qquad \text{for all } x \in \Omega;$$

hence $\mathbf{D}\varphi(x)$ is injective and we have proved the following.

5.4 Proposition. *Let $\varphi : \Omega \subset \mathbb{R}^r \to \mathbb{R}^n$ be a diffeomorphism from an open set $\Omega \subset \mathbb{R}^r$ into $\mathbb{R}^n$, then $r \leq n$, and for every $x \in \Omega$ the linear map $\mathbf{D}\varphi(x)$ is injective or, equivalently, $\mathbf{D}\varphi(x)$ has maximal rank r.*

5.5 ¶. Let $\varphi : \Omega \subset \mathbb{R}^r \to \mathbb{R}^n$, Ω open, be a diffeomorphism. Prove that, if $\varphi(\Omega)$ is open in $\mathbb{R}^n$, then $r = n$ and $\mathbf{D}f(x)$ is nonsingular.

A typical diffeomorphism from an open set of $\mathbb{R}^r$ is the projection of a graph.

5.6 Proposition. *Let $\Omega \subset \mathbb{R}^r$ be an open set and let $f : \Omega \to \mathbb{R}^m$ be a map of class C^1. Then the map $\varphi : \Omega \to \mathbb{R}^r \times \mathbb{R}^m$ given by $\varphi(x) := (x, f(x))$ is a diffemorphism from Ω onto the graph of f*

$$G_f := \Big\{ (x, y) \in \mathbb{R}^n \times \mathbb{R}^m \,\Big|\, x \in \Omega, \ y = f(x) \Big\}.$$

Proof. In fact, φ is injective, $\mathrm{Im}\,\varphi = G_f$, and the inverse of $\varphi : G_f \to \Omega$ has a C^∞ extension as, for instance, the orthogonal projection on the first factor $\Omega \times \mathbb{R}^n \to \Omega$ defined by $(x, y) \to x$. $\qquad \square$

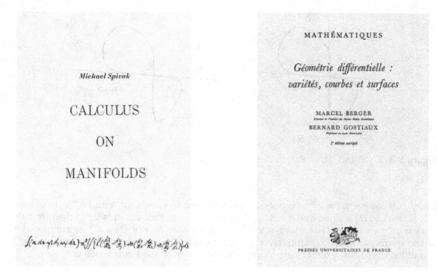

Figure 5.2. Two books on surfaces.

a. Tangent vectors

5.7 Definition. *Let $\Omega \subset \mathbb{R}^r$ be an open set and let $\varphi : \Omega \to \mathbb{R}^n$ be a diffeomorphism with $1 \leq r < n$. For $x_0 \in \varphi(\Omega)$, let $u_0 \in \Omega$ be the unique point with $\varphi(u_0) = x_0$. The* tangent space *to $\varphi(\Omega)$ at $x_0 \in \varphi(\Omega)$ is the linear subspace of $\mathbb{R}^n$ image of the tangent map of φ at u_0,*

$$\operatorname{Tan}_{x_0}\varphi(\Omega) := \operatorname{Im}\left(\mathbf{D}\varphi(u_0)\right).$$

Since φ is a diffeomorphism, $\mathbf{D}\varphi(u_0)$ has maximal rank r, hence $\operatorname{Tan}_{x_0}\varphi(\Omega)$ has dimension r.

As it is defined, the tangent space depends on the parameterization φ and not just on its image $\varphi(\Omega)$, meaning that there may exist another diffeomorphism $\psi : \Delta \to \mathbb{R}^n$ defined on another open set $\Delta \subset \mathbb{R}^s$ with $\psi(\Delta) = \varphi(\Omega)$ such that $\operatorname{Im}\left(\mathbf{D}\psi(x_0)\right) \neq \operatorname{Im}\left(\mathbf{D}\varphi(u_0)\right)$ at some point $x_0 = \varphi(u_0) = \psi(v_0)$. But this cannot happen. In fact, from Proposition 5.8 below we have $r = s$ and $\psi = \varphi \circ h$ for a diffeomorphism $h : \Delta \to \Omega$, thus $h(v_0) = u_0$ and

$$\mathbf{D}\psi(v_0) = \mathbf{D}\varphi(x_0)\mathbf{D}h(v_0).$$

Since h is nonsingular, we conclude $\operatorname{Im}\left(\mathbf{D}\psi(v_0)\right) = \operatorname{Im}\left(\mathbf{D}\varphi(u_0)\right)$. This also shows that the dimension of a parameterized surface depends only on the surface and not on the particular parameterization.

5.8 Proposition. *Let Ω and Δ be two open sets respectively in $\mathbb{R}^r$ and $\mathbb{R}^s$, and let $\varphi : \Omega \to \mathbb{R}^n$, $\psi : \Delta \to \mathbb{R}^n$ be two diffeomorphisms. If $\varphi(\Omega) = \psi(\Delta)$, then there exists a diffeomorphism $h : \Delta \to \Omega$ onto Ω such that $\psi = \varphi \circ h$. In particular $r = s$.*

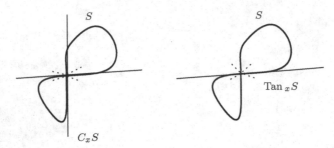

Figure 5.3. $C_x S$ and $\mathrm{Tan}\,_x S$.

Proof. Of course, $h := \varphi^{-1} \circ \psi$ does it: we only need to show that h and h^{-1} are of class C^1. If $W \supset \varphi(\Omega)$ is open and $f : W \to \Omega$ is the C^1-map that extends φ^{-1}, then we have $h(z) = \varphi^{-1} \circ \psi(z) = f(\psi(z))\ \forall z \in \psi^{-1}(W)$. Therefore h is of class C^1 as composition of two maps of class C^1. Similarly, one shows that h^{-1} is of class C^1. In conclusion, h is a diffeomorphism from Ω onto Δ, in particular $r = s$, see Proposition 5.4. $\qquad\square$

We conclude with a few remarks on the tangent space to a surface. Let $S \subset \mathbb{R}^n$ be a set. We say that a vector $v \in \mathbb{R}^n$ is *tangent* to S at $x \in S$ if there exists a curve $s :] - \delta, \delta[\to S$ of class C^1 with $s(0) = 0$ and $s'(0) = v$. As λv, $\lambda \in \mathbb{R}_+$ is tangent to S at x if v is tangent to S at x, the set of tangent vectors to S at x form a cone with vertex at 0 denoted

$$C_x S.$$

5.9 Proposition. *Let Ω be an open set in $\mathbb{R}^r$. Let $\varphi : \Omega \subset \mathbb{R}^r \to \mathbb{R}^n$, Ω open, be a diffeomorphism. Then the tangent cone to $\varphi(\Omega)$ at $x_0 := \varphi(u_0)$ is the tangent space to $\varphi(\Omega)$ at x_0,*

$$C_{x_0}\varphi(\Omega) = \mathrm{Tan}\,_{x_0}\varphi(\Omega).$$

Proof. Let $r(t) := \nu t + u_0$, $\nu \in \mathbb{R}^r$, be the line in $\mathbb{R}^r$ through u_0. The curve $s(t) := \varphi(r(t))$ is well defined for t near zero, lies in $\varphi(\Omega)$ and is of class C^1; moreover $s(0) = x_0$ and $s'(0) = \mathbf{D}\varphi(x_0)\nu$. Since $\nu \in \mathbb{R}^r$ is arbitrary, we then infer

$$\mathrm{Im}\,(\mathbf{D}\varphi(u_0)) \subset C_{x_0}\varphi(\Omega).$$

Let $s :] - \delta, \delta[\to S$ be a curve of class C^1 with trajectory in $\varphi(\Omega)$ with $s(0) = x_0$ and let $r(t) := \varphi^{-1}(s(t))$ be the corresponding curve in Ω. Since φ is a diffeomorphism, $r(t)$ is of class C^1 in Ω, and we have $r(0) = u_0$ and $s'(0) = \mathbf{D}\varphi(u_0)(r'(0))$, therefore

$$C_{x_0}\varphi(\Omega) \subset \mathrm{Im}\,(\mathbf{D}\varphi(u_0)).$$

$\qquad\square$

5.1.2 r-dimensional surfaces in $\mathbb{R}^n$

The image of an open set Ω of $\mathbb{R}^r$ by a diffeomorphism $\varphi : \Omega \to \mathbb{R}^n$ into $\mathbb{R}^n$, $1 \leq r < n$, realizes only partially the intuitive idea of a surface, since several surfaces cannot be parameterized on a *open* set of $\mathbb{R}^r$, as, for instance, the circle $S^1 := \{x^2 + y^2 = 1\}$ in $\mathbb{R}^2$, which is not homeomorphic to any interval of $\mathbb{R}$.

Roughly an r-dimensional surface in $\mathbb{R}^n$, $1 \leq r \leq n$, is a subset that is *locally* diffeomorphic to an open set of $\mathbb{R}^r$. However, there are two possible ways of localizing: in the space of parameters or in the target space. This leads to two notions of surface, both useful.

a. Submanifolds

Localizing the definition of diffeomorphism in the target space yields the following.

5.10 Definition. *A r-submanifold of $\mathbb{R}^n$ of class C^k is a set $M \subset \mathbb{R}^n$ all points of which have an open neighborhood $W \subset \mathbb{R}^n$ such that $M \cap W$ is C^k-diffeomorphic to an open set of C^r.*

Of course, if $\varphi : \Omega \subset \mathbb{R}^r$, Ω open, is a diffeomorphism, then $\varphi(\Omega)$ is trivially a r-submanifold of $\mathbb{R}^n$. We say that $\varphi(\Omega)$ is an *embedded submanifold* of $\mathbb{R}^n$ of dimension r. In particular, the graph of a map $f : \Omega \subset \mathbb{R}^r \to \mathbb{R}^m$ of class C^1 is an r-dimensional embedded submanifold of $\mathbb{R}^{r+m}$.

As a consequence of Proposition 5.9 we easily get

5.11 Corollary. *Let M be an r-submanifold of $\mathbb{R}^n$. Then $\operatorname{Tan}_x M$ has dimension r and $C_x M = \operatorname{Tan}_x M \; \forall x \in M$.*

5.12 ¶. Show the following.
   (i) In $\mathbb{R}^2$, the hyperbolas $x^2 - y^2 = 1$, the parabola $y = x^2$, the ellipse $x^2 + 2y^2 = 1$ define 1-dimensional submanifolds of class C^∞ of $\mathbb{R}^2$.
   (ii) The unit sphere $\mathbb{R}^n$

$$S^{n-1} := \left\{ x \in \mathbb{R}^n \; \middle| \; |x|^2 = 1 \right\}$$

   is a $(n-1)$-submanifold of $\mathbb{R}^n$ of class C^∞.
   (iii) The set $\{(x, y) \in \mathbb{R}^2 \mid x^2 = y^2\}$ is not an r-submanifold of $\mathbb{R}^n$.
   (iv) An r-submanifold of $\mathbb{R}^n$ is locally homeomorphic to an open set of $\mathbb{R}^r$.

b. Immersions

Localizing the definition of diffeomorphism in the space of parameters we instead set the following.

5.13 Definition. *An immersion is a map $\varphi : \Omega \to \mathbb{R}^n$, where $\Omega \subset \mathbb{R}^r$ is open, $1 \leq r < n$, that is locally a diffeomorphism, i.e., any $u_0 \in \Omega$ has an open neighborhood U_{x_0} such that $\varphi_{|U_{u_0}}$ is a diffeomorphism from U_{u_0} onto $\varphi(U_{u_0})$. An r-dimensional immersed submanifold in $\mathbb{R}^n$ is the image $\varphi(\Omega)$ of an immersion $\varphi : \Omega \subset \mathbb{R}^r \to \mathbb{R}^n$.*

Figure 5.4. From the left: (i) and (ii) are two injective immersed 1-surfaces in $\mathbb{R}^2$, and (iii) is a 1-submanifold in $\mathbb{R}^2$.

Notice that, if $\varphi : \Omega \to \mathbb{R}^n$ is an immersion, then $\mathbf{D}\varphi(u)$ has maximal rank r at every $u \in \Omega$.

Let $\varphi : \Omega \subset \mathbb{R}^r \to \mathbb{R}^n$ be an immersion. Of course, the noninjectivity of φ is an obstruction for φ to be a diffeomorphism from Ω onto $\varphi(\Omega)$. Moreover, an injective immersion is not yet a diffeomorphism, see Figure 5.5. Also the tangent cone at a point is in general a real cone and not a plane, see Figure 5.3. However, the obstruction for φ to be a diffeomorphism is purely topological. We in fact have the following.

5.14 Theorem. *Let $\varphi : \Omega \subset \mathbb{R}^r \to \mathbb{R}^n$, $1 \le r < n$ be an injective immersion. The following claims are equivalent.*

(i) *φ is open, equivalently, $\varphi^{-1} : \varphi(\Omega) \to \Omega$ is continuous.*
(ii) *$\varphi(\Omega)$ is an r-dimensional embedded submanifold of $\mathbb{R}^n$.*
(iii) *φ is a diffeomorphism.*

Proof. Trivially (iii) implies (i) and (ii).

Let us prove that (i) implies (ii) and (iii). Let $x_0 \in \varphi(\Omega)$, $u_0 \in \Omega$ be such that $\varphi(u_0) = x_0$ and let U_0 be an open neighborhood of u_0 such that $\varphi_{|U_0}$ is a diffeomorphism. Since φ is open, we have $\varphi(U_0) = W \cap \varphi(\Omega)$ for an open set $W \subset \mathbb{R}^n$ that contains x_0, thus x_0 has an open neighborhood W such that $W \cap \varphi(\Omega)$ is diffeomorphic to U_{u_0}. Since x_0 is arbitrary, (ii) holds. Moreover, on account of Theorem 2.95 we have a locally finite covering $\{B_i\}$ of $\varphi(\Omega)$ by open balls, and corresponding maps $\psi_i : \mathbb{R}^n \to U_i \subset \mathbb{R}^r$ of class C^1 such that $\psi_i(\varphi(u)) = u \,\forall u \in U_i$. Let $\{\alpha_i\}$ be an associated partition of unity to $\{B_i\}$, see Theorem 2.97. Set $\Delta := \cup_i B_i$ and let $\psi : \Delta \to \mathbb{R}^r$ be defined by

$$\psi(x) = \sum_{i=1}^{\infty} \alpha_i(x)\psi(x), \qquad x \in \Delta,$$

where we think of the ψ's as defined on the whole space. Trivially, Δ is open, $\Delta \supset \varphi(\Omega)$, ψ is of class C^1 and for all $u \in \Omega$ we have

$$\psi(\varphi(u)) = \sum_{i=1}^{\infty} \alpha_i(\varphi(u))\psi_i(\varphi(u)) = \sum_{\{i \,|\, \varphi(u) \in B_i\}} \alpha_i(\varphi(u))\psi_i(\varphi(u))$$

$$= \sum_{\{i \,|\, \varphi(u) \in B_i\}} \alpha_i(\varphi(u))u = \sum_{\{i \,|\, \varphi(u) \in B_i\}} \Big(\sum_{i=1}^{\infty} \alpha_i(\varphi(u))\Big) u = u.$$

It remains to prove that (ii) implies (i). For that, fix $x_0 \in \Omega$, let u_0 be such that $\varphi(u_0) = x_0$, and let $W \subset \mathbb{R}^n$ be an open neighborhood of x_0, let $A \subset \mathbb{R}^r$, let $h : A \to W \cap \varphi(\Omega)$ be a diffeomorphism and $k : W \to \mathbb{R}^r$ be such that $k(h(y)) = y \,\forall y \in A$. Then the map

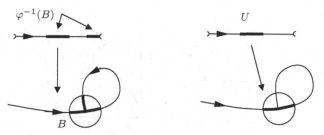

Figure 5.5. On the left an injective immersion that is not a homeomorphism since $\varphi(\Omega) \cap B$ is connected whereas $\varphi^{-1}(B)$ is not; nevertheless, $\varphi_{|U}$ is a diffeomorphism as shown on the right.

$g := k \circ \varphi : \varphi^{-1}(W) \to A$ is injective, of class C^1 and trivially, $\varphi^{-1} = g^{-1} \circ h$. Now, by the chain rule $\mathbf{D}g(u) = \mathbf{D}k(\varphi(u))\mathbf{D}\varphi(u)$, hence $\mathbf{D}g(u)$ is nonsingular. It then follows from the local invertibility theorem that g has a C^1 inverse, in particular, $\varphi^{-1} = g^{-1} \circ h$ is continuous. $\qquad \square$

5.1.3 Parameterizations of maximal rank

5.15 Theorem. *Let $\varphi : \Omega \subset \mathbb{R}^r \to \mathbb{R}^n$ be a function of class C^k, $k \geq 1$, where Ω is an open set in $\mathbb{R}^r$, $1 \leq r < n$. If $\mathbf{D}\varphi(u_0)$ has maximal rank r at $u_0 \in \Omega$, then there exists an open neighborhood U of u_0 such that $\varphi_{|U}$ is a C^k-diffeomorphism. Moreover, $\varphi(U)$ is the graph of a map of class C^k defined on an open set of a coordinate r-plane of $\mathbb{R}^n$.*

Proof. Let $\varphi = (\varphi^1, \varphi^2, \ldots, \varphi^r, \ldots, \varphi^n)$. By reordering the coordinates we may assume that

$$\det \frac{\partial(\varphi^1, \ldots, \varphi^r)}{\partial u}(u_0) \neq 0.$$

Split $\mathbb{R}^n$ as $\mathbb{R}^n = \mathbb{R}^r \times \mathbb{R}^{n-r}$, and denote by (x, y), $x \in \mathbb{R}^r$, $y \in \mathbb{R}^{n-r}$, its coordinates; finally, set $\varphi^{(1)} := (\varphi^1, \varphi^2, \ldots, \varphi^r)$ and $\varphi^{(2)} := (\varphi^{r+1}, \ldots, \varphi^n)$. Since the Jacobian of the map $\varphi^{(1)} : \Omega \to \mathbb{R}^r$ is nonzero at u_0, the local invertibility theorem yields an open neighborhood $U \subset \mathbb{R}^r$ of u_0 such that the restriction $\gamma := \varphi^{(1)}_{|U}$ is an open map with inverse $h : \gamma(U) \to U$ of class C^k, i.e.,

$$\begin{cases} x = \varphi^{(1)}(u), \\ y = \varphi^{(2)}(u), \qquad \text{if and only if} \\ u \in U \end{cases} \qquad \begin{cases} u = h(x), \\ y = \varphi^{(2)}(h(x)), \\ x \in \gamma(U). \end{cases}$$

In other words, the map $\psi : \gamma(U) \times \mathbb{R}^{n-r} \to U$, $\psi(x, y) := h(x)$ is of class C^k and inverts $\varphi_{|U}$; finally $\varphi(U)$ is the graph of the function $k(x) := \varphi^{(2)}(h(x))$, $x \in \gamma(U)$. $\qquad \square$

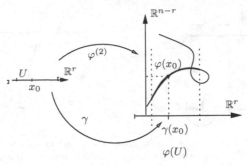

Figure 5.6. Illustration of the proof of Theorem 5.15.

5.16 Definition. *Let $\varphi : \Omega \to \mathbb{R}^n$ be a map of class C^1 defined on an open set $\Omega \subset \mathbb{R}^r$, $1 \le r < n$. If $\mathbf{D}\varphi(u_0)$ has maximal rank r at every point $u_0 \in \Omega$, we say that φ is a* regular parameterization *of $\varphi(\Omega)$.*

Theorem 5.15 then yields the following.

5.17 Corollary. *Let $\varphi : \Omega \subset \mathbb{R}^r \to \mathbb{R}^n$, Ω open, $1 \le r < n$, be a map of class C^k, $k \ge 1$. The following claims are equivalent.*

(i)   *φ is a regular parameterization of $\varphi(\Omega)$,*
(ii) *φ is an immersion,*
(iii) *for every $u_0 \in \Omega$ there exists a neighborhood U of u_0 such that $\varphi(U)$ is the graph of a C^k-map $k : W \to \mathbb{R}^{n-r}$ defined on an open set W of a r-dimensional coordinate plane.*

5.18 ¶. Let N be an r-dimensional linear subspace of $\mathbb{R}^n$, $1 \le r < n$. We say that $\Sigma \subset \mathbb{R}^n$ is a *graph with respect to N* or *over N* if there exist an open set $W \subset \mathbb{R}^r$ and a map $k : W \to \mathbb{R}^{n-r}$ such that

$$R(\Sigma) = \Big\{ (u,v) \in \mathbb{R}^r \times \mathbb{R}^{n-r} \,\Big|\, u \in W, \ v = k(u) \Big\}$$

where $R : \mathbb{R}^n \to \mathbb{R}^n$ maps linearly N onto $\{(u,v) \in \mathbb{R}^r \times \mathbb{R}^{n-r} \,|\, v = 0\}$.

Show that each of the claims in Corollary 5.17 is actually equivalent to saying that for every $u_0 \in \Omega$ there exists a neighborhood U of u_0 such that $\varphi(U)$ is a graph of a C^k-map over $\operatorname{Tan}_{\varphi(u_0)}\varphi(U)$. [*Hint:* Observe that $\mathbf{D}\varphi(u_0)$ and $\mathbf{D}(R \circ \varphi)(u_0)$ have the same rank and use Theorem 5.15.]

The next theorem is a consequence, actually is equivalent to Theorem 5.15. It claims that if $\varphi : \Omega \subset \mathbb{R}^r \to \mathbb{R}^n$ has maximal rank at some point $u_0 \in \Omega$, then we can deform the target space with a diffeomorphism in such a way that $\varphi(\Omega)$ becomes flat. Here we present a direct proof: Instead of factorizing φ, we apply the local invertibility theorem to an extension of φ.

5.19 Theorem. *Let $\Omega \subset \mathbb{R}^r$ be an open set and let $\varphi : \Omega \to \mathbb{R}^n$, $r < n$, be a regular C^k-map, $k \ge 1$. For every $x_0 \in \Omega$ there exist a neighborhood*

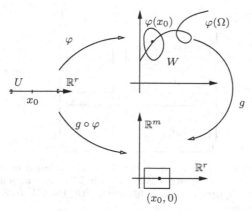

Figure 5.7. Illustration of the proof of Theorem 5.19.

U of x_0 in $\mathbb{R}^r$, a ball $B(0,\delta) \subset \mathbb{R}^{n-r}$, an open neighborhood W of $\varphi(x_0)$ in $\mathbb{R}^n$ and a diffeomorphism $g : W \subset \mathbb{R}^n \to U \times B(0,r) \subset \mathbb{R}^n$ of class C^k such that

$$g(\varphi(x)) = (x, 0) \qquad \forall x \in U. \tag{5.1}$$

Therefore, the maps $h := (g^1, \ldots, g^r)$ and $k := (g^{r+1}, \ldots, g^n)$ defined on W are of class C^k and

$$\begin{cases} h(\varphi(x)) = x, \\ k(\varphi(x)) = 0, \end{cases} \qquad \forall x \in U.$$

Moreover, the map h extends $(\varphi_{|U})^{-1}$ to the open set W, $\varphi(U)$ is the zero-set of k,

$$\varphi(U) = \Big\{ w \in W \,\Big|\, k(w) = 0 \Big\}, \tag{5.2}$$

and, finally, the Jacobian matrices of h and k have maximal rank.

Proof. By reordering the variables of $\mathbb{R}^n$, we can assume

$$\det \frac{\partial(\varphi^1, \ldots, \varphi^r)}{\partial(x^1, \ldots, x^r)}(x) \neq 0.$$

Let f be the extension of φ to $\Omega \times \mathbb{R}^{n-r}$, $f = f(x^1 \ldots, x^r, t_1, \ldots, t_{n-r})$ given by $f = (f^1, f^2, \ldots, f^n)$ with

$$\begin{cases} f^i(x,t) := \varphi^i(x) & \text{if } i \leq n - r, \\ f^i(x,t) := \varphi^i(x) + t_{i-r} & \text{if } r < i \leq n. \end{cases} \tag{5.3}$$

We have

$$\det \mathbf{D}f(x,0) = \det \begin{pmatrix} \dfrac{\partial(\varphi^1\ldots,\varphi^r)}{\partial(x^1\ldots,x^r)} & 0 \\[2ex] \dfrac{\partial(\varphi^{r+1}\ldots,\varphi^n)}{\partial(x^1\ldots,x^r)} & \mathrm{Id}_{n-r} \end{pmatrix} = \det \frac{\partial(\varphi^1\ldots,\varphi^r)}{\partial(x^1\ldots,x^r)}(x),$$

hence $\det \mathbf{D}f(x_0,0) \neq 0$. By Theorem 5.15 there exist a neighborhood $Z := U \times B(0,\delta)$ of $(x_0,0)$ in $\mathbb{R}^r \times \mathbb{R}^{n-r}$, an open set $W := f(Z)$ in $\mathbb{R}^n$, and a map $g : W \to Z$ of class C^k that inverts $f_{|Z}$. Therefore, g is a *diffeomorphism from W to Z* and for all $x \in U$ and $t \in B(0,r)$ we have

$$g(f(x,t)) = (x,t) \qquad \forall x \in U, \ \forall t \in B(0,\delta); \tag{5.4}$$

in particular $g(f(x,0)) = (x,0)$, i.e., (5.1).

The relation (5.1) can be written as

$$\begin{cases} h(\varphi(x)) = x, \\ k(\varphi(x)) = 0 \end{cases} \qquad \forall x \in U$$

if $g =: (h,k)$. The first equality says that $h : W \to \mathbb{R}^r$ extends $(\varphi_{|U})^{-1}$ to W. The second relation implies that $\varphi(U) \subset \{w \in W \,|\, k(w) = 0\}$. On the other hand, if $w \in W$, there exists $(x,t) \in U \times B(U,\delta)$ such that $f(x,t) = w$. If $k(w) = 0$, then $t = 0$ and in conclusion there exists $x \in U$ such that $\varphi(x) = f(x,0) = w$, that is,

$$\left\{ w \in W \,\Big|\, k(w) = 0 \right\} \subset \varphi(U),$$

and (5.2) is proved.

Observing that $\mathbf{D}f(x_0,0)$ is the block matrix

$$\mathbf{D}f(x_0,0) = \begin{pmatrix} \mathbf{A} & 0 \\[1ex] \mathbf{B} & \mathrm{Id} \end{pmatrix}$$

with $\det \mathbf{A} \neq 0$, we deduce that $\mathbf{D}g(\varphi(x_0))$ has the form

$$\mathbf{D}g(\varphi(x_0)) = \begin{pmatrix} \mathbf{A}^{-1} & 0 \\[1ex] -\mathbf{B}\mathbf{A}^{-1} & \mathrm{Id} \end{pmatrix}.$$

Consequently, the Jacobian matrices of the maps h and g at $\varphi(x_0)$ are given by the two matrices of maximal rank

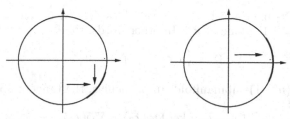

Figure 5.8. From the left: (a) $y = -\sqrt{1-x^2}$ or $x = \sqrt{1-y^2}$, and (b) $x = \sqrt{1-y^2}$.

$$\mathbf{D}h = \left(\begin{array}{c|c} \mathbf{A}^{-1} & \mathbf{0} \end{array}\right), \quad \text{and} \quad \mathbf{D}k = \left(\begin{array}{c|c} -\mathbf{B}\mathbf{A}^{-1} & \text{Id} \end{array}\right). \quad (5.5)$$

$\square$

5.20 ¶. Deduce Theorem 5.19 from Theorem 5.15.

5.2 Implicit Function Theorem

5.2.1 Implicit functions

The linear implicit equation in $\mathbb{R}^2$

$$ax + by = 0$$

rewrites as $y = (-a/b)x$ if $b \neq 0$, i.e., its solutions are the points of the graph of a real function of one variable. If we consider nonlinear implicit equations as for instance

$$\phi(x, y) = x^2 + y^2 - 1,$$

the situation is more complex. For instance, the solutions of $x^2 + y^2 - 1 = 0$ are the points of the unit circle of $\mathbb{R}^2$ which is not a graph. However, pieces of it are graphs. We say that the circle is *locally* a graph, see Figure 5.8.

A theorem due to Ulisse Dini (1845–1918) states that the solutions of a generic nonlinear implicit equation or system are locally a graph provided some qualitative conditions hold. Let

$$\phi(x) = 0, \qquad x = (x^1, x^2, \ldots, x^n) \in \mathbb{R}^n$$

be a nonlinear equation, $\phi : \mathbb{R}^n \to \mathbb{R}$ of class C^1 and let x_0 be a solution. If $\mathbf{D}\phi(x_0) \neq 0$, then the solutions of $\phi(x) = 0$ are locally the points of the graph of a C^1-function of $(n-1)$ variables defined on an open set of a coordinate $(n-1)$-plane of $\mathbb{R}^n$. In other words, the set of solutions

$$\Gamma := \left\{ x \in \mathbb{R}^n \,\middle|\, \phi(x) = 0 \right\}$$

is near x_0 a $(n-1)$-submanifold. In particular, the tangent space to Γ at x_0 exists and

$$\operatorname{Tan}_{x_0}\Gamma = \ker \mathbf{D}\phi(x_0) = \nabla\phi(x_0)^\perp.$$

As stated the theorem also holds for systems of implicit equations

$$\begin{cases} \phi^1(x^1, \ldots, x^n) = 0, \\ \ldots \\ \phi^m(x^1, \ldots, x^n) = 0 \end{cases} \tag{5.6}$$

and it is known as the *implicit function theorem*. Again the linear case is illuminating. Let $C \in M_{m,n}(\mathbb{R})$ be a matrix of maximal rank. By reordering the variables we may assume

$$\mathbf{C} = \left(\boxed{\mathbf{A}}\ \boxed{\mathbf{B}}\right)$$

where $\mathbf{A} \in M_{m,n-m}(\mathbb{R})$ and $\mathbf{B} \in M_{m,m}(\mathbb{R})$ and $\det \mathbf{B} \neq 0$. Denoting by $z = (x, y)$, $x \in \mathbb{R}^{n-m}$, $y \in \mathbb{R}^m$, the coordinates in $\mathbb{R}^n$, the system $\mathbf{C}z = 0$ rewrites as

$$\left(\boxed{\mathbf{A}}\ \boxed{\mathbf{B}}\right)\begin{pmatrix} x \\ y \end{pmatrix} = \mathbf{A}x + \mathbf{B}y = 0.$$

Since $\det \mathbf{B} \neq 0$, we may write the m-variable y as function of the $n-m$-variable x

$$y = -\mathbf{B}^{-1}\mathbf{A}x.$$

Therefore, *if $\mathbf{C}$ has maximal rank n, the solutions of $\mathbf{C}z = 0$ are the points on the (linear) graph of $y = -\mathbf{B}^{-1}\mathbf{A}x$.*

The implicit function theorem extends the previous claim to the case of a system of nonlinear equations.

5.21 Theorem (Implicit function theorem, I). *Let $x_0 \in \mathbb{R}^n$ be a solution of $\phi(x_0) = 0$ where $\phi := (\phi^1, \phi^2, \ldots, \phi^m)$, $m < n$, is of class C^1 in an open neighborhood $U \subset \mathbb{R}^n$ of x_0. Suppose that $\mathbf{D}\phi(x_0)$ has maximal rank m. Then the zero set of ϕ, $\Gamma := \{x \in U \,|\, \phi(x) = 0\}$ is near x_0 the graph of a function of class C^1 of $n-m$ coordinate variables defined on an open set of an $(n-m)$-dimensional coordinate plane of $\mathbb{R}^n$. In particular, the tangent space at x_0 to Γ is the kernel of the linear tangent map of ϕ at x_0*

$$\operatorname{Tan}_{x_0}\Gamma = \ker \mathbf{D}\phi(x_0) = \operatorname{Span}\left\{\nabla\phi^1(x_0), \nabla\phi^2(x_0), \ldots, \nabla\phi^m(x_0)\right\}^\perp.$$

Actually, a more precise statement holds. We present it in the next paragraph together with a few examples.

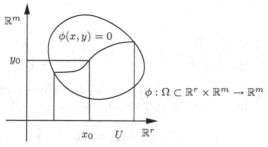

Figure 5.9. The implicit function theorem.

a. The theorem

Let $\phi : \Omega \subset \mathbb{R}^n \to \mathbb{R}^m$, $m < n$, $\phi = (\phi^1, \ldots, \phi^m)$ be of class C^k, $k \geq 1$, and let us consider the system

$$\phi(x^1, x^2, \ldots, x^n) = 0. \tag{5.7}$$

Suppose that the variables in $\mathbb{R}^n$ are split in two groups of respectively r and m variables $r + m = n$, which we reorder in such a way that $x = (x^1, \ldots, x^r) \in \mathbb{R}^r$ and $y = (x^{r+1}, \ldots, x^n) \in \mathbb{R}^m$. Denote now by

$$\frac{\partial \phi}{\partial x}(x, y) \in M_{m,r}(\mathbb{R}) \qquad \text{and} \qquad \frac{\partial \phi}{\partial y}(x, y) \in M_{m,m}(\mathbb{R})$$

respectively the $m \times r$ submatrix of the first r columns and the $m \times m$-submatrix of the remaining m-columns of $\mathbf{D}\phi$ at (x, y). We have

$$\mathbf{D}\phi = \left(\begin{array}{c|c} \dfrac{\partial \phi}{\partial x} & \dfrac{\partial \phi}{\partial y} \end{array} \right).$$

5.22 Theorem (Implicit function theorem, II). *Let $\phi : \Omega \subset \mathbb{R}^n \to \mathbb{R}^m$ be a map of class $C^k(\Omega)$, $k \geq 1$, where Ω is open. With the previous notations, suppose that $\mathbb{R}^n = \mathbb{R}^r \times \mathbb{R}^m$,*

$$\phi(x_0, y_0) = 0 \qquad \text{and} \qquad \det \frac{\partial \phi}{\partial y}(x_0, y_0) \neq 0$$

at some $(x_0, y_0) \in \Omega$. Then there exist open neighborhoods W of (x_0, y_0) in $\mathbb{R}^r \times \mathbb{R}^m$, U of x_0 in $\mathbb{R}^r$ and a map $\varphi : U \to \mathbb{R}^m$ such that

(i) *$\phi_{|W} : W \to \mathbb{R}^m$ is open,*
(ii) *φ is open and of class C^k,*

(iii) *finally,*

$$\begin{cases} (x,y) \in W, \\ \phi(x,y) = 0 \end{cases} \qquad \text{if and only if} \qquad \begin{cases} x \in U, \\ y = \varphi(x). \end{cases}$$

We again postpone its proof until we state an apparently more general version of it, Theorem 5.29, and we instead discuss a few consequences. Set

$$\Gamma := \Big\{ (x,y) \in \Omega \,\Big|\, \phi(x,y) = 0 \Big\}$$

for the zero level set of ϕ.

5.23 Corollary. *Under the previous hypotheses and notations, the following holds.*

(i) *$\Gamma \cap W$ is the graph of the map $\varphi : U \to \mathbb{R}^m$,*

$$\Gamma \cap W = \Big\{ (x,y) \in W \,\Big|\, \phi(x,y) = 0 \Big\} = \Big\{ (x,y) \,\Big|\, x \in U, \ y = \varphi(x) \Big\}.$$

(ii) *By differentiating the system $\phi(x,\varphi(x)) = 0 \ \forall x \in U$, we get*

$$\mathbf{D}\varphi(x) = -\Big(\frac{\partial \phi}{\partial x}\Big)^{-1}(x,\varphi(x))\frac{\partial \phi}{\partial y}(x,\varphi(x)) \qquad \forall x \in U. \qquad (5.8)$$

and for every $z \in \Gamma \cap W$

$$\mathrm{Tan}\,_z\Gamma = \ker \mathbf{D}\phi(z) \ = \mathrm{Span}\,\Big\{\nabla\phi^1(z),\dots,\nabla\phi^m(z)\Big\}^{\perp}. \qquad (5.9)$$

Proof. (i) It is a rewriting of Theorem 5.22 (iii).

(ii) On account of (i), we have $\phi(x,\varphi(x)) = 0 \ \forall x \in U$. Differentiating (recall the chain rule) we get

$$\mathbf{D}\phi(x,\varphi(x)) \begin{bmatrix} \mathrm{Id} \\ \mathbf{D}\varphi(x) \end{bmatrix} = 0, \qquad (5.10)$$

that one can rewrite as

$$\frac{\partial \phi}{\partial x}(x,\varphi(x)) + \frac{\partial \phi}{\partial y}(x,\varphi(x))\mathbf{D}\varphi(x) = 0,$$

i.e., (5.8). Moreover, the tangent space to $\Gamma \cap W$ at $z = (x,\varphi(x)) \in W$ is the tangent space to the graph of φ at $(x,\varphi(x))$, which is spanned by the linearly independent columns of the right matrix in (5.10), hence by (5.10)

$$\mathrm{Tan}\,_{(x,\varphi(x))}\Gamma \subset \ker \mathbf{D}\phi(x,\varphi(x));$$

(5.9) then follows since $\mathrm{Tan}\,_{(x,\varphi(x))}\Gamma$ and $\ker \mathbf{D}\phi(x,\varphi(x))$ have the same dimension $n - m$. $\qquad \square$

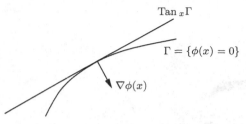

Figure 5.10. The tangent and normal planes to Γ.

5.24 Remark. We emphasize that, in the scalar case, $m = 1$, i.e., $\phi : \Omega \to \mathbb{R}$, for every $z \in \Gamma \cap W$, the tangent space $\mathrm{Tan}\,_z\Gamma$ is the perpendicular hyperplane to $\nabla\phi(z)$, as we recall, $\ker \mathbf{D}\phi(z) = \mathrm{Span}\,\nabla\phi(z)^{\perp}$.

5.25 Example. Trivially, the point $(x_0, y_0) := (1/\sqrt{2}, 1/\sqrt{2})$ is on the unit circle

$$\phi(x, y) = x^2 + y^2 - 1 = 0 \qquad \text{in } \mathbb{R}^2,$$

and $(x - x_0) + (y - y_0) = 1$ and $x + y = 1$ are respectively the tangent line and the tangent space to the circle at (x_0, y_0).

Let us discuss this simple situation in terms of the implicit function theorem. We have $\mathbf{D}\phi(x, y) = (2x, 2y)$, hence $\mathbf{D}\phi(x_0, y_0) = \sqrt{2}(1, 1)$. Therefore, near (x_0, y_0), the circle is both the graph of a function $y = \varphi(x)$ and $x = \psi(y)$ of class C^∞, and the tangent space is

$$\ker \mathbf{D}\phi(x_0, y_0) = \left\{ (x, y) \,\Big|\, \sqrt{2}(1, 1) \begin{pmatrix} x \\ y \end{pmatrix} = 0 \right\},$$

i.e., $x + y = 0$.

Once we assume that $\phi(x, y) = 0$ is the graph of a C^1 map $y = \varphi(x)$ near $1/\sqrt{2}$, in order to compute the tangent space, we may also write

$$x^2 + \varphi^2(x) = 1 \qquad \text{for } x \text{ near to } 1/\sqrt{2}.$$

Differentiating with respect to x we get $2x + 2\varphi(x)\varphi'(x) = 0$, hence $\varphi'(x_0) = -1$; in other words, $(1, -1)$ is the velocity vector of the curve $x \to \varphi(x)$, thus $(1, -1)$ spans the tangent space to the circle at (x_0, y_0).

5.26 Example. Let $\Gamma \subset \mathbb{R}^3$ be the set of solutions of

$$\begin{cases} x + \log y + 2z - 2 = 0, \\ 2x + y^2 + e^z - 1 - e = 0 \end{cases}$$

and let $\phi(x, y, z) := (x + \log y + 2z - 2, 2x + y^2 + e^z - 1 - e)$. We have $(0, 1, 1) \in \Gamma$ and

$$\mathbf{D}\phi(0, 1, 1) = \begin{pmatrix} 1 & 1 & 2 \\ 2 & 2 & e \end{pmatrix}.$$

Since $\det \begin{pmatrix} 1 & 2 \\ 2 & e \end{pmatrix} = e - 4 \neq 0$, there is an open neighborhood $W \subset \mathbb{R}^3$ of $(0, 1, 1)$ such that $\Gamma \cap W$ writes as

Figure 5.11. Ulisse Dini (1845–1918) and a page of his *Lezioni di Analisi Infinitesimale*, Pisa, 1909.

$$\Gamma \cap W = \left\{ (x, y, z) \,\middle|\, y = \alpha(x), z = \beta(x) \right\} \tag{5.11}$$

where $\varphi := (\alpha, \beta)$ is of class C^∞, i.e., $\Gamma \cap W$ is the trajectory of the parameterized curve $x \to (x, \alpha(x), \beta(x))$. Moreover, the tangent space (i.e., the tangent line) to the curve is the kernel of $\mathbf{D}\phi(0, 1, 1)$, i.e., the solutions of

$$\begin{cases} x + 2y = 0, \\ 2x + 2y + ez = 0. \end{cases}$$

Notice that these solutions are orthogonal to the two rows $(1, 1, 2)$ and $(2, 2, e)$ of $\mathbf{D}\varphi(0, 0, 1)$.

Alternatively, once we assume (5.11), i.e.,

$$\begin{cases} x + \log \alpha(x) + 2\beta(x) - 2 = 0, \\ 2x + \alpha^2(x) + e^{\beta(x)} - 1 - e = 0 \end{cases}$$

near zero, differentiating and taking into account that $\alpha(0) = \beta(0) = 1$, we find

$$\begin{cases} 2 + 2\alpha'(0) + 4\beta'(0) = 0, \\ 2 + 2\alpha'(0) + e\beta'(0) = 0, \end{cases}$$

i.e., $\alpha'(0) = -1$, $\beta'(0) = 0$. As $(1, \alpha'(0), \beta'(0))$ is the velocity vector of the curve $x \to (x, \alpha(x), \beta(x))$, we conclude that the parametric equation of the tangent line to Γ at $(0, 1, 1)$ is

$$\begin{pmatrix} x \\ y \\ z \end{pmatrix} = t \begin{pmatrix} 1 \\ -1 \\ 0 \end{pmatrix} + \begin{pmatrix} 0 \\ 1 \\ 1 \end{pmatrix}.$$

Notice that $(1, -1, 0)$ is perpendicular both to $(1, 1, 2)$ and $(2, 2, e)$.

Finally, notice that the tangent line to Γ is the intersection of the tangent planes at $(0, 1, 1)$ to the surfaces of equations

$$x + \log y + 2z - 2 = 0 \qquad \text{and} \qquad 2x + y^2 + e^z - 1 - e = 0$$

corresponding to the rows of ϕ.

5.27 ¶. Prove that the level set $xe^y + ye^x = 1$ is the graph of a smooth function $y = \varphi(x)$ in a neighborhood of $(1, 0)$. Compute $\varphi''(0)$.

5.28 ¶. Prove that the subset of $\mathbb{R}^3$ of solutions of

$$\begin{cases} x^2 + y^2 + z^2 = 1, \\ xe^y + y \cos z + xz = 1 \end{cases}$$

is the graph of a curve $x = x(z), y = y(z)$ in a neighborhood of $(1, 0, 0)$, and compute its acceleration at $(1, 0, 0)$.

b. Foliations

Let $\phi : \Omega \to \mathbb{R}^m$ be a function of class C^1 defined on an open set $\Omega \subset \mathbb{R}^r \times \mathbb{R}^m$. For every $c \in \mathbb{R}^m$ the c *level set of* ϕ is the set

$$\Gamma_c := \Big\{ (x, y) \in \Omega \,\Big|\, \phi(x, y) = c \Big\},$$

with evident analogy with the level lines of a geographic map in which ϕ models the altitude over the sea. The implicit function theorem states that, if

$$\phi(x_0, y_0) = 0 \qquad \text{and} \qquad \det \frac{\partial \phi}{\partial y}(x_0, y_0) \neq 0,$$

then the zero level line is the graph of a function $y = \varphi(x)$ of class C^1 near (x_0, y_0). What can we say about the close level lines, i.e., about

$$\Gamma_c := \Big\{ (x, y) \,\Big|\, \phi(x, y) = c \Big\}, \qquad c \in \mathbb{R}^m, \ |c| \text{ small?}$$

Of course, $\det \frac{\partial \phi}{\partial y}(x, y) \neq 0$ in a neighborhood of (x_0, y_0). Since ϕ is of class C^1, the implicit function theorem applies to close level sets. Actually, there is a C^1-map $(x, c) \to \varphi(x, c)$ such that for c close to zero, the c level line is near (x_0, y_0) the graph of $x \to \varphi(x, c)$, see Figure 5.12, as it is stated in the following.

5.29 Theorem (Implicit function theorem, III). *Let $\phi : \Omega \to \mathbb{R}^m$ be a function of class C^k, $k \geq 1$, where $\Omega \subset \mathbb{R}^r \times \mathbb{R}^m$ is an open set, $m < n$ and $r := n - m$. Suppose that at $(x_0, y_0) \in \Omega$ we have*

$$\phi(x_0, y_0) = 0 \qquad \text{and} \qquad \det \frac{\partial \phi}{\partial y}(x_0, y_0) \neq 0.$$

Then there exist an open connected neighborhood U of $x_0 \in \mathbb{R}^r$ a ball $B(0, \delta)$ in $\mathbb{R}^m$, an open set $W \subset \mathbb{R}^{r+m}$ and a function $\varphi : U \times B(0, \delta) \to \mathbb{R}^m$ such that

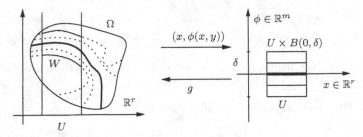

Figure 5.12. Illustration of the implicit function theorem.

(i) $\phi_{|W}$ *is open,*
(ii) φ *is an open map of class* C^k,
(iii) *finally,*

$$
\begin{cases}
(x, y) \in W, \\
c \in B(0, \delta), \qquad\quad \text{if and only if} \\
\phi(x, y) = c,
\end{cases}
\qquad
\begin{cases}
x \in U, \\
c \in B(0, \delta), \\
y = \varphi(x).
\end{cases}
$$

Consequently, we have the following.

5.30 Corollary. *Under the hypotheses and with the notation of Theorem 5.29, we have*

(i) *for all* $c \in B(0, \delta)$, *the c level line* Γ_c *of* ϕ *is the graph of the map* $x \to \varphi(x, c)$, $x \in U$,

$$
\Gamma_c \cap W = \Big\{ (x, y) \,\Big|\, x \in U, \; y = \varphi(x, c) \Big\}, \qquad (5.12)
$$

in particular $\Gamma_c \cap W$ *is a* $(n - m)$-*submanifold in* $\mathbb{R}^n$.
(ii) *We have* $\phi(x, \varphi(x, c)) = c \; \forall (x, c) \in U \times B(0, \delta)$, *and, differentiating, we get*

$$
\begin{cases}
\dfrac{\partial \varphi}{\partial x}(x, c) = -\Big[\dfrac{\partial \phi}{\partial y}\big(x, \varphi(x, c)\big)\Big]^{-1} \dfrac{\partial \phi}{\partial x}(x, \varphi(x, c)), \\[4mm]
\dfrac{\partial \varphi}{\partial c}(x, c) = \Big[\dfrac{\partial \phi}{\partial y}\big(x, \varphi(x, c)\big)\Big]^{-1}.
\end{cases}
\qquad (5.13)
$$

In particular, for every $z \in \Gamma \cap W$

$$
\text{Tan}_z \Gamma \cap W = \ker D\phi(z) = \text{Span} \Big\{ \nabla\phi^1(z), \dots, \nabla\phi^m(z) \Big\}^{\perp}.
$$

Proof. (i) is a rewriting of Theorem 5.29 (iii); this yields also that $\phi(x, \varphi(x, c)) = c$ $\forall (x, c) \in U \times B(0, \delta)$. Thus, differentiating the previous identity we get

$$\mathbf{D}\phi(x, \varphi(x, c)) \begin{pmatrix} \mathrm{Id}_r & 0 \\ \dfrac{\partial \phi}{\partial x} & \dfrac{\partial \phi}{\partial c} \end{pmatrix} = \begin{pmatrix} 0 & \mathrm{Id}_m \end{pmatrix}$$

i.e., (5.13) and the last part of the claim. $\qquad\qquad\qquad\qquad\qquad\qquad\square$

Proof of Theorem 5.29. The function $f(x, y) := (x, \phi(x, y))$, $(x, y) \in \Omega$, is of class C^k and its Jacobian $(r + m) \times (r + m)$-matrix is given by

$$\mathbf{D}f = \begin{pmatrix} \mathrm{Id}_r & 0 \\ \dfrac{\partial \phi}{\partial x} & \dfrac{\partial \phi}{\partial y} \end{pmatrix}.$$

We have $\det \mathbf{D}f = \det \frac{\partial \phi}{\partial y}$, hence $\det \mathbf{D}f$ is nonzero at (x_0, y_0). From the local invertibility theorem, Theorem 1.89, there exists an open neighborhood V of (x_0, y_0) such that $Z := f(V)$ is open and $f_{|V}$ is a diffeomorphism of class C^k from V onto Z. Now we choose an open connected neighborhood U of x_0 in $\mathbb{R}^r$ and a ball $B(0, \delta)$ in $\mathbb{R}^m$ such that $U \times B(0, \delta) \subset Z$. If $W := f^{-1}(U \times B(0, \delta))$, then f is also a diffeomorphism of class C^k from W onto $U \times B(0, \delta)$. In particular, $f_{|W}$ is open, hence $\phi_{|W}$ is open as composition of f with the orthogonal projection onto the second factor. The inverse map $g : U \times B(0, \delta) \to W$ of $f_{|W}$ is of class C^k and open. Let us write g as $g =: (\psi, \varphi)$ where

$$\psi := (g^1, g^2, \dots, g^r) \qquad \text{and} \qquad \varphi := (g^{r+1}, \dots, g^{r+m}).$$

The maps ψ, φ are clearly of class C^k and open since they are the compositions of g with an orthogonal projection. This concludes the proof of (i) and (ii).

(iii) then readily follows. We only notice that the identities

$$\begin{cases} f(g(x, c)) = (x, c) & \forall (x, c) \in U \times B(0, \delta), \\ g(f(x, y)) = (x, y) & \forall (x, y) \in W \end{cases}$$

are clearly equivalent to

$$\begin{cases} \psi(x, c) = x & \forall (x, c) \in U \times B(0, \delta), \\ \phi(\psi(x, c), \varphi(x, c)) = c & \forall (x, c) \in U \times B(0, \delta), \qquad\qquad (5.14) \\ \varphi(x, \phi(x, y)) = y & \forall (x, y) \in W \end{cases}$$

because of the form of f and g, and $x \in U$ if $(x, y) \in W$ and $\phi(x, y) = c$. $\qquad\square$

5.31 Remark. We have in fact proved that $f : W \to Z = U \times B(0, \delta)$ is a diffeomorphism from the ambient space $\mathbb{R}^n$ into $\mathbb{R}^n$ that maps each level set $\Gamma_c = \{(x, y) \in W \mid \phi(x, y) = c\}$ of ϕ onto the planar surface $\{(x, c), x \in U\}$ if $c \in B(0, \delta)$.

5.32 ¶. Theorem 5.29 generalizes Theorem 5.22. Prove that the two theorems are actually equivalent. [*Hint:* Under the assumptions of Theorem 5.29, apply Theorem 5.22 to the function

$$\psi : \mathbb{R}^r \times \mathbb{R}^m \times \mathbb{R}^m \to \mathbb{R}^m, \qquad \psi(x, c, y) := \phi(x, y) - c$$

to get the conclusion of Theorem 5.29.]

5.33 Example. The hypotheses of Theorem 5.22 are only *sufficient* to prove the smoothness of the zero level set. In fact, if $\varphi : U \times \mathbb{R}^m$ is of class C^1 and $\phi : U \times \mathbb{R} \to \mathbb{R}$ is defined by $\phi(x, y) := y - \varphi(x)$, we have $\det \frac{\partial \phi}{\partial y}(x, y) = 1$ and $\det \frac{\partial \phi^2}{\partial y}(x, y) = 0$ while the zero level set of both functions ϕ and ϕ^2 is given by the graph $y = \varphi(x)$.

5.34 An algorithm for the level sets. Let Ω be open in $\mathbb{R}^r \times \mathbb{R}^m$ and let $\phi : \Omega \to \mathbb{R}^m$ be a map of class C^1 with

$$\phi(0,0) = 0 \qquad \text{and} \qquad \det \frac{\partial \phi}{\partial y}(0,0) \neq 0.$$

Near $(0,0)$ the zero level set is the graph of a C^1-function $y = \varphi(x)$. Thinking of the proofs of the inverse and the implicit function theorems, we may design an algorithm for the approximation of φ. Let

$$L = \frac{\partial \phi}{\partial x}(0,0), \qquad M := \frac{\partial \phi}{\partial y}(0,0).$$

From the first-order Taylor formula, we infer

$$0 = f(x,y) - f(0,0) = Lx + My + R(x,y)$$

with $|R(x,y)| = o(|x| + |y|)$; therefore, since M is invertible,

$$y = -M^{-1}Lx - M^{-1}R(x,y).$$

Hence, for a given x, $\varphi(x)$ is a fixed point of

$$y \to F(y) := -M^{-1}Lx - M^{-1}R(x,y).$$

For $|x|$ sufficiently small, F is a contraction on a suitable defined Banach space. Therefore, an approximation scheme for $\varphi(x)$ is given by

$$\begin{cases} y_0 := F(0), \\ y_{k+1} := F(y_k) \end{cases}$$

as $k \to \infty$, and one sees that, for $|x|$ sufficiently small, $\{y_k\}$ converges to $\varphi(x)$ at least exponentially.

5.35 ¶. Prove Theorem 5.22 in the case $m = 1$ and $n = 2$ developing the following steps that use only the one-dimensional calculus.

    (i) Show that there exists $\epsilon > 0$ such that $\phi(x_0, y) < 0$ if $y_0 - \epsilon \leq y < y_0$ and $\phi(x_0, y) > 0$ if $y_0 < y \leq y_0 + \epsilon$.

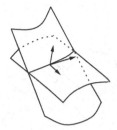

Figure 5.13. Intersection of two transversal surfaces.

(ii) Show that there exists $\delta > 0$ such that $\phi(x, y_0 - \epsilon) < 0$ and $\phi(x, y_0 + \epsilon) > 0$ if $|x - x_0| < \delta$.

(iii) Set $R := \{(x, y) \mid |x - x_0| < \delta, \ |y - y_0| < \epsilon\}$, and observe that one can choose ϵ and δ in such a way that $\phi_y(x, y) > 0 \ \forall \ (x, y) \in R$. Then show that for $|x_1 - x_0| < \delta$ the equation $\phi(x_1, y) = 0$ has a unique solution y_1 with $(x_1, y_1) \in R$: This defines a function $y = \varphi(x)$ such that $\phi(x, \varphi(x)) = 0$ for all $x \in (x_0 - \delta, x_0 + \delta)$.

(d) Finally, show that φ is of class C^1 and

$$\phi_x(x, \varphi(x)) + \phi_y(x, \varphi(x))\varphi'(x) = 0.$$

Extend the previous proof to the case $m = 1$, $n \geq 2$.

5.36 ¶. Assume that the implicit function theorem, Theorem 5.22, holds true. Infer from it the local invertibility theorem, Theorem 1.89. Conclude that the statements of Theorems 1.89 and 5.22 are equivalent. [*Hint:* Consider the equation $y - \phi(x) = 0$.]

c. Submersions

5.37 Definition. *Let Ω be an open set of $\mathbb{R}^n$. A map $\phi : \Omega \to \mathbb{R}^m$ of class C^1 for which $m < n$ and $\operatorname{Rank} D\phi(x) = m$ for all $x \in \Omega$ is called a submersion.*

If $\phi : \Omega \subset \mathbb{R}^n \to \mathbb{R}^m$ is a submersion, at every point $x \in \Omega$ an $m \times m$ submatrix of the Jacobian matrix has nonzero determinant. After eventually reordering the variables, we can apply the implicit function theorem to state the following.

5.38 Theorem. *Let $\phi : \Omega \subset \mathbb{R}^n \to \mathbb{R}^m$ be a submersion of class C^k, $k \geq 1$. Then ϕ is an open map, and every level line of it is a $(n - m)$-submanifold of $\mathbb{R}^n$ of class C^1; moreover, $\operatorname{Tan}_x \Gamma_{\phi(x)} = \ker D\phi(x) \ \forall x \in \Omega \cap \Gamma_{\phi(x)}$ where $\Gamma_t := \{y \mid \phi(y) = t\}$ is the t level set of ϕ.*

For future use, we explicitly restate the case $m = 1$.

5.39 Proposition. *Let $\phi : \Omega \subset \mathbb{R}^n \to \mathbb{R}$ be a submersion of class C^1. Then for every $x \in \Omega$, the level set of ϕ through x,*

$$\Gamma_x := \left\{ y \in \Omega \mid \phi(y) = \phi(x) \right\}$$

Figure 5.14. From the left: (a) $x^2 - y^2 = 0$, (b) $\rho^2 - 2a^2 \cos \theta = 0$ and (c) $\rho = \sin \theta \cos 2\theta$.

is an $(n-1)$-submanifold of $\mathbb{R}^n$ and

$$\operatorname{Tan}_x \Gamma_x = \ker \mathbf{D}\phi(x) = \nabla \phi(x)^\perp \qquad \forall x \in \Omega.$$

5.40 ¶ Intersections of submanifolds. Let M and N be two submanifolds of dimension 2 of $\mathbb{R}^3$ with $N \cap M \neq \emptyset$. Show that if the tangent planes of M and N at each point do not agree, then $M \cap N$ is a C^1-curve.

In general, prove that, if M and N are submanifolds of $\mathbb{R}^n$ of dimension respectively r and s, with $r+s > n$, and if $\operatorname{Tan}_{x_0} M \cap \operatorname{Tan}_{x_0} N$ has dimension $r+s-n$ at $x_0 \in M \cap N$, then there exists an open neighborhood W of x_0 such that $M \cap N \cap W$ is an $r+s-n$-submanifold of $\mathbb{R}^n$. [*Hint:* Write M and N locally as level sets and apply the implicit function theorem.]

5.2.2 Irregular level sets

The study and actually the behavior of the level set of a function near a point at which its Jacobian matrix does not have maximal rank is more complicated. A theorem that goes in this direction is in Section 5.3.8. Here we confine ourselves to showing a few examples of irregular level sets in two variables

$$\left\{ (x, y) \in \mathbb{R}^2 \,\middle|\, \phi(x, y) = \phi(x_0, y_0) \right\}$$

in a neighborhood of a *critical point* (x_0, y_0).

5.41 Example. For $\phi(x, y) = x^2 + y^2$, the zero level set $\phi(x, y) = 0$ is just the point $(0, 0)$.

5.42 Example. Let $\phi(x, y) := x^2 - y^2$. The zero level set is the union of two lines $y = \pm x$, see Figure 5.14. Near the origin we have a similar situation for *Bernoulli's lemniscate*

$$(x^2 + y^2)^2 - 2a^2(x^2 - y^2) = 0 \qquad a > 0,$$

i.e., in polar coordinates, $\rho^2 - 2a^2 \cos \theta = 0$, see Figure 5.14.

5.43 Example. The zero level set of
$$\phi(x, y) := (x^2 + y^2)^2 - y(x^2 - y^2),$$
that in polar coordinates is the graph of $\rho = \sin \theta \cos 2\theta$ has a double knot at the origin, see Figure 5.14.

5.44 Example. The *cissoid of Diocles* is the zero level set of the function
$$\phi(x, y) = x(x^2 + y^2) - 2y^2.$$
It has a cusp at the origin with the positive x-axis as tangent cone, Figure 5.15.

5.45 Example. The zero level set of
$$(x^2 + y^2)^2 - 3xy^2 = 0,$$
has a triple point at the origin, see Figure 5.15.

5.46 Example. A famous example is the family of algebraic curves
$$y^2 = x^2(x + \lambda)$$
that depend continuously on the parameter λ. They have the form in Figure 5.16. When λ moves from positive to negative values, we are in the presence of a drastic change of form, a "catastrophe".

5.3 Some Applications

The local invertibility theorem and the implicit function theorems are useful tools in several contexts. In this section we shall illustrate some of their applications.

5.3.1 Small perturbations

This is a typical situation that appears quite often: Quadratic terms are smaller than linear terms if the variable is small.

Figure 5.15. From the left: (a) The cissoid of Diocles and (b) $(x^2 + y^2)^2 - 3xy^2 = 0$.

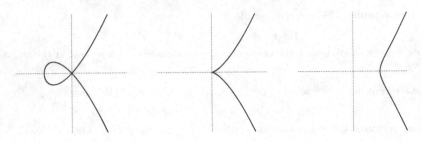

Figure 5.16. $y^2 = x^2(x + \lambda)$: From the left (a) $\lambda > 0$, (b) $\lambda = 0$ and (c) $\lambda < 0$.

a. Quadratic systems

Suppose we want to solve the following system of n equations in n unknowns

$$\sum_{ij=1}^{n} a_{ij}^h x_i x_j + \sum_{i=1}^{n} b_i^h x_i + c^h = 0, \qquad h = 1, n \qquad (5.15)$$

or in brief

$$X^T \mathbf{A} X + \mathbf{B} X + C = 0$$

where $X \in \mathbb{R}^n$, $\mathbf{A} \in M_{n,n}(\mathbb{R}^n)$, $\mathbf{A}_{ij} := (a_{ij}^h)$, $\mathbf{B} \in M_{n,n}$, and $C \in \mathbb{R}^n$. The map $\phi : \mathbb{R}^n \to \mathbb{R}^n$ defined by $\phi(X) := X^T \mathbf{A} X + \mathbf{B} X$ is of class C^∞, moreover, $\phi(0) = 0$, and

$$d\phi_X(v)^h = \sum_{i,j=1}^{n} (a_{ij}^h + a_{ji}^h) X^i v^j + \sum_{i=1}^{n} \mathbf{B}_i^h v^i$$

so that $d\phi_0(v) = \mathbf{B}v$. If $\mathbf{B}$ is invertible, the local invertibility theorem says that there are two neighborhoods U and V of $0 \in \mathbb{R}^n$ so that (5.15) is uniquely solvable in U for every $C \in V$.

b. Nonlinear Cauchy problem

Suppose we want to solve the Cauchy problem

$$\begin{cases} u'(t) = F(u(t)), & t \in [0, T], \\ u(0) = u_0 \end{cases} \qquad (5.16)$$

where $F : \Omega \subset \mathbb{R}^n \to \mathbb{R}^n$ is of class C^1, $F(0) = 0$, and $(0, u_0) \in \Omega$. Observe that if $u_0 = 0$, then $u(t) = 0$ is the unique solution to the problem.

Consider the transformation $\phi : X \to Y$, $f(u) := (u' - F(u), u(0))$, between the Banach spaces $X := C^1([0, b])$ and $Y := C^0([0, b]) \times \mathbb{R}$. It is easily seen that $\phi(0) = (0, 0)$, ϕ is of class C^1 with differential at 0 given by the linear map $d\phi_0 : X \to Y$, $d\phi_0(v(t)) = (v'(t) - \mathbf{D}F(0)v(t), v(0))$. The theory of *linear* systems yields a unique solution for the Cauchy problem

$$\begin{cases} v'(t) - \mathbf{D}F(0)v(t) = f(t), & t \in [0, T] \\ v(0) = v_0 \end{cases}$$

for any $f \in C^0([0, T])$ and $v_0 \in \mathbb{R}$ for which

$$||v||_{C^1([0,T])} \leq C\Big(||f||_{\infty,[0,T]}, v_0\Big)$$

see [GM3], or, in other words the map $d\phi_0 : X \to Y$ is an isomorphism of Banach spaces. The local invertibility theorem (in Banach spaces) then yields the unique solvability of (5.16) in a neighborhood of 0 in $C^1([a, b])$ provided u_0 is sufficiently small.

c. A boundary value problem

Suppose we want to solve the following boundary value problem

$$\begin{cases} u''(t) + F(u(t)) = \varphi(t), & t \in [0, 1], \\ u(0) = u_0, & \\ u(1) = u_1 \end{cases} \tag{5.17}$$

where $F : \Omega \subset \mathbb{R}^n \to \mathbb{R}^n$ is of class C^1, $F(0) = \mathbf{D}F(0) = 0$ and u_0, u_1 are given real numbers. We consider the transformation $\phi : X \to Y$ from the Banach spaces $X := C^2([0, 1])$ and $Y := C^0([0, 1]) \times \mathbb{R} \times \mathbb{R}$ given by $\phi(u(t)) := (u''(t) + F(u(t)), u(0), u(1))$. It is not difficult to check that $\phi(0) = (0, 0, 0)$, ϕ is differentiable at every $u \in X$ with differential given by

$$df_u(v(t)) - \Big(v''(t) + \mathbf{D}F(u(t))v(t), v(0), v(1)\Big),$$

in particular, $df_0(v(t)) = (v''(t), v(0), v(1))$. Since $df_0 : X \to Y$ is invertible, the local invertibility theorem tell us that, for u_0, u_1 small enough and for $\varphi(t)$ close to zero in the uniform norm, there is a unique function close to zero in norm C^2 that solves (5.17).

d. C^1-dependence on initial data

Let $F(t, y)$ be a function of class C^1 in a neighborhood of $(t_0, y_0) \in \mathbb{R} \times \mathbb{R}^n$. As we know, see, e.g., [GM3, Chapter 3], the Cauchy problem

$$\begin{cases} y'(t) = F(t, y(t)), \\ y(t_0) = y_0 \end{cases} \tag{5.18}$$

has a unique local solution in a neighborhood of (t_0, y_0); moreover, the solution $y(t; y_0)$ depends continuously on the initial datum y_0. Fix y_0 and shorten $y(t; y_0)$ as $\overline{y} :]t_0 - \delta, t_0 + \delta[\to \mathbb{R}^n$.

Consider the map $\phi : X \to Y$ between the Banach spaces $X := C^1(]t_0 - \delta, t_0 + \delta[, \mathbb{R}^n)$ and $Y := C^0(]t_0 - \delta, t_0 + \delta[) \times \mathbb{R}^n$ given by

$\phi(u) := (u' - F(t,u), u(t_0))$. Trivially, $\phi(\overline{y}) = (0, y_0)$; one also checks that ϕ is differentiable with differential $d\phi_{\overline{y}} : X \to Y$ given by

$$d\phi_{\overline{y}}(v(t)) = \Big(v'(t) - F(t, \overline{y}(t))v(t), v(t_0)\Big).$$

The theory of linear systems of ODE says that $d\phi_{\overline{y}} : X \to Y$ is an isomorphism of Banach spaces. Then the local invertibility theorem states that ϕ has an inverse ϕ^{-1} of class C^1. In particular, for every choice $(\epsilon(t), \xi)$, where $\epsilon(t)$ is a continuous function sufficiently close in the uniform norm to 0 and ξ a vector close to y_0, the problem

$$\begin{cases} y'(t) = F(t, y(t)) + \epsilon(t), \\ y(t_0) = \xi \end{cases} \tag{5.19}$$

has a unique solution close to $\overline{y}(t)$ in X. Such a solution $y(t, \xi) := \phi^{-1}(\epsilon(t), \xi)$ has a C^1-dependence on ξ. In particular, we have proved the following.

5.47 Theorem. *Let $F(t, y)$ be a function of class C^1 in a neighborhood of (t_0, y_0) in $\mathbb{R} \times \mathbb{R}^n$. There exist $\delta > 0$ and $\rho > 0$ such that for every $\xi \in B(y_0, \rho)$ there exists a solution $y = y(t, \xi) \in C^1(]t_0 - \delta, t_0 + \delta[\times B(y_0, \rho))$ that is of class C^1 in the variables (t, ξ).*

5.48 ¶. Let $F(t, y, \lambda)$ be of class C^1 in a neighborhood of (t_0, y_0, λ_0). Prove that the local solution of

$$\begin{cases} y(t, \lambda) = F(t, y(t, \lambda), \lambda), \\ y(t_0, \lambda) = y_0 \end{cases}$$

has a C^1-dependence on the parameter λ, and infer Theorem 5.47 from this. [*Hint:* Consider the local solutions to

$$\begin{cases} \frac{\partial}{\partial t} z(t, \lambda) = \frac{\partial F}{\partial y}(t, y(t, \lambda), \lambda) z(t, \lambda) + \frac{\partial F}{\partial \lambda_j}(t, y(t, \lambda), \lambda) \\ z(t_0, \lambda) = 0 \end{cases}$$

and the differential increments $u_{h,j} := h^{-1}\Big(z(t, \lambda + he_j) - z(t, \lambda)\Big)$, and prove by means of Gronwall's lemma that $u_{h,j} \to z$ as $h \to 0$.]

5.3.2 Rectifiability theorem for vector fields

Let Ω be an open set of $\mathbb{R}^n$ and let $a(x) := (a^1(x), \ldots, a^n(x))$, $x \in \Omega$, be a vector field of class C^1. In a neighborhood of every point of Ω the *local* (in time) *flow* of a, denoted by $\varphi_t(x)$, is defined as the unique solution of the Cauchy problem

$$\begin{cases} \frac{d}{dt}\varphi_t(x) = a(\varphi_t(x)), \\ \varphi_0(x) = x. \end{cases}$$

For t sufficiently small, the system being autonomous, we have

$$\varphi_s \circ \varphi_t(x) = \varphi_{t+s}(x) = \varphi_t \circ \varphi_s(x).$$

5.49 Definition. *Two vector fields $a \in C^1(\Omega, \mathbb{R}^n)$ and $b \in C^1(\Omega^*, \mathbb{R}^n)$ are said to be* equivalent *if there is a diffeomorphism $u : \Omega^* \to \Omega$ such that $b(y) = (\mathbf{D}u)^{-1}(y)a(u(y)) \ \forall y \in \Omega^*$.*

The relation of equivalence among vector fields is symmetric, reflexive, and transitive. Moreover, a and b are equivalent if and only if there is a diffeomorphism that transforms the flow of a into the flow of b. In fact, if $b(y) = (\mathbf{D}u)^{-1}(u(y))a(u(y))$, and $\varphi_t(x)$, $\psi_t(y)$ are the flows respectively of a and b, then

$$\psi_t(y) = u^{-1} \circ \varphi_t \circ u(y),$$

since for the map $v_t(y) := u^{-1} \circ \varphi_t \circ u(y)$, we compute $v_0(y) = y$ and

$$\frac{dv_t}{dt}(y) = \mathbf{D}u^{-1}(\varphi_t(y))\frac{d\varphi_t}{dt}(u(y)) = (\mathbf{D}u)^{-1}(v_t(y))a(\varphi_t(u(y)) = b(v_t(y)),$$

thus $\psi_t = v_t$ because of the uniqueness of the Cauchy problem. Conversely, differentiating $\psi_t = u^{-1} \circ \varphi_t \circ u$ we see that a and b are equivalent.

5.50 Definition. *We say that x_0 is a* singular point, *or a point in equilibrium or a stagnation point for the vector field $a(x)$ if $a(x_0) = 0$.*

5.51 Theorem (Rectifiability of vector fields). *Let a be a vector field of class C^1 in Ω. Suppose $a(x_0) \neq 0$. Then in a neighborhood of x_0, the vector field $a(x)$ is equivalent to the parallel vector field $e_1 := (1, 0, \ldots, 0)$. Consequently, two vector fields without stagnation points are locally equivalent.*

Proof. After an affine transformation, we may assume $x_0 = 0$, and even $a(x_0) = e_1$. Let φ_t be the flow generated by a. In a neighborhood of $y = 0$, the map $x = u(y)$ defined as

$$u(y) := \varphi_{y^1}(0, y^2, \ldots, y^n)$$

is a local diffeomorphism since $\det \mathbf{D}u(0) = 1$; moreover, from

$$\varphi_t(u(y)) = \varphi_t \circ \varphi_{y^1}(0, y^2, \ldots, y^n) = \varphi_{y^1+t}(0, y^2, \ldots, y^n)$$
$$= u(t + y^1, y^2, \ldots, y^n),$$

we conclude

$$u \circ \varphi_t \circ u(y) = y + te_1.$$

$\square$

5.3.3 Critical points and critical values: Sard's lemma

Let Ω be an open set in $\mathbb{R}^n$ and let $f : \Omega \to \mathbb{R}^m$ be a map of class C^1.

5.52 Definition. *We say that $x \in \Omega$ is a* critical point *for f if the rank of $\mathbf{D}f(x)$ is not maximal, i.e., $\operatorname{Rank}\mathbf{D}f(x) < m$. A point $y \in \mathbb{R}^m$, whose counterimage $f^{-1}(y)$ contains a critical point, is called a* critical value *for f.*

The set of critical points of f is closed in Ω. In terms of the new terminology we can state the following.

5.53 Theorem (Local invertibility theorem). *Let $f : \Omega \subset \mathbb{R}^n \to \mathbb{R}^n$ be a map of class C^1. If $y \in \mathbb{R}^n$ is not a critical value for f, then the level set $f^{-1}(y)$ is a discrete set.*

5.54 Theorem (Implicit function theorem). *Let $f : \Omega \subset \mathbb{R}^n \to \mathbb{R}^m$, $m < n$, be a map of class C^1. If $y \in \mathbb{R}^m$ is not a critical value for f, then the level set $f^{-1}(y)$ is a $(n-m)$-submanifold of $\mathbb{R}^n$.*

A natural question to ask is how large the set of critical values can be. If $m > n$, then all points of Ω are critical for f, hence all points of $f(\Omega)$ are critical values for f. Instead, if $m \leq n$, critical values are rare.

5.55 Theorem. *Let $f : \Omega \subset \mathbb{R}^n \to \mathbb{R}^m$ be a function of class C^1.*

(i) *If $n \leq m$, then the set of critical values has zero Lebesgue measure in $\mathbb{R}^m$.*

(ii) *(SARD) If $m < n$ and f is sufficiently smooth, for instance $f \in C^k$, $k \geq n-m+1$, then the set $f(A)$ has zero Lebesgue measure in $\mathbb{R}^m$.*

In particular, in both cases the set of noncritical values is dense in the image of f.

The proof of Theorem 5.55 goes beyond the goals of this book. However, we notice that the result (i) is contained in the *area formula*, and actually, it is part of its proof, compare Chapter 2 and [GMS, Vol. 5]. The claim (ii) is instead more delicate.[1] Nontrivial examples show that the smoothness of f is essential. For the reader's convenience we prove Theorem 5.55 (i) only in the case $n = m$.

---

[1] The proof, see, e.g., J. MILNOR, *Topology from the Differentiable Viewpoint*, Princeton Univ. Press, 1965, uses both the implicit function theorem and Fubini's theorem and consists in showing, by induction on n, that, if C_i denotes the set of points at which the partial derivatives of f or order $\leq i$ vanish, then successively the sets $f(\Omega \setminus C_1)$, $f(C_{i+1} \setminus C_i)$, and $f(C_k)$, k large, have zero measure.

Proof. The open set Ω is a denumerable union of bounded closed cubes, see, e.g., [GM2]; therefore we can assume that $Q = [0,1] \times \cdots \times [0,1]$ is contained in Ω and that $f(A)$ has zero measure for

$$A := \left\{ x \in Q \,\middle|\, x \text{ is a critical value for } f \right\}.$$

Set $M := \sup_{x \in Q} \|\mathbf{D}f(x)\|$, so that

$$|f(x) - f(y)| \le M \,|x - y| \qquad \forall x, y \in Q,$$

consequently, $f(y) \in B(f(x), M\,|x - y|)$ for all $x, y \in Q$. Also notice that, by Taylor's formula, we have

$$|f(x) - f(y) - \mathbf{D}f(y)(x - y)| \le \omega(|x - y|)\,|x - y|$$

with $\omega(t)$ nondecreasing, $0 \le \omega(t) \le M$ and $\omega(t) \to 0$ as $t \to 0^+$.

Let a be a critical point for f. The image of the linear tangent map is a linear subspace of dimension strictly less than n. If $x \in Q$, then

$$\mathrm{dist}\,(f(x), f(a) + \mathbf{D}f(a)(\mathbb{R}^n)) \le |f(x) - f(a) - \mathbf{D}f(a)(x - a)| \le \omega(|x - a|)\,|x - a|$$

i.e., if $H := f(a) + \mathbf{D}f(a)(\mathbb{R}^n)$,

$$\mathrm{dist}\,(f(x), H) \le \omega(|x - a|)\,|x - a|.$$

We then infer for $\eta > 0$ and $x \in Q \cap B(a, \eta)$

$$\mathrm{dist}\,(f(x), H) \le \omega(\eta)\,\eta$$

while $|f(x) - f(a)| \le M\eta$. Therefore $f(Q \cap B(a, \eta))$ is contained in the intersection of the ball $B(f(a), \eta)$ with the strip parallel to H of width $2\,\omega(\eta)\,\eta$, thus in a parallelepiped of volume at most

$$2^{n-1}\,M^{n-1}\eta^{n-1}\,2\,\omega(\eta)\,\eta = 2^{n}\,M^{n-1}\eta^{n}\,\omega(\eta).$$

Now divide Q in equal cubes of side $1/k$. If one of these cubes contains a critical point, then it is contained in the ball $B(a, \sqrt{n}/k)$ and its image lies in a parallelepiped of volume at most

$$2^{n}\,M^{n-1}n^{n/2}k^{-n}\omega\left(\frac{\sqrt{n}}{k}\right).$$

Since there are at most k^n cubes in Q, we have covered $f(A)$ with a finite union of cubes of total measure at most

$$2^{n}\,M^{n-1}n^{n/2}\omega\left(\frac{\sqrt{n}}{k}\right).$$

The result follows since $\omega(\frac{\sqrt{n}}{k}) \to 0$ as $k \to \infty$. $\square$

5.56 Corollary. *Let $f : \Omega \subset \mathbb{R}^n \to \mathbb{R}^n$ be a map of class C^1. The set of noncritical values of f is dense in $f(\Omega)$. In particular, the counterimage $f^{-1}(y)$ of y is made of a finite set of points for a.e. $y \in f(\Omega)$.*

5.3.4 Morse lemma

5.57 Proposition (Morse lemma). *Let f be a smooth function defined in an open set of $\mathbb{R}^n$. Suppose that $0 \in \Omega$ is a nondegenerate critical point of f, i.e., $\mathbf{D}f(0) = 0$, $\det \mathbf{H}f(0) \neq 0$. Then there exist an open neighborhood U of 0 and a diffeomorphism $\varphi : U \to \mathbb{R}^n$ such that*

$$f(\varphi(\xi)) = -(\xi^1)^2 - \cdots - (\xi^k)^2 + (\xi^{k+1})^2 + \cdots + (\xi^n)^2$$

where k is the dimension of the largest eigenspace on which the quadratic form associated to the Hessian matrix $\mathbf{H}f(0)\xi \bullet \xi$ is negative.

Proof. We recall that by Hadamard's lemma we have

$$f(x) = \sum_{i,j=1}^{n} h_{ij}(x)x^i x^j$$

in a ball around 0.

Let us prove the claim by induction on the dimension. After a linear transformation we may assume that $h_{11}(0) = 1$. Consider now the trasnformation $y := \gamma(x)$ defined by

$$\begin{cases} y^i = x^i & \text{if } i \neq 1, \\ y^1 = x^1 + \sum_{i \geq 1} h_{1i}(x)x^i. \end{cases}$$

Since $\mathbf{D}\gamma(0) \neq 0$, γ is invertible in a neighborhood of zero and it is not difficult to show that

$$f(\gamma^{-1}(y)) = \pm(y^1)^2 + \sum_{i,j>1} h_{ij}^{(1)}(y)y^i y^j.$$

Now, assume that in a neighborhood of 0 we have in suitable coordinates

$$f(y) = \pm(y^1)^2 \pm (y^2)^2 \pm \cdots \pm (y^{r-1})^2 + \sum_{i,j=r}^{n} H_{ij}(x)y^i y^j,$$

then by a linear change of coordinates we may achieve $H_{rr}(0) \neq 0$; the transform

$$\begin{cases} y^i = x^i & \text{if } i \neq r, \\ y^r = x^r + \sum_{i>r} H_{ir}(x)x^i \end{cases}$$

is then invertible near zero and again

$$f(y) = \sum_{i \leq r} \pm(y^i)^2 + \sum_{i,j>r} H_{ij}(y)y^i y^j.$$

$\square$

5.58 Corollary. *The nondegenerate critical points of a C^2-function are isolated.*

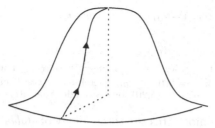

Figure 5.17. The gradient flow.

5.3.5 Gradient flow

Let A be an open set in $\mathbb{R}^n$ and let $f : A \to \mathbb{R}$ be a function of class $C^2(A)$. For every initial position that is not critical for f the (local in time) unique solution of

$$\begin{cases} x'(t) = \nabla f(x(t)), \\ x(0) = \overline{x}, \end{cases}$$

is called a *trajectory of the gradient flow* through $\overline{x}$. We have:

(i) The trajectories of the gradient flow are orthogonal to the level sets of f.
(ii) f is increasing along the trajectories of the gradient flow of f, in fact,

$$\frac{d}{dt} f(x(t)) = \nabla f(x(t)) \bullet x'(t) = |\nabla f(x(t))|^2.$$

5.59 Proposition. *If a trajectory of the gradient flow exists for all times $t > 0$ and has a limit x_0 when $t \to \infty$, then x_0 is a critical point for f.*

Proof. On the contrary, suppose $\nabla f(x_0) \neq 0$. Then there exists a bounded neighborhood U of x_0 such that $|\nabla f(x)| \geq m > 0 \ \forall x \in U$ for some $m > 0$. We may also assume that $|f(x)| \leq M$ for some $M > 0$ and that for $t > \overline{t}$ the trajectory lies in U, $x(t) \in U$. For all $t > \overline{t}$ we then infer

$$f(x(t)) - f(x(\overline{t})) = \int_{\overline{t}}^{t} \frac{d}{ds} f(x(s)) \, ds \geq m^2 (t - \overline{t}),$$

i.e., $f(x(t)) \to +\infty$, in contradiction with $f(x(t)) \to f(x_0)$. $\square$

5.60 Proposition. *Let x_0 be a critical point for f with $\mathbf{H}f(x_0) < 0$. Then there exists a neighborhood U of x_0 such that every trajectory of the gradient flow that begins at time $t = 0$ in U ends at x_0.*

Proof. We have

$$\nabla f(x) \bullet (x - x_0) = \sum_{i,j=1}^{n} f_{x_i x_j}(x_0 + s(x - x_0))(x - x_0)_i (x - x_0)_j \leq -\frac{m}{2} |x - x_0|^2$$

for all x in a ball $B(x_0, \delta)$ in which $\mathbf{H}f(x)v \bullet v < -(m/2)|v|^2$. If we set $\psi(t) := |x(t) - x_0|^2$, we find

$$\psi'(t) = 2\,(x(t) - x_0) \bullet x'(t) \; \le \; -m|x(t) - x_0|^2 = -m\,\psi(t),$$

and, after integration,

$$\log(\psi(t)) - \log(\psi(0)) \le -mt \tag{5.20}$$

for all times for which the trajectory is contained in $B(x_0, \delta)$. Thus the trajectory $x(t)$ is contained in $\overline{B(x_0, r)}$, $r := |x(0) - x_0|$. Therefore, $x(t)$ is defined for all $t \ge 0$ and, passing to the limit as $t \to \infty$ in (5.20), one concludes that $x(t) \to x_0$. $\qquad\square$

Proposition 5.60 motivates the terminology of *stable critical point* for a point for which $\nabla f(x_0) = 0$ and $\mathbf{H}f(x_0) < 0$. The above provides us with a method for finding critical points of f. It is called the *gradient method*, or the *method of steepest descent* (*ascent*, in our case).

5.3.6 Constrained critical points: the multiplier rule

In this section, we discuss some necessary conditions for the extremals of a smooth function in presence of *constraints* for the independent variables.

Let Ω be an open set in $\mathbb{R}^n$ and let $S \subset \mathbb{R}^n$ be a set. We say that $x_0 \in S$ is a *relative* or *local maximum* (*minimum*) *point* for f *constrained* to S if there exists an open neighborhood W of x_0 in $\mathbb{R}^n$ such that $f(x_0) \ge f(x)$ ($f(x_0) \le f(x)$) for all $x \in S \cap W$. Maximum and minimum points are also called *extremal points*.

We shall only discuss extremals relative to *smooth bilateral constraints*, i.e., submanifolds. For more general situations, see, e.g., [GM5].

Our considerations are purely local. Thus, we assume that $\varphi : U \subset \mathbb{R}^r \to \mathbb{R}^n$ is a diffeomorphism and the constraint is given by the embedded submanifold $\varphi(U)$. Let $\Omega \supset \varphi(U)$ be an open set and let $f : \Omega \to \mathbb{R}$ be a C^1-map. If $x_0 = \varphi(u_0)$ is a relative maximum or minimum point constrained to $\varphi(U)$, then u_0 is a relative internal maximum or minimum point for $f \circ \varphi : U \to \mathbb{R}$. By Fermat's theorem

$$\mathbf{D}(f \circ \varphi)(u_0) = 0$$

i.e., using the chain rule,

$$\mathbf{D}f(x_0)\mathbf{D}\varphi(u_0) = 0.$$

Since the columns of the Jacobian matrix $\mathbf{D}\varphi(u_0)$ span the tangent space to $\varphi(U)$ at x_0, the derivatives in the tangential directions of Γ vanish,

$$\mathbf{D}f(x_0)(v) = 0 \qquad \forall v \in \mathrm{Tan}\,_{x_0}\Gamma$$

or, equivalently,

$$\nabla f(x_0) \perp \mathrm{Tan}\,_{x_0}\varphi(U).$$

The above motivates the following definition.

5.61 Definition. *Let Ω be an open set in $\mathbb{R}^n$, let $f : \Omega \to \mathbb{R}$ be a C^1-function, and let Γ be an r-dimensional submanifold, $1 \leq r < n$. We say that $x_0 \in \Gamma$ is a* critical point *of f constrained (or relative to) Γ if*

$$\nabla f(x_0) \perp \operatorname{Tan}_{x_0} \Gamma. \tag{5.21}$$

In other words, x_0 is a critical point of f relative to Γ if the derivatives in the tangential directions of Γ vanish,

$$\frac{\partial f}{\partial v}(x_0) = \mathbf{D}f(x_0)(v) = \nabla f(x_0) \bullet v = 0 \qquad \forall v \in \operatorname{Tan}_{x_0} \Gamma.$$

5.62 Theorem (Lagrange multiplier rule). *Let $\phi : \Omega \to \mathbb{R}^m$, $\phi = (\phi^1, \phi^2, \ldots, \phi^m)$ be a function of class C^1 where Ω is an open set of $\mathbb{R}^n$ and $m < n$. Let $x_0 \in \Omega$ be a point for which $\operatorname{Rank} \mathbf{D}\phi(x_0) = m$ and let $\Gamma := \{x \in \Omega \,|\, \phi(x) = \phi(x_0)\}$. Finally let $f : \Omega \to \mathbb{R}$ be a function of class C^1. The following claims are equivalent*

(i) *x_0 is a critical point for f constrained on Γ.*
(ii) *There exist constants $\lambda_1^0, \ldots, \lambda_m^0$ such that*

$$\mathbf{D}f(x_0) = \sum_{i=1}^{m} \lambda_i^0 \mathbf{D}\phi^i(x_0). \tag{5.22}$$

(iii) *There exist constants $\lambda_1^0, \ldots, \lambda_m^0$ such that*

$$\nabla f(x_0) = \sum_{i=1}^{m} \lambda_i^0 \nabla \phi^i(x_0). \tag{5.23}$$

(iv) *There exists $\lambda^0 \in \mathbb{R}^r$ such that (x_0, λ^0) is an unconstrained critical point for the function $F : \Omega \times \mathbb{R}^r \to \mathbb{R}$,*

$$F(x, \lambda) := f(x) - \sum_{i=1}^{m} \lambda_i \phi^i(x).$$

(v) *We have $\nabla f(x_0) \in \operatorname{Im} \mathbf{D}\phi(x_0)^*$, where $\mathbf{D}\phi(x_0)^*$ denotes the adjoint matrix of $\mathbf{D}\phi(x_0)$ computed using an arbitrary scalar product in $\mathbb{R}^m$.*

The numbers $\lambda_1^0, \ldots, \lambda_m^0$ are called the *Lagrange multipliers*.

Proof. (i) $\Leftrightarrow$ (ii). We have

$$\operatorname{Tan}_{x_0} \Gamma = \ker \mathbf{D}\phi(x_0)$$

and

$$\mathbf{D}f(x_0)(v) = 0 \qquad \forall v \text{ such that } \mathbf{D}\phi(x_0)(v) = 0.$$

If we introduce the $(n - r + 1) \times n$ matrix

$$\mathbf{G} := \begin{pmatrix} \boxed{\mathbf{D}\phi(x_0)} \\ \mathbf{D}f(x_0) \end{pmatrix},$$

then (i) holds if and only if $\dim \ker \mathbf{G} = \dim \ker \mathbf{D}\phi(x_0) = r$. Since the rows of $\mathbf{D}\phi(x_0)$ are linearly independent, (i) holds if and only if $\mathbf{D}f(x_0)$ is a linear combination of the rows of $\mathbf{D}\phi(x_0)$.

(ii) $\Leftrightarrow$ (iii) is trivial, since for any map $f : \Omega \to \mathbb{R}$ $\mathbf{D}f(v) = \nabla f \bullet v$ $\forall v \in \mathbb{R}^n$.

(ii) $\Leftrightarrow$ (iv) since $\ker \mathbf{D}\phi(x_0)^{\perp} = \operatorname{Im} \mathbf{D}\phi(x_0)^*$ by the alternative theorem.

(ii) $\Leftrightarrow$ (v). The vanishing of the gradient of $F(x, \lambda)$ at (x_0, λ^0) is in fact equivalent to (ii),

$$\begin{cases} \mathbf{D}f(x_0) = \sum_{i=1}^{r} \lambda_i^0 \mathbf{D}\phi^i(x_0), \\ \phi(x_0) = 0. \end{cases}$$

$\square$

5.63 ¶. Notice that in Theorem 5.62 the assumption that Γ is a submanifold is essential. If $f(x, y) = y$ in $\mathbb{R}^2$ and $\Gamma := \{(x, y) \,|\, y^3 - x^2 = 0\}$, show that $(0, 0)$ is an absolute minimizer for f constrained to Γ; however, $F(x, y, \lambda) := y + \lambda(y^3 - x^2)$ has no critical point.

5.3.7 Some applications

Several relevant inequalities, which are obtained by linear algebra methods or by Jensen's inequality, can also be proved by means of Lagrange's multiplier rule. We conclude this section with a few examples.

a. Orthogonal projection and eigenvectors

5.64 Orthogonal projection. Let S be a linear subspace of $\mathbb{R}^n$, $(\,|\,)$ an inner product on $\mathbb{R}^n$, $||x||^2 := (x|x)$ the induced norm, and $b \in \mathbb{R}^n$. The function $f(x) := ||x - b||^2$, $x \in S$, is continuous and nonnegative; moreover, $f(||x||) \to +\infty$ as $||x|| \to +\infty$, $x \in S$. Therefore, f has at least a minimum point x_0 constrained to S for which

$$\nabla(||x_0 - b||^2) = 2(x_0 - b) \perp \operatorname{Tan}_{x_0} S = S,$$

i.e.,

$$2(x_0 - b|v) = 0 \qquad \forall v \in S.$$

In other words, x_0 is the foot of the perpendicular to S through b.

5.65 The method of least squares. In the least mean square method and in the linear regression, we need to find the minimizers of the function $x \to ||\mathbf{A}x - y||^2$ where $y \in \mathbb{R}^N$ and $\mathbf{A} \in M_{k,N}(\mathbb{R})$. As we have seen, using linear algebra x is such that $y - \mathbf{A}x \perp \operatorname{Im} \mathbf{A}$, i.e., once we fix an inner product also in $\mathbb{R}^k$, $\mathbf{A}^*(\mathbf{A}x - y) = 0$. This last equation is easily found using Fermat's theorem since

$$||\mathbf{A}x - y||^2 = ||y||^2 - 2\,y \bullet \mathbf{A}x + ||\mathbf{A}x||^2,$$
$$\nabla(\,y \bullet \mathbf{A}x\,) = \nabla(\,\mathbf{A}^* y \bullet x\,) = \mathbf{A}^* y,$$
$$\nabla||\mathbf{A}x||^2 = \nabla(\,\mathbf{A}^*\mathbf{A}x \bullet x\,) = 2\mathbf{A}^*\mathbf{A}x$$

hence

$$\nabla||\mathbf{A}x - y||^2 = \mathbf{A}^*(\mathbf{A}x - y).$$

Of course, we may proceed similarly if the data model is nonlinear as for instance if the data model is built upon a diffeomorphism $\phi : \Omega \subset \mathbb{R}^k \to \mathbb{R}^n$ of class C^1. In this case we minimize

$$x \to ||\phi(x) - b|| = \sqrt{(\phi(x) - b) \bullet (\phi(x) - b)},$$

and by Fermat's theorem, we get

$$0 = \frac{\partial ||\phi(x) - b||^2}{\partial x^i}(x_0) = 2 \sum_{i=1}^{N} (\phi(x_0) - b)^i \mathbf{D}\phi^i(x_0),$$

i.e., $\phi(x) - y \perp \operatorname{Tan}_{\varphi(x)}\varphi(\Omega)$.

5.66 Eigenvalues of a self-adjoint matrix. Let $\mathbf{A} \in M_{n,n}(\mathbb{R}^n)$ be a self-adjoint matrix. In linear algebra, see, e.g., [GM3], one sees that for the largest eigenvalue L one has

$$L = \max_{|x|=1} \mathbf{A}x \bullet x$$

and the maximum is attained at the corresponding eigenvectors. We can find again such a result by means of the Lagrange multiplier rule applied to the problem

$$\begin{cases} \text{Maximize } \mathbf{A}x \bullet x \\ \text{on the constraint } |x|^2 = 1. \end{cases}$$

Since $S := \{x \,|\, |x| = 1\}$ is compact, $\mathbf{A}x \bullet x$ attains its maximum at some $x_0 \in S$. Since S is a submanifold of $\mathbb{R}^n$,

$$\nabla(\mathbf{A}x \bullet x)(x_0) = 2\mathbf{A}x_0 \perp \operatorname{Tan}_{x_0}S, \quad \text{and} \quad |x|^2 - 1 = 0$$

i.e.,

$$\begin{cases} 2\mathbf{A}x_0 = 2\lambda\, x_0, \\ |x_0| = 1 \end{cases}$$

for some $\lambda \in \mathbb{R}$. In other words, x_0 is an eigenvector, and λ is the corresponding eigenvalue. Since $\mathbf{A}x_0 \bullet x_0 = \lambda |x_0|^2 = \lambda$, and x_0 maximizes $\mathbf{A}x \bullet x$ on S, we conclude that λ is the largest eigenvalue and x_0 a corresponding eigenvector of norm one.

5.67 Least distance between two surfaces. Let S and T be two compact and boundary-less submanifolds of $\mathbb{R}^n$. We want to describe the minimizers of $|x - y|$ with the constraint $x \in S$ and $y \in T$. Since the function $|x - y|$, $(x, y) \in \mathbb{R}^n \times \mathbb{R}^n$, is continuous, and $S \times T \subset \mathbb{R}^n \times \mathbb{R}^n$ is compact, S and T being compact, the existence of a couple (x_0, y_0), $x_0 \in S$ and $y_0 \in T$, which realizes the least distance between S and T follows at once from Weierstrass's theorem. Clearly, $S \cap T \neq \emptyset$ if and only if the least distance between S and T is zero, and, in this case, all points (x, x), $x \in S \cap T$, are minimum points. Otherwise, i.e., if $S \cap T = \emptyset$, if (x_0, y_0), $x_0 \neq y_0$, is a point of least distance, then $y \to |x_0 - y|$, $y \in T$, has a minimum at y_0, and $x \to |x - y_0|$, $x \in S$ has a minimum at x_0. Therefore, see 5.64, the vector $x_0 - y_0$ is both perpendicular to $\operatorname{Tan}_{x_0}S$ and $\operatorname{Tan}_{y_0}T$.

5.68 ¶ Linear programming. Let $\{c_j\}$, $\{a_{ij}\}$ and $\{b_i\}$, $i = 1, \ldots, m$, $j = 1, \ldots, n$ be given constants. Consider the problems

$$\begin{cases} \text{Minimize } \sum_{j=1}^{n} c_j x_j \to \min \\ \text{under the constraints } \sum_{j=1}^{n} a_{ij} x_j = b_i, \ i = 1, \ldots, m, \end{cases} \tag{5.24}$$

and

$$\begin{cases} \text{Maximize } \sum_{i=1}^{m} b_i w_i \to \max \\ \text{under the constraints } \sum_{i=1}^{m} a_{ij} w_i = c_j, \ j = 1, \ldots, n. \end{cases} \tag{5.25}$$

Prove that (5.24) is solvable if and only if (5.25) is solvable.

5.69 Boltzmann's distribution. Let E be the energy of a system of N particles, let $E_1, E_2, \ldots E_k$ be the possible energetic levels of the particles, and let n_i be the number of particles with E_i as energy. The most probable distribution is the one that maximizes the quantity

$$\binom{N}{n_1}\binom{N-n_1}{n_2}\cdots\binom{N-n_1-n_2\cdots-n_{k-2}}{n_k-1} = \frac{N!}{n_1!n_2!\ldots n_k!}$$

under the constraints

$$N = \sum_{i=1}^{k} n_i, \qquad E = \sum_{i=1}^{k} n_i E_i. \tag{5.26}$$

Assuming N and n_i, $i = 1, \ldots, k$ large enough and using Stirling's formula to approximate the form of the distribution energy, we need to minimize

$$\sum_{i=1}^{k} \left(n_i(\log n_i - 1) + \frac{1}{2}\log n_i \right)$$

under the constraints (5.26). If we further simplify, assuming that the n_i are continuous variables, and set

$$F(n_1, n_2, \ldots, n_k, \lambda, \mu)$$
$$:= \sum_{i=1}^{k} \left(n_i(\log n_i - 1) + \frac{1}{2}\log n_i \right) + \lambda\left(\sum_{i=1}^{k} n_i - N \right) + \mu\left(\sum_{i=1}^{k} n_i E_i - E \right),$$

differentiating we get

$$n_i \frac{1}{n_i} + \log n_i - 1 + \frac{1}{2n_i} + \lambda + \mu E_i = 0, \qquad i = 1, \ldots, k$$

and, neglecting the term $1/n_i$,

$$\log n_i = \mu E - \lambda,$$

concluding that the *Boltzmann distribution* is given by

$$n_i = Ce^{-\mu E_i}$$

where C and μ are constants to be found so that (5.26) holds.

5.70 ¶. Taking the asymptotic development of $\log \Gamma$ and of its derivative, substantiate the procedure we followed.

b. Inequalities

5.71 Young's inequality. If $a, b > 0$ and $p^{-1} + q^{-1} = 1$, then $ab \le \frac{a^p}{p} + \frac{b^q}{q}$, see [GM1]. To prove it we can also proceed by minimizing the function $(a, b) \to f(a, b) := \frac{a^p}{p} + \frac{b^q}{q}$ under the constraints $ab = 1$, $a > 0$, $b > 0$. Since when $ab = 1$, for

$$\varphi(a) := f(a, b) = f(a, 1/a) = \frac{a^p}{p} + \frac{a^{-q}}{q}$$

we have $\varphi(t) \to +\infty$ as $t \to 0^+$ and $t \to +\infty$, the function f has at least a minimizer (x, y) in $S = \{ab = 1, a > 0, b > 0\}$. The multiplier rule says that

$$\begin{cases} \nabla f(x, y) = \lambda \nabla(xy - 1), \\ xy = 1 \end{cases}$$

for some $\lambda \in \mathbb{R}$, i.e.,

$$\begin{cases} x^{p-1} = \lambda y, \\ y^{q-1} = \lambda x, \\ xy = 1, \ x > 0, \ y > 0. \end{cases}$$

This system has $(x, y, \lambda) = (1, 1, 1)$ as unique solution, thus the unique constrained minimizer is $(1, 1)$, and $f(1, 1) = 1$.

5.72 Arithmetic and geometric means. Recall that, for positive numbers $x_1, x_2,$ $\ldots, x_n$ we have

$$\sqrt[n]{x_1 x_2 \cdots x_n} \le \frac{1}{n} \sum_{i=1}^{n} x_i$$

with equality if and only if $x_1 = x_2 = \cdots = x_n$, see [GM1]. We may find it again by minimizing the function

$$f(x_1, x_2, \ldots, x_n) := \frac{1}{n} \sum_{i=1}^{n} x_i$$

constrained to

$$S := \left\{ x = (x_1, x_2, \ldots, x_n) \,\middle|\, x_1 x_2 \cdots x_n = 1, \ x_i > 0 \right\}$$

and proving that the unique minimizer is $\bar{x} := (1, 1, \ldots, 1)$.

First, S is a submanifold. Next, it is easy to see that $f(x) \to +\infty$ when x tends to a boundary point of S. Therefore, Weierstrass's theorem implies the existence of a minimizer $x = (x^1, x^2, \ldots, x^n)$ for f. The multiplier rule yields

$$\begin{cases} \nabla f(x) = \lambda \nabla (x_1 x_2 \cdots x_n - 1), \\ x_1 x_2 \cdots x_n = 1, \ x_i > 0, \end{cases}$$

for some $\lambda \in \mathbb{R}$, i.e.,

$$\begin{cases} x_1 = n\lambda x_2 x_3 \cdots x_n, \\ x_2 = n\lambda x_1 x_3 \cdots x_n, \\ \cdots \\ x_n = n\lambda x_1 x_2 \cdots x_{n-1}. \end{cases}$$

Multiplying the ith row by x_i, we find $x_i^2 = n\lambda$ for all i, hence $x_1^2 = x_2^2 = \cdots = x_n^2$, thus $x_1 = 1$ for all i. Finally, $\lambda = 1/n$.

The inequality between geometric and arithmetic means follows at once. In fact, if $x = (x_1, x_2, \ldots, x_n) \in \mathbb{R}^n$ and $x_1 x_2 \ldots x_n = k^n$, then $x/k \in S$, hence

$$\frac{1}{n} \sum_{i=1}^{n} \frac{x_i}{k} \ge 1,$$

that yields

$$\frac{1}{n} \sum_{i=1}^{n} x_i \ge k = \sqrt[n]{x_1 x_2 \cdots x_n}.$$

5.73 Hadamard's inequality. Let $\mathbf{A} \in M_{n,n}(\mathbb{R})$ and let $A_1, A_2, \ldots A_n$ be its columns. *The Hadamard inequality*

$$(\det \mathbf{A})^2 \le \prod_{i=1}^{n} |A_i|^2 \tag{5.27}$$

holds with equality if and only if the columns of $\mathbf{A}$ are orthogonal.

The following proof uses only linear algebra and the inequality between geometric and arithmetic means.

First, suppose that $\mathbf{A}$ is symmetric with nonnegative eigenvalues. $\lambda_1, \lambda_2, \lambda_n$ so that $\det \mathbf{A} = \lambda_1 \lambda_2 \cdots \lambda_n$ and $\operatorname{tr} \mathbf{A} = \sum_{i=1}^{n} \lambda_i$. From the inequality between geometric and arithmetic means we get

$$\det \mathbf{A} = \lambda_1 \lambda_2 \cdots \lambda_n \leq \left(\frac{1}{n} \sum_{i=1}^{n} \lambda_i \right)^n = \left(\frac{\operatorname{tr} \mathbf{A}}{n} \right)^n \tag{5.28}$$

with equality if and only if $\mathbf{A} = \lambda \operatorname{Id}$, $\lambda \geq 0$.

Now, we observe that (5.27) is trivial if one of the columns is zero, otherwise, we can assume that each column has length 1, $|A_i| = 1$. In this case it suffices to prove that

$$(\det \mathbf{A})^2 \leq 1,$$

with equality if and only if $A_i \bullet A_j = \delta_{ij}$. Since $\mathbf{A}^T \mathbf{A}$ is symmetric, with nonnegative eigenvalues, we get from (5.28)

$$(\det \mathbf{A})^2 = \det \mathbf{A}^T \mathbf{A} \leq \left(\frac{\operatorname{tr} \mathbf{A}^T \mathbf{A}}{n} \right)^n = 1$$

with equality if and only if $\mathbf{A}^T \mathbf{A} = \operatorname{Id}$, i.e., $A_i \bullet A_j = \delta_{ij}$.

An alternative proof of Hadamard's inequality can be done using the multiplier rule. It suffices to show that the function

$$f : M_{n,n}(\mathbb{R}) \to \mathbb{R}_+, \qquad f(\mathbf{A}) = \det \mathbf{A},$$

constrained to

$$\Sigma := \left\{ \mathbf{A} \in M_{n,n} \,\middle|\, |A_i| = 1 \; \forall i = 1, n \right\},$$

A_i being the columns of $\mathbf{A}$, has a constrained maximum point, and all such constrained maximum points $\mathbf{X}$ satisfy $\mathbf{X}^T \mathbf{X} = \operatorname{Id}$.

Since Σ is compact and f is continuous, both maximizers and minimizers exist in Σ. Σ is defined by the equations

$$\begin{cases} a_{11}^2 + a_{12}^2 + \cdots + a_{1n}^2 = 1, \\ a_{21}^2 + a_{22}^2 + \cdots + a_{2n}^2 = 1, \\ \cdots \\ a_{n1}^2 + a_{n2}^2 + \cdots + a_{nn}^2 = 1. \end{cases}$$

Its Jacobian matrix is the $n \times n^2$-matrix given by

$$2 \begin{pmatrix} a_{11} & \cdots & a_{1n} & 0 & \cdots & \cdots & \cdots & \cdots & 0 \\ 0 & \cdots & 0 & a_{21} & \cdots & a_{2n} & 0 & \cdots & 0 \\ \vdots & \vdots & \vdots & \vdots & \vdots & \vdots & \vdots & \vdots & \vdots \\ 0 & \cdots & \cdots & \cdots & \cdots & 0 & a_{n1} & \cdots & a_{nn} \end{pmatrix}.$$

Thus it has maximal rank n since its column vectors are nonzero. Consequently, at points $\mathbf{X} = [x_{ij}]$ of constrained maximum or minimum for f we have

$$\frac{\partial \det \mathbf{X}}{\partial x_{ij}} = \sum_{h=1}^{n} \lambda_h x_{ij}$$

for some $(\lambda_1, \lambda_2, \ldots, \lambda_n) \in \mathbb{R}^n$. Since

$$\mathbf{X} \operatorname{cof}(\mathbf{X}) = \det \mathbf{X} \operatorname{Id}, \tag{5.29}$$

we infer $\frac{\partial \det \mathbf{X}}{\partial x_{ij}} = \text{cof}\,(\mathbf{X})_{ji}$, hence the stationarity equations become

$$\begin{cases} \text{cof}\,(\mathbf{X})_{ji} = 2\lambda_i x_{ij}, & \text{if } i, j = 1, \ldots, n, \\ (\mathbf{X}^T\mathbf{X})_{ii} = \sum_{j=1}^n x_{ji}x_{ji} = 1 \end{cases}$$

for some $(\lambda_1, \lambda_2, \ldots, \lambda_n)$. Multiplying the first equation for x_{kj}, $k \neq i$, and summing on j, we find, using again (5.29),

$$\begin{cases} 0 = (\mathbf{X}\text{cof}\,(\mathbf{X}))_{ki} = 2\lambda_i \sum_{j=1}^n x_{kj}x_{ij} = 2\lambda_i(\mathbf{X}\mathbf{X}^T)_{ki}, \\ (\mathbf{X}^T\mathbf{X})_{ii} = 1. \end{cases}$$

Now, we claim that the multipliers are nonzero. In fact, if one of the λ_i's is zero, then $\text{cof}\,(\mathbf{X})_{ji} = 0$ for all j, hence $\det \mathbf{X} = 0$ by (5.29): a contradiction, if $\mathbf{X}$ is a maximum or a minimum point of $\mathbf{A} \to \det \mathbf{A}$ restricted to Σ.

The stationarity equations then say that $\mathbf{X}^T\mathbf{X} = \text{Id}$, and we conclude that, if $\mathbf{X}$ is a maximizer or a minimizer for $\mathbf{A} \to \det \mathbf{A}$ constrained to Σ, then $\mathbf{X}^T\mathbf{X} = \text{Id}$. For these matrices we finally have

$$(\det \mathbf{X})^2 = \det \mathbf{X}^T\mathbf{X} = 1$$

hence $|\det \mathbf{A}| \leq |\det \mathbf{X}| = 1 \; \forall \mathbf{A} \in \Sigma$, with equality if and only if $\mathbf{A}^T\mathbf{A} = \text{Id}$.

5.74 Isoperimetric inequality. As a consequence of Hadamard's formula, using again the geometric-arithmetic inequality,

$$(\det \mathbf{A})^2 \leq \prod_{i=1}^n |A_i|^2 \leq \Big(\frac{1}{n}\sum_{i=1}^n |A_i|^2\Big)^n = n^{-n}||\mathbf{A}||_2^{2n}, \qquad ||\mathbf{A}||_2^2 := \sum_{ij}\mathbf{A}_{ij}^2,$$

i.e, the *isoperimetric inequality for matrices*

$$|\det \mathbf{A}| \leq n^{-n/2}||\mathbf{A}||_2,$$

with equality if and only if $\mathbf{A}^T\mathbf{A} = \text{Id}$.

5.3.8 Lyapunov–Schmidt procedure

We can make use of the implicit function theorem for the study of nonlinear equations

$$\phi(x) = 0$$

in a neighborhood of a solution x_0 also when the linear tangent map of ϕ is not of maximal rank at x_0.

Let $\phi : \Omega \subset \mathbb{R}^n \to \mathbb{R}^q$ be a given map of class C^1. We assume that we can split the coordinates of the domain space $\mathbb{R}^n$ and the components of ϕ into two groups, respectively $x := (x^1, \ldots, x^r)$, $y := (x^{r+1}, \ldots, x^n)$ and $\phi^{(1)} := (\phi^1, \ldots, \phi^{q-m})$, $\phi^{(2)} := (\phi^{q-m+1}, \ldots, \phi^q)$, in such a way that $r = n - m$ and

$$\phi(x_0, y_0) = 0, \qquad \det \frac{\partial \phi^{(2)}}{\partial y}(x_0, y_0) \neq 0.$$

Though $\mathbf{D}\phi$ is not of maximal rank if $q > m$, we can apply Theorem 5.22 to $\phi^{(2)} : \Omega \to \mathbb{R}^m$ that is of maximal rank to find the following formulation of the implicit function theorem.

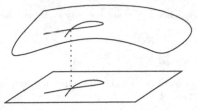

Figure 5.18. The level set of ϕ is a graph over the level set of g.

5.75 Theorem (Implicit function theorem, IV). *Let* $\phi : \Omega \subset \mathbb{R}^r \times \mathbb{R}^m \to \mathbb{R}^q$ *be of class* C^k, $k \geq 1$, *where* $\Omega \subset \mathbb{R}^{r+m}$ *is open and* $q > m$. *Set*

$$\phi^{(1)} := (\phi^1, \dots, \phi^{q-m}), \qquad \phi^{(2)} := (\phi^{q-m+1}, \dots, \phi^q).$$

If at $(x_0, y_0) \in \Omega \subset \mathbb{R}^r \times \mathbb{R}^m$ *we have*

$$\phi(x_0, y_0) = 0, \qquad \det \frac{\partial \phi^{(2)}}{\partial y}(x_0, y_0) \neq 0,$$

then there exist an open neighborhood U *of* x_0, *an open set* $W \subset \mathbb{R}^r \times \mathbb{R}^m$, *and a map* $\varphi : U \to \mathbb{R}^m$ *that is open and of class* $C^k(U)$ *such that*

$$\begin{cases} (x, y) \in W, \\ \phi(x, y) = 0 \end{cases} \quad \text{if and only if} \quad \begin{cases} x \in U, \\ \phi^{(1)}(x, \varphi(x)) = 0, \\ y = \varphi(x). \end{cases}$$

In other words, if $\Gamma = \{(x, y) \in \Omega \,|\, \phi(x, y) = 0\}$ is the zero level set of ϕ, then

$$\Gamma \cap W = \Big\{ x \in U \,\Big|\, \phi^{(1)}(x, \varphi(x)) = 0, y = \varphi(x) \Big\}$$

i.e., $\Gamma \cap W$ *is a graph over the zero level set* Δ,

$$\Delta := \Big\{ x \in U \,\Big|\, \phi^{(1)}(x, \varphi(x)) = 0 \Big\},$$

of the function $\psi(x) := \phi^{(1)}(x, \varphi(x))$. This way, the analysis of the level set $\Gamma \subset \mathbb{R}^{r+m}$ is reduced to the analysis of the level set $\Delta \subset \mathbb{R}^r$ defined with less equations and less unknowns, thus potentially easier. Notice that $\Gamma \cap W$ and Δ are diffeomorphic.

5.76 Remark. Notice that the number of equations and independent variables of $\psi := \phi^{(1)}(x, \varphi(x))$ is smaller and at most m. Therefore, the best we can do is choose coordinates so that $m = \operatorname{Rank} \mathbf{D}\phi(x_0)$. In this case the residual implicit equation

$$\phi^{(1)}(x, \varphi(x)) = 0, \qquad x \in U \subset \mathbb{R}^r, \quad r := \dim \ker \mathbf{D}\phi(x_0, y_0)$$

is called the *bifurcation equation* of the level $\phi(x, y) = 0$. Of course, when $\mathbf{D}\phi(x_0, y_0)$ has maximal rank, we have no bifurcation equation, and Theorem 5.75 reduces to Theorem 5.22.

Theorem 5.75 is the finite dimensional version in coordinates of the so-called Lyapunov–Schmidt reduction procedure that has its natural context in Banach spaces. First, we recall that we can decompose the domain and the range of a linear map $L : X \to Y$ between vector spaces of finite dimension into supplementary spaces

$$X := \ker L \oplus S, \qquad Y := \operatorname{Im} L \oplus T$$

so that all $x \in X$, $y \in Y$ uniquely decompose as

$$x = x_1 + x_2, \; x_1 \in \ker L, \; x_2 \in S,$$
$$y = y_1 + y_2, \; y_1 \in \operatorname{Im} L, \; y_2 \in T.$$

and the map $L_{|S} : S \to \operatorname{Im} L$ is one-to-one and onto.

If we choose coordinates in X in such a way that the first group describes $\ker L$ and the remaining y the supplementing space S, and we choose the coordinates in Y in such a way that the first q describe $\operatorname{Im} L$, then $q = \dim S$ and $\det \frac{\partial L}{\partial y} \neq 0$. Moreover, the previous decomposition identifies $L_{|S}$ as the *invertible factor* of L.

Secondly, we recall that the previous construction can be done for every linear continuous map $L : X \to Y$ between Banach spaces X and Y provided L *has closed range* (this is always true if Y is finite dimensional, see [GM3]). The implicit function theorem takes then the following form.

5.77 Theorem (Lyapunov–Schmidt procedure). *Let $\phi : X \to Y$ be a map of class C^k, $k \geq 1$ with $\phi(0) = 0$ between the Banach spaces X and Y. Suppose that the Frechét differential at 0 of $L := \phi'(0)$ has closed image. Then $X = \ker L \oplus S$, i.e., every $x \in X$ uniquely decomposes as $x = x_1 + x_2$, $x_1 \in \ker L$, $x_2 \in S$; moreover, there are two open sets $U, W \in X$ and two maps $\varphi : U \cap \ker L \to X$ and $\psi : U \cap \ker L \to Y$ such that*

$$\begin{cases} x \in W, \\ \phi(x) = 0 \end{cases} \qquad \text{if and only if} \qquad \begin{cases} x = x_1 + x_2, x_1 \in U \cap \ker L, \\ \psi(x_1) = 0, \\ x_2 = \varphi(x_1). \end{cases}$$

$$(5.30)$$

In other words, for

$$\Gamma = \Big\{ x \in X \, \Big| \, \phi(x) = 0 \Big\},$$

the piece $\Gamma \cap W$ of the level set Γ is the graph of the C^k-map φ over the level set $\{ x_1 \in \ker L \mid \psi(x_1) = 0 \}$. The equation $\psi(x_1) = 0$, $x_1 \in U \cap \ker L$, is called the *bifurcation equation* of $\Gamma \cap W$.

Since $\Gamma \cap W$ and $\Delta = \{ x \in U \cap \ker L \mid \psi(x) = 0 \}$ are diffeomorphic, the invariants under diffeomorphisms of $\Gamma \cap W$ are described by the invariants under diffeomorphisms of $\Delta \subset \ker L$.

Proof. Step 1. As already stated, since the Fréchet differential of ϕ at 0 has closed range, there exist closed subspaces $S \subset X$ and $T \subset Y$ supplementing respectively, $\ker L$ and $\operatorname{Im} L$,

$$X := \ker L \oplus S, \qquad Y := \operatorname{Im} L \oplus T,$$

so that every $x \in X$ and $y \in Y$ uniquely decompose as

$$x = x_1 + x_2, \ x_1 \in \ker L, \ x_2 \in S,$$
$$y = y_1 + y_2, \ y_1 \in \operatorname{Im} L, \ y_2 \in T$$

and the restriction $L_{|S}$ is one-to-one, onto, continuous as map form S to $\operatorname{Im} L$, and its inverse $(L_{|S})^{-1} : \operatorname{Im} L \to S$ is continuous by Banach's theorem, see [GM3]. Now we extend $(L_{|S})^{-1}$ to a continuous linear map $M : Y \to S$ by $M(y_1 + y_2) := L^{-1}(y_1)$. By construction $\ker M = T$ and $\operatorname{Im} M = S$.

Now we repeat the proof of the implicit function theorem. Let $F : X \to X$ be the map defined by

$$F(x) = x_1 + M\phi(x).$$

For $v \in X$ we have

$$F'(0)(v) = v_1 + ML(v) = v_1 + ML(v_1 + v_2) = x_1 + ML(v_2) = v_1 + v_2 = v.$$

Then, the local invertibility theorem implies that F is locally invertible near 0. There exists an open neighborhood W of 0 such that $F_{|W}$ is a diffeomorphism with inverse $G : U \to W$, $U = F(W)$ of class C^k. Finally, we set for $x_1 \in U \cap \ker L$

$$\varphi(x_1) := G(x_1)_2, \qquad \psi(x_1) = \phi(G(x_1))_2. \qquad (5.31)$$

Step 2. We have

$$G(x_1)_1 = x_1 \qquad \text{and} \qquad x_1 + \varphi(x_1) = G(x_1) \qquad (5.32)$$

for all $x_1 \in U \cap \ker L$. In fact, for $x_1 \in U$, we have $G(x_1) \in W$ and $x_1 = F(G(x_1)) = G(x_1) + M\phi(G(x_1))$. Since the image of M is S, we have $(My)_1 = 0 \ \forall y$, thus

$$x_1 = (x_1)_1 = G(x_1)_1 \qquad \text{hence} \qquad G(x_1) = G(x_1)_1 + G(x_1)_2 = x_1 + \varphi(x_1).$$

Step 3. Finally, let us prove (5.30). If $x \in W$ and $\phi(x) = 0$, then

$$x = G(F(x)) = G(x_1 + M\phi(x)) = G(x_1),$$

in particular, $x_1 \in U$. From the first equality of (5.32) we infer $x_1 \in \ker L$, and from the second one of (5.32) that $x_1 + x_2 = x = G(x_1) = x_1 + \varphi(x_1)$, i.e., $x_2 = \varphi(x_1)$. Moreover,

$$\psi(x_1) = \phi(G(x_1))_2 = \phi(x)_2 = 0_2 = 0.$$

Conversely, from

$$\begin{cases} x = x_1 + x_2, \ x \in U \cap \ker L, \\ \psi(x_1) = 0, \\ x_2 = \varphi(x_1) \end{cases}$$

and the second equality of (5.32) we get $x = x_1 + x_2 = x_1 + \varphi(x_1) = G(x_1)$. In particular, $x \in G(U) = W$. Also $F(x) = x_1 + M\phi(x) = x_1$, thus $M\phi(x) = 0$ and, M being invertible, $\phi(x)_1 = 0$. Since $\phi(x)_2 = \phi(G(x_1))_2 = \psi(x_1) = 0$, we conclude

$$\phi(x) = \phi(x)_1 + \phi(x)_2 = 0.$$

$\square$

5.78 Remark. We conclude this section with some remarks. We refer to the notation in the statement and the proof of Theorem 5.77.

(i) Suppose $\phi'(0)$ is surjective. We then have $T = \{0\}$, and the bifurca-
     tion equation $\psi(x_1) = 0$ holds for every $x_1 \in \ker L$ by definition of
     ψ. Therefore, in this case, the level set $\{x \in W \mid \phi(x) = 0\}$ is a graph
     over $\ker L$.

(ii) An interesting case in which the Lyapunov–Schmidt procedure ap-
     plies is that of maps whose Fréchet differential $L = \phi'(x_0)$ is a *Fred-
     holm operator*. Recall that L is a Fredholm operator if $\ker L$ and
     the supplementary space to $\operatorname{Im} L$ are both finite-dimensional; Fred-
     holm operators have closed range. In this case the set of solutions
     of $\phi(x) = 0$ is, near x_0, a graph over a level set contained in $\ker L$,
     hence of finite dimension. This way, the study of a large class of
     infinite-dimensional equations transforms into the study of a system
     of finitely many equations in a finite number of unknowns.

(iii) A special case arises when $X = Y = H$ is a Hilbert space and $L :=$
      $\phi'(0)$ is a compact perturbation of the identity, see [GM3], as L has
      closed range because of the alternative theorem. In this case, it is
      convenient to choose $S := \ker L^{\perp}$, $T := \operatorname{Im} L^{\perp}$ and to repeat the proof
      of the Lyapunov–Schmidt theorem by choosing as M the adjoint L^*
      of L instead of $L_{|S}^{-1}$, since $\ker L^* = \operatorname{Im} L^{\perp}$, $\ker L = \operatorname{Im} L^{*\perp}$ and

$$F(x) := x_1 + L^*\phi(x)$$

is also a local diffeomorphism. This last claim deserves an expla-
nation. The function $L^*L : \ker L^{\perp} \to \ker L^{\perp}$ is an isomorphism of
Hilbert spaces by the alternative theorem. It follows that $F'(0)(v) =$
$v_1 + L^*L(v) = v_1 + L^*L(v_2)$ is an isomorphism of H onto itself and
$F(x) := x_1 + L^*\phi(x)$ is a local diffeomorphism because of the local
invertibility theorem.

(iv) Since the inverse G of $F(x) = x_1 + M\phi(x)$ is obtained via local
     invertibility, we can set for G a scheme of successive approximations
     that allow us to work in quite an explicit way on the bifurcation
     equation $\psi(x_1) = \phi(G(x_1))_2 = 0$.

(v) The critical points of a map $V : \Omega \subset \mathbb{R}^n \to \mathbb{R}$ are the solutions
     of the system of n equations $\phi(x) := \mathbf{D}V(x) = 0$ in n unkowns;
     assume 0 is one of the critical points. We have $\mathbf{D}\phi(0) = \mathbf{D}^2V(0)$ and
     the bifurcation equation $\psi(x_1) = [\mathbf{D}V(G(x_1))]_2 = 0$ is defined on
     $\ker \mathbf{D}\phi(0)$, i.e., on the null space of $\mathbf{D}^2V(0)$.

5.3.9 Maps with locally constant rank and functional dependence

Suppose that the $n \times n$ matrix of the linear system

$$\mathbf{A}x = 0$$

has rank smaller than n. Then some of the equations are linear combinations of the others and therefore they can be eliminated as irrelevant.

In the spirit of the implicit function theorem, a similar result holds for nonlinear equations $\phi(x) = 0$ when $\mathbf{D}\phi$ has constant rank.

5.79 Theorem. *Let $\phi : \Omega \subset \mathbb{R}^r \times \mathbb{R}^m \to \mathbb{R}^q$ be a function of class C^k, $k \geq 1$ where Ω is open and $q > m$. Set $\phi := (\phi^{(1)}, \phi^{(2)})$ where*

$$\phi^{(1)} := (\phi^1, \dots, \phi^{q-m}), \qquad \phi^{(2)} := (\phi^{q-m+1}, \dots, \phi^q)$$

and suppose that at $(x_0, y_0) \in \Omega \subset \mathbb{R}^r \times \mathbb{R}^m$ we have

$$\phi(x_0, y_0) = 0, \qquad \text{Rank}\, \frac{\partial \phi^{(2)}}{\partial y}(x_0, y_0) = m,$$

and, moreover, that $\text{Rank}\, \mathbf{D}\phi(x, y)$ is m in all points of a neighborhood of (x_0, y_0). Then there exist an open neighborhood W of $(x_0, y_0) \in \mathbb{R}^{r+m}$, a ball $B(0, \delta) \subset \mathbb{R}^m$, an open set $Z \subset \mathbb{R}^m$, and a map $k : Z \to \mathbb{R}^{q-m}$ of class C^k such that $\phi(W)$ is the graph of k over Z,

$$\phi(W) = \Big\{ (u, v) \in \mathbb{R}^q \,\Big|\, u \in \mathbb{R}^{q-m},\, v \in Z,\, u = k(v) \Big\}. \tag{5.33}$$

In particular,

$$\phi^{(1)}(x, y) = k(\phi^{(2)}(x, y)) \qquad \forall (x, y) \in W. \tag{5.34}$$

In particular, (5.33) states that $\phi(W)$ is an m-submanifold of $\mathbb{R}^q$, and (5.34) states the *functional dependence* of the first components $\phi^{(1)}$ from the second $\phi^{(2)}$ according to the following.

5.80 Definition. *Let $\phi^1, \dots, \phi^q$ be q functions of class C^1 defined in the open set $\Omega \subset \mathbb{R}^n$, and let $\phi : \Omega \to \mathbb{R}^q$, $\phi := (\phi^1, \dots, \phi^q)$. We say that $\phi^1, \dots, \phi^q$ are* functionally dependent *if there exists a submersion $F : \Delta \to \mathbb{R}^p$, where Δ is open, $\phi(\Omega) \subset \Delta \subset \mathbb{R}^q$, and $p < q$, such that*

$$F(\phi^1(x), \phi^2(x), \dots, \phi^q(x)) = 0 \qquad \forall x \in \Omega.$$

Proof of Theorem 5.79. Step 1. We repeat once again the proof of the implicit function theorem. The function $f(x, y) := (x, \phi^{(2)}(x, y))$ is a local diffeomorphism near (x_0, y_0) since $\det \mathbf{D}f(x_0, y_0) \neq 0$. Therefore, there exist an open connected neighborhood U of x_0 and a ball $B(0, \delta) \subset \mathbb{R}^m$ such that for $W := f^{-1}(U \times B(0, \delta))$, the map $f_{|W}$ is invertible with inverse $g : U \times B(0, \delta) \to W$ of class C^k. The map $\varphi : U \times B(0, \delta) \to \mathbb{R}^m$ defined by $g(x, c) =: (x, \varphi(x, c))$, is then open, of class C^k, and

$$\begin{cases} (x, y) \in W, \\ c \in B(0, \delta), \\ \phi^{(2)}(x, y) = c \end{cases} \quad \text{if and only if} \quad \begin{cases} x \in U, \\ c \in B(0, \delta), \\ y = \varphi(x, c). \end{cases}$$

In particular, $\phi^{(2)}(x, \varphi(x, c)) = c\ \forall (x, c) \in U \times B(0, \delta)$. Differentiating in x, we get

$$\mathbf{D}\phi^{(2)}(x, \varphi(x, c))\mathbf{T}(x, c) = 0 \qquad \forall x \in U \tag{5.35}$$

where

$$\mathbf{T}(x) := \begin{pmatrix} \text{Id} \\ \frac{\partial \varphi}{\partial x}(x,c) \end{pmatrix}.$$

Step 2. Since Rank $\mathbf{D}\phi$ is constantly m near x_0 and det $\frac{\partial \phi^{(2)}}{\partial y}(x,y) \neq 0$ near (x_0, y_0), we may assume, possibly with smaller U and $B(0, \delta)$, that the rows of $\mathbf{D}\phi^{(1)}(x,y)$ linearly depend on the rows of $\mathbf{D}\phi^{(2)}(x,y)$ at every point of W. Consequently, there is $\mathbf{R}(x,y) \in M_{m-r,m}$ such that

$$\mathbf{D}\phi^{(1)}(x, \varphi(x,c)) = \mathbf{R}(x, \varphi(x,c))\, \mathbf{D}\phi^{(2)}(x, \varphi(x,c)), \qquad \forall x \in U, \ \forall c \in B(0, \delta). \quad (5.36)$$

By differentiating the function $\psi(x,c) := \phi^{(1)}(x, \varphi(x,c))$ with respect to x, we infer from (5.36) and (5.35)

$$\mathbf{D}\psi(x) = \mathbf{D}\phi^{(1)}(x, \varphi(x,c))\, \mathbf{T}(x,c) = \mathbf{R}(x, \varphi(x,c))\, \mathbf{D}\phi^{(2)}(x, \varphi(x,c))\, \mathbf{T}(x,c) = 0.$$

Therefore ψ is constant in x,

$$\psi(x,c) = k(c) = \phi^{(2)}(x, \varphi(x,c)) \qquad \forall (x,c) \in U \times B(0, \delta),$$

consequently, if $Z := \phi^{(2)}(W) \subset \mathbb{R}^m$, we have that Z is open, k is of class $C^k(Z)$, and

$$\phi^{(1)}(x, \varphi(x,c)) = k\Big(\phi^{(2)}(x, \varphi(x,c))\Big) \qquad \forall (x,c) \in U \times B(0, \delta),$$

or, equivalently, $\phi^{(1)}(z) = k\Big(\phi^{(2)}(z)\Big) \ \forall z \in W$. $\qquad\qquad\square$

5.4 Curvature of Curves and Surfaces

5.4.1 Curvature of a curve in $\mathbb{R}^n$

Let $\gamma : [0, L] \to \mathbb{R}^n$ be a curve of class C^2 parameterized by its arc length so that $|\gamma'(s)| = 1$. The unit vector $\vec{t}(s) := \gamma'(s)$ is tangent to the trajectory of γ at s, and the acceleration vector, $\gamma''(s)$, i.e., the variation of the tangent unit vector $\vec{t}(s)$ to γ, is perpendicular to $\vec{t}$ since

$$0 = \frac{d}{ds}|\gamma'(s)|^2 = 2(\gamma'(s)|\gamma''(s)).$$

The vector

$$\vec{k}(s) := \frac{d\vec{t}}{ds}(s) = \gamma''(s)$$

is also called the *curvature vector* of γ. When $\vec{k}(s)$ is nonzero, the vector

$$\vec{n}(s) := \frac{\vec{k}(s)}{|\vec{k}(s)|} = \frac{\gamma''(s)}{|\gamma''(s)|}$$

is called the *principal normal* to γ at s, the nonnegative number $|\vec{k}(s)|$ is called the *scalar curvature* of γ, its inverse, $\rho(s) := 1/|\vec{k}(s)|$ is called the *radius of curvature* of γ at s, and, finally, the *circle* of center $\gamma(s)+\rho(s)\vec{n}(s)$ and radius $\rho(s)$ in the plane generated by $(\vec{t}(s), \vec{n}(s))$ is called the *osculating circle* to γ at s.

5.81 ¶. Prove that a circle of radius R in $\mathbb{R}^2$ has $1/R$ and R respectively, as curvature and curvature radius.   •

Writing Taylor's formula $\gamma(s+h) = \gamma(s) + \gamma'(s)h + \gamma''(s)h^2/2 + o(h^2)$ as

$$\gamma(s+h) = \gamma(s) + h\,\vec{t}(s) + \frac{h^2}{2}\vec{k}(s) + o(|h|^2) \qquad \text{as } h \to 0;$$

we see that the curvature vector points in the direction in which the curve turns, and its modulus is a measure (up to second order) of how regularly the curve deflects from its tangent line in the plane spanned by $\vec{t}(s)$ and $\vec{n}(s)$.

5.82 ¶. Prove that Taylor's developments of γ and its osculating circle (both parameterized by the arc length) agree up to second terms included.

Let $\gamma : [a,b] \to \mathbb{R}^n$ be a regular and *simple* curve of class C^1 with $\gamma(a) \neq \gamma(b)$. At each of its points p there are two unit tangent vectors $\vec{t}(s)$ and $-\vec{t}(s)$ corresponding to the parameterization of γ by the arc length and to the opposite reparameterization respectively, $\delta : [0, L] \to \mathbb{R}^n$, $\delta(s) := \int_0^s |\gamma'(t)|\,dt$, and $\delta_1 : [0, L] \to \mathbb{R}^n$, $\delta_1(s) = \delta(L - s)$. In both cases the curvature vector at $p = \delta(s)$ depends only on the trajectory of the curve (independently of its parameterization). Therefore, we may also refer to the curvature vector, the scalar curvature, and the osculating circle of γ at a point $p = \delta(s)$ of its trajectory and write

$$\vec{k}(p) := \vec{k}(s) \qquad \text{if} \qquad p = \gamma(s).$$

5.83 Curvature for a parameterized curve. Let $\gamma : [a,b] \to \mathbb{R}^n$ be a curve of class C^2 parameterized by an arbitrary parameter t. Then,

$$\vec{t}(t) := \frac{\gamma'(t)}{|\gamma'(t)|}$$

hence

$$\vec{k}(t) := \frac{d\vec{t}}{ds} = \frac{d\vec{t}}{dt}\frac{dt}{ds} = \frac{d}{dt}\left(\frac{\gamma'(t)}{|\gamma'(t)|}\right)\frac{1}{|\gamma'(t)|} = \frac{1}{|\gamma'|^2}\left(\gamma'' - \frac{\gamma''\bullet\gamma'}{|\gamma'|^2}\gamma'\right).$$

By denoting with $[v]^N$, $v \in \mathbb{R}^n$, the orthogonal projection of v into the normal space to $\vec{t}$,

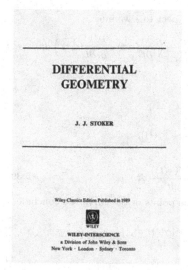

Figure 5.19. Frontispieces of two books on curves and surfaces.

$$[v]^N := v - (v \bullet \vec{t}) \, \vec{t},$$

we also have

$$\vec{k}(t) = \frac{1}{|\gamma'(t)|^2} \left[\gamma''(t) \right]^N. \tag{5.37}$$

The curvature is strongly related to the *first variation* of the length with respect to deformations.

5.84 Proposition (First variation of the length). *Let* $\gamma : [a, b] \to \mathbb{R}^n$ *be a regular curve of class* C^2 *and let* $\phi : [-1, 1] \times \mathbb{R}^n \to \mathbb{R}^n$ *be a deformation of* γ *for which the end-points remain fixed, i.e., a map of class* C^2 *with* $\phi(0, x) = x$ $\forall x$ *and*

$$\phi(\epsilon, \gamma(a)) = \gamma(a), \qquad \phi(\epsilon, \gamma(b)) = \gamma(b) \qquad \forall \epsilon.$$

If L_ϵ *denotes the length of the curve* $\gamma_\epsilon(t) := \phi(\epsilon, t)$, *then*

$$\frac{dL_\epsilon}{d\epsilon}\bigg|_{\epsilon=0} = - \int_\gamma \vec{k} \bullet V \, d\mathcal{H}^1 \tag{5.38}$$

where $V(x) := \frac{\partial \phi}{\partial \epsilon}(0, x)$ *is the velocity of the flow* $\phi(\epsilon, x)$ *at* $\epsilon = 0$.

Proof. Without loss of generality, we assume that γ is parameterized by the arc length, $\gamma : [0, L] \to \mathbb{R}^n$, $L =$ length of γ and $|\gamma'(t)| = 1$, so that $\vec{t}(t) = \gamma'(t)$ and $\vec{k}(t) := \gamma''(t)$. We set

$$h(\epsilon, t) := \gamma_\epsilon(t) = \phi(\epsilon, \gamma(t))$$

and notice $h \in C^2(] - 1, 1[\times [a, b])$. For every ϵ we have

$$L_\epsilon = \int_0^L \left| \frac{\partial h}{\partial t}(\epsilon, t) \right| dt$$

and, differentiating under the integral sign, with respect to ϵ we infer

$$\frac{dL_\epsilon}{d\epsilon}(\epsilon) = \int_0^L \frac{\partial}{\partial\epsilon}\left(\left|\frac{\partial h}{\partial t}(\epsilon,t)\right|\right)dt = \int_0^L \frac{1}{\left|\frac{\partial h}{\partial t}(\epsilon,t)\right|}\left(\frac{\partial h}{\partial t}(\epsilon,t)\bullet\frac{\partial^2 h}{\partial\epsilon\partial t}(\epsilon,t)\right)dt.$$

Since

$$\frac{\partial h}{\partial t}(0,t) = \gamma'(t), \qquad |\gamma'(t)| = 1,$$

$$\frac{\partial h}{\partial\epsilon}(0,t) = \frac{\partial\phi}{\partial\epsilon}(0,\gamma(t)) = V(\gamma(t)),$$

we find

$$\frac{dL_\epsilon}{d\epsilon}\Big|_{\epsilon=0} = \int_0^L \gamma'(t)\bullet\frac{dV(\gamma(t))}{dt}\,dt. \tag{5.39}$$

Integrating by parts, since V is zero at the extreme points of the curve, we conclude

$$\frac{dL_\epsilon}{d\epsilon}\Big|_{\epsilon=0} = -\int_0^L \gamma''(t)\bullet V(\gamma(t))\,dt,$$

i.e., (5.38). $\square$

a. Moving frame for a planar curve

5.85 Moving frame and oriented curvature. For planar curves the following alternative presentation can be useful. Let $\gamma : [0, L] \to \mathbb{R}^2$ be a simple plane curve of class C^2 parameterized by the arc length and let $\vec{t}(s)$ be its velocity vector. We choose the unit vector $\vec{n}(s)$ perpendicular to $\vec{t}(s)$ so that $(\vec{t}(s), \vec{n}(s))$ is *positively oriented*, meaning that $\det[\vec{t}(s)|\vec{n}(s)] = 1$. In coordinates, if $\vec{t}(s) = (x(s), y(s))$, set $\vec{n}(s) := (-y(s), x(s))$. Thus,

$$\vec{k}(s) = k(s)\vec{n}(s) \qquad \text{where} \qquad k(s) := \vec{k}(s)\bullet\vec{n}(s) = \gamma''(s)\bullet\vec{n}(s);$$

the sign of $k(s)$ depends on the choice of $\vec{n}$, accordingly $k(s)$ is called the *oriented curvature* of γ, and the vectors $(\vec{t}(s), \vec{n}(s))$ are called the *moving frame along γ*.

5.86 ¶. Show that $k > 0$ if γ "turns left".

5.87 ¶. Prove that the oriented curvature of the graph of $f : [a, b] \to \mathbb{R}$, $x \to (x, f(x))$, is

$$k(x) = \frac{f''(x)}{(1 + |f'(x)|^2)^{3/2}} = \left(\frac{f'}{\sqrt{1 + f'^2}}\right)'(x).$$

In particular, $k(x) \geq 0\ \forall x$ if and only if f is convex.

Formula (5.37) writes with respect to the moving frame as the *Huygens formula*

$$\gamma''(t) = (\gamma''(t)\bullet\vec{t}(t))\,\vec{t}(t) + |\gamma'(t)|^2\,k(t)\,\vec{n}(t)$$

from which we deduce

$$k(t) = \frac{1}{|\gamma'(t)|^2}\gamma''(t)\bullet\vec{n}(\gamma(t)) = \frac{\det[\gamma'(t)|\gamma''(t)]}{|\gamma'|^3} \tag{5.40}$$

or, in terms of the components $(x(t), y(t))$ of $\gamma(t)$,

$$k(t) = \frac{x'(t)y''(t) - y'(t)x''(t)}{(x'(t)^2 + y'(t)^2)^{3/2}}. \tag{5.41}$$

5.88 Serret–Frenet formulas. Let γ be a curve in $\mathbb{R}^2$ of class C^2, and let $(\vec{t}(s), \vec{n}(s))$ be its moving frame. The curvature vector of γ is a multiple of $\vec{n}$ and by the definition of oriented curvature

$$\vec{t}'(s) = \vec{k}(s) = k(s)\,\vec{n}(s).$$

On the other hand, since $\vec{n}(s)$ and $\vec{n}'(s)$ are perpendicular and $n = 2$, we have $\vec{n}'(s) = \alpha(s)\,\vec{t}(s)$. We may compute the proportionality coefficient $\alpha(s)$ from

$$0 = \frac{d}{ds}\,\vec{t}(s) \bullet \vec{n}(s) = (k(s)\vec{n}(s))\vec{n}(s) + (\vec{t}(s))\alpha(s)\vec{t}(s) = k(s) + \alpha(s),$$

hence

$$\vec{n}'(s) = -\kappa(s)\vec{t}(s).$$

In conclusion, the *moving frame* $(\vec{t}(s), \vec{n}(s))$ *along* γ and the oriented curvature $k(s)$ are related by the *Serret–Frenet formulas* for planar curves

$$\begin{cases} \vec{t}'(s) = k(s)\,\vec{n}(s), \\ \vec{n}'(s) = -k(s)\,\vec{t}(s), \end{cases} \tag{5.42}$$

which can be written as the system of first-order differential equations

$$\left[\vec{t}'(s)\,\middle|\,\vec{n}'(s)\right] = \mathbf{A}(s)\left[\vec{t}(s)\,\middle|\,\vec{n}(s)\right], \qquad \mathbf{A}(s) = \begin{pmatrix} 0 & k(s) \\ -k(s) & 0 \end{pmatrix}.$$

Integrating these equations twice, we see that the curvature vector of a curve determines the curve apart from an isometry. More interesting is the fact that the *oriented curvature* of a curve suffices to determine the curve (modulus rigid motions of the plane).

5.89 Theorem. *For any given continuous function* $h : [0, L] \to \mathbb{R}$ *there exists a curve* $\gamma : [0, L] \to \mathbb{R}^2$ *parameterized by the arc length with oriented curvature* h; *moreover,* γ *is unique modulus isometries of the plane.*

Proof. Uniqueness. Suppose $\gamma_1, \gamma_2 : [0, L] \to \mathbb{R}^2$ with respectively, $(\vec{t}_1, \vec{n}_1)$, $(\vec{t}_2, \vec{n}_2)$ as moving frames, have the same oriented curvature. Let $\mathbf{R}$ be the rotation that moves the vectors $\vec{t}_1(0)$ and $\vec{n}_1(0)$ respectively into $\vec{t}_2(0)$ and $\vec{n}_2(0)$, and let

$$\gamma_3(s) := \mathbf{R}(\gamma_2(s) - \gamma_2(0)) + \gamma_1(0), \qquad \forall s \in [0, 1]$$

be the curve (obtained by a rigid motion of γ_2) so that its moving frame agrees at $s = 0$ with the moving frame of γ_1. We claim that $\gamma_3(s) = \gamma_1(s)\ \forall s$.

It is easily seen that

$$\begin{cases} \vec{t_2}\,'(s) = k(s)\,\vec{n_2}(s), \\ \vec{n_2}\,'(s) = -k(s))\,\vec{t_2}(s) \end{cases} \quad \text{implies} \quad \begin{cases} \vec{t_3}\,'(s) = k(s)\,\vec{n_3}(s), \\ \vec{n_3}\,'(s) = -k(s))\,\vec{t_3}(s). \end{cases}$$

Therefore the matrices $\mathbf{X_1}(s) = \left[\vec{t_1}(s)|\vec{n_1}(s)\right]$ and $\mathbf{X_3}(s) = \left[\vec{t_3}(s)|\vec{n_3}(s)\right]$ both solve the Cauchy problem

$$\begin{cases} \mathbf{X}'(s) = \mathbf{A}(s)\mathbf{X}(s), \\ \mathbf{X}(0) = \mathbf{X_1}(0), \end{cases} \qquad \mathbf{A}(s) := \begin{pmatrix} 0 & h(s) \\ -h(s) & 0 \end{pmatrix}$$

hence agree for all s. In particular, $\gamma_3'(s) = \vec{t_3}(s) = \vec{t_1}(s) = \gamma_1'(s)$ $\forall s$ and, since $\gamma_3(0) = \gamma_1(0)$, we conclude $\gamma_3(s) = \gamma_1(s)$ $\forall s$.

Existence. Let $\mathbf{A} := \begin{pmatrix} 0 & h \\ -h & 0 \end{pmatrix}$. The Cauchy problem for $\mathbf{X}(t) \in M_{2,2}(\mathbb{R})$

$$\begin{cases} \mathbf{X}'(s) = \mathbf{A}(s)\mathbf{X}(s), \\ \mathbf{X}(0) = \mathrm{Id} \end{cases}$$

has a unique global solution $\mathbf{X}(s)$, $s \in [0, L]$, see [GM3]. Moreover, since $\mathbf{A}$ is antisymmetric, we have

$$(\mathbf{X}^T\mathbf{X})' = \mathbf{X}^T\mathbf{A}^T\mathbf{X} + \mathbf{X}^T\mathbf{A}\mathbf{X} = \mathbf{X}^T(\mathbf{A}^T + \mathbf{A})\mathbf{X} = 0,$$

hence $\mathbf{X}^T\mathbf{X}$ is constant. Since $\mathbf{X}(0) = \mathrm{Id}$, we conclude that the columns u, v of $\mathbf{X} = [u|v]$ are orthonormal $\det[u|v] = 1$. If we define

$$\gamma(s) := \int_0^s u(\tau)\,d\tau, \qquad s \in [0, L],$$

γ is parameterized by its arc length with (u, v) as moving frame and oriented curvature. $\square$

b. Moving frame of a curve in $\mathbb{R}^3$

We can proceed similarly for curves in $\mathbb{R}^n$. For the sake of brevity, we only deal with curves in $\mathbb{R}^3$.

5.90 Moving frame and torsion. Let $\gamma : I \to \mathbb{R}^3$ be a curve of $\mathbb{R}^3$ of class C^3 parameterized by the arc length with $\gamma'(s) \neq 0$ and $\gamma''(s) \neq 0$. The unit vectors $\vec{t}(s) = \gamma'(s)$ and $\vec{n}(s) := \gamma''(s)/|\gamma''(s)|$ are orthonormal, and we call $\gamma''(s)$ the (vector) curvature of γ at s, and denote by $k(s) := |\gamma''(s)|$ the scalar curvature so that $\gamma''(s) = k(s)\,\vec{n}(s)$.

We now choose a third unit vector $\vec{b}(s)$, which is called the *binormal vector*, perpendicular to both $\vec{t}(s)$ and $\vec{n}(s)$ such that $\det\left[\vec{t}(s)|\vec{n}(s)|\vec{b}(s)\right] = 1$, or, in terms of *cross product*,

$$\vec{b}(s) := \vec{t}(s) \times \vec{n}(s).$$

The triplet $(\vec{t}(s), \vec{n}(s), \vec{b}(s))$ is called the *moving frame* along γ and the plane generated by $\vec{t}(s)$ and $\vec{n}(s)$ is called the *osculating plane* to γ at $\gamma(s)$. Trivially,

$$\vec{n}(s) \times \vec{b}(s) = \vec{t}(s), \qquad \vec{b}(s) \times \vec{t}(s) = \vec{n}(s).$$

Since $|\vec{n}(s)| = 1$ and $\vec{n}(s)$ and $\vec{n}'(s)$ are perpendicular,

$$\vec{n}'(s) = \alpha(s)\vec{t}(s) - \tau(s)\vec{b}(s)$$

where α can be computed as

$$\alpha(s) = \vec{n}'(s) \bullet \vec{t}(s) = \frac{d}{ds}\,\vec{n}(s) \bullet \vec{t}(s) - \vec{n}(s) \bullet \vec{t}\,'(s) = -k(s)|\vec{n}(s)|^2 = -k(s),$$

while the function $\tau(s) := -\beta(s) = -\vec{n}'(s) \bullet \vec{b}(s)$ is called the *torsion* of γ at $\gamma(s)$.

5.91 Serret–Frenet formulas. Since

$$\vec{b}'(s) = (\vec{t}(s) \times \vec{n}(s))' = k(s)\vec{n}(s) \times \vec{n}(s) + \vec{t}(s) \times \vec{n}'(s)$$

$$= \beta(s)\vec{t}(s) \times \vec{b}(s) = -\beta(s)\vec{n}(s),$$

the moving frame $(\vec{t}, \vec{n}, \vec{b})$ along γ satisfies the system of Serret–Frenet ordinary differential equations

$$\begin{cases} \vec{t}\,'(s) &= -k(s)\vec{n}(s), \\ \vec{n}'(s) &= -k(s)\vec{t}(s) + \tau(s)\vec{b}(s), \\ \vec{b}'(s) &= -\tau(s)\vec{n}(s), \end{cases}$$

equivalently,

$$\left[\vec{t}\,\middle|\,\vec{n}\,\middle|\,\vec{n}\right]'(s) = \mathbf{A}(s)\left[\vec{t}\,\middle|\,\vec{n}\,\middle|\,\vec{n}\right](s), \qquad \mathbf{A}(s) = \begin{pmatrix} 0 & k(s) & 0 \\ -k(s) & 0 & \tau(s) \\ 0 & -\tau(s) & 0 \end{pmatrix}.$$

We notice that the torsion of a planar curve is null. Moreover, if $\gamma \in C^3$, since

$$\begin{cases} \gamma'(s) = \vec{t}(s), \\ \gamma''(s) = k(s)\vec{n}(s), \\ \gamma'''(s) = (k(s)\vec{n}(s))' = k'(s)\vec{n}(s) - k^2(s)\vec{t}(s) + k(s)\tau(s)\vec{b}(s) \end{cases}$$

Taylor's formula of third order writes as

$$\gamma(s+h) = \gamma(s) + \left(h - \frac{k^2}{6}h^3\right)\vec{t}(s) + \left(\frac{k(s)}{2}h^2 + \frac{k'(s)}{6}h^3\right)\vec{n}(s)$$

$$+ \frac{k(s)\tau(s)}{6}h^3\,\vec{b}(s) + o(h^3) \qquad \text{per } h \to 0.$$

Therefore we can state: $k(s)$ measures the deviation of the curve at $\gamma(s)$ (up to second-order terms) from the tangent direction $\vec{t}(s)$ in the osculating plane, while the torsion, together with the curvature, measures the deviation of the curve (up to third order) from the osculating plane.

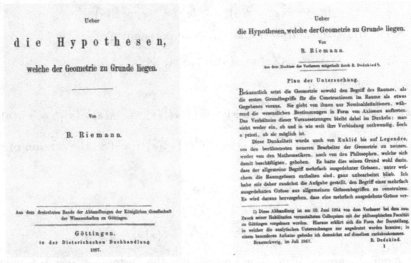

Figure 5.20. The frontispiece and the first page of a celebrated paper by G. F. Bernhard Riemann (1826–1866) on geometries.

5.92 ¶. Prove that the curvature and the torsion of a curve $\gamma : I \to \mathbb{R}^3$ parameterized by a generic parameter t are given by

$$k(t) = \frac{|\gamma'(t) \times \gamma''(t)|}{|\gamma'(t)|^3},$$

$$\tau(t) = \frac{\det \left[\gamma'(t) \,\middle|\, \gamma''(t) \,\middle|\, \gamma'''(t)\right]}{|\gamma'(t) \times \gamma''(t)|^2}.$$

The curvature and the torsion form a *complete set of invariants* for curves in $\mathbb{R}^3$ according to the following *fundamental theorem of the local theory of curves*.

5.93 Theorem. *For any couple of given continuous functions $k(s) > 0$ and $\tau(s)$, $s \in [0, L]$, there exists a curve $\gamma : [0, L] \to \mathbb{R}^3$ parameterized by the arc length with curvature $k(s)$ and torsion $\tau(s)$ at s; moreover, such a curve is unique modulus rigid motions of $\mathbb{R}^3$.*

5.94 ¶. Prove Theorem 5.89. [*Hint:* Repeat the argument for planar curves.]

5.4.2 Curvature of a submanifold of $\mathbb{R}^n$

In this section, we define the curvature of an m-submanifold M in $\mathbb{R}^n$ and discuss some basic facts related to it. All our considerations will be of local nature, therefore it is not restrictive to assume that M is an embedded submanifold, i.e., the image of a diffeomorphism $\varphi : B \to \mathbb{R}^n$, $B =$

$B(0,1) \subset \mathbb{R}^m$, $m < n$, of class C^1. We shall denote with $u^1, u^2, \ldots, u^m$ the coordinates in B and with $x^1, x^2, \ldots, x^n$ the coordinates in $\mathbb{R}^n$. As we know, the tangent space T_pM to M at $p = \varphi(u)$ is the image of $\mathbf{D}\varphi(u)$ and, since $\mathbf{D}\varphi(u)$ has maximal rank m, $\left(\frac{\partial \varphi}{\partial u^1}(u), \ldots, \frac{\partial \varphi}{\partial u^m}(u) \right)$ is a basis for $T_p(M)$. Finally, we shall denote N_pM the normal subspace to T_pM in $\mathbb{R}^n$.

a. First fundamental form

Let M be an embedded submanifold of $\mathbb{R}^n$, let $\varphi : \Omega \to M$, Ω open in $\mathbb{R}^m$, be a diffeomorphism, and let $p = \varphi(u) \in M$. The norm of $\mathbb{R}^n$ induces by restriction a quadratic form on T_pM

$$\mathrm{I}_p(a) := |a|^2 \qquad \forall a \in T_pM$$

called the *first fundamental form* of M at p. In coordinates, if $p = \varphi(u)$ and $a = \sum_{\alpha=1}^m \xi^\alpha \frac{\partial \varphi}{\partial u^\alpha}(u) \in T_pM$, then

$$\mathrm{I}_p(a) = \sum_{\alpha,\beta=1}^m g_{\alpha\beta}(p)\xi^\alpha\xi^\beta = \xi^T \mathbf{G}\xi$$

where the matrix $\mathbf{G} = (\mathbf{G}_{\alpha\beta}) := (g_{\alpha\beta})$,

$$g_{\alpha\beta}(p) := \frac{\partial \varphi}{\partial u^i}(u) \bullet \frac{\partial \varphi}{\partial u^j}(u).$$

The matrix $\mathbf{G}(u) := (g_{\alpha\beta}(u)) \in M_{m,m}(\mathbb{R})$ is called the *metric tensor* of the parameterization φ at u. The metric $\mathbf{G}(u)$ is symmetric, moreover

$$\mathbf{G}(u) = \mathbf{D}\varphi(u)^T \mathbf{D}\varphi(u).$$

Since $\mathbf{D}\varphi(u)$ is injective, all eigenvalues of $\mathbf{G}(u)$ are positive, and $\mathbf{G}(u)$ and $\mathbf{G}(u)^{1/2}$ are invertible with symmetric inverses. The entries of $\mathbf{G}(u)^{-1}$ are denoted as $\mathbf{G}^{-1} = (g^{\alpha\beta})$.

The metric and the first fundamental form appear in the calculus of the area of M and of the length of curves on M. Indeed, the area formula of Chapter 2 states that

$$\mathcal{H}^m(M) = \int_B \sqrt{g} \, d\mathcal{L}^m(u), \qquad g := \det \mathbf{G}.$$

Moreover, if $\gamma : [a,b] \to M$ is a C^1-curve in M, then

$$|\gamma'(t)|^2 = \mathrm{I}_{\gamma(t)}(\gamma'(t))$$

and

$$L(\gamma) = \int_a^b \sqrt{\mathrm{I}_{\gamma(t)}(\gamma'(t))} \, dt.$$

5.95 Orthonormal bases of the tangent plane. Let $X_1, \ldots, X_m$ be m vectors in $\mathbb{R}^m$ and for every $i = 1, \ldots, m$, let $a_i := \mathbf{D}\varphi(u)X_i$. Denote by $\mathbf{X}$ the $m \times m$ matrix $\mathbf{X} := [X_1 | X_2 | \ldots | X_m]$. Then

$$(a_i \bullet a_j)_{\mathbb{R}^n} = \mathbf{D}\varphi(u)X_i \bullet \mathbf{D}\varphi(u)X_j = (\mathbf{X}^T \mathbf{G} \mathbf{X})_{ij}.$$

Therefore the following claims are equivalent.

  (i) $(a_1, a_2, \ldots, a_m)$ is an orthonormal base in $\mathbb{R}^n$,
 (ii) $\mathbf{X}^T \mathbf{G} \mathbf{X} = \mathrm{Id}$,
(iii) $\mathbf{G}^{1/2} \mathbf{X}$ is an isometry of $\mathbb{R}^m$,
(iv) $\mathbf{X}\mathbf{X}^T = \mathbf{G}^{-1}$.

b. Second fundamental form

Let M be an embedded submanifold of $\mathbb{R}^n$, let $\varphi : \Omega \to M$, Ω open in $\mathbb{R}^m$, be a diffeomorphism, and let $p = \varphi(u) \in M$. For any vector $v \in \mathbb{R}^n$ denote by v^N the orthogonal projection of v on $N_p M$.

The *second fundamental form* of M at p is the map $\mathbb{II}_p : T_p M \to N_p M$ defined for $a = \sum_{\alpha=1}^m \xi^\alpha \frac{\partial \varphi}{\partial u^\alpha}(u) \in T_p M$ by

$$\mathbb{II}_p(a) := \sum_{\alpha,\beta=1}^m \left[\frac{\partial^2 \varphi}{\partial u^\alpha \partial u^\beta}(0) \right]^N \xi^\alpha \xi^\beta. \tag{5.43}$$

If we introduce the matrix $\Phi \in M_{m,m}(\mathbb{R}^n)$ whose entries are vectors in $\mathbb{R}^n$

$$\Phi = (\Phi_{\alpha\beta}), \qquad \Phi_{\alpha\beta} := \left[\frac{\partial^2 \varphi}{\partial u^\alpha \partial u^\beta}(0) \right]^N, \tag{5.44}$$

we shorten (5.43) as

$$\mathbb{II}_p(a) = \xi^T \Phi \xi, \qquad \xi = (\xi^1, \xi^2, \ldots, \xi^m).$$

A priori, $\mathbb{II}_p$ depends on the parameterization of M as we have used the components of a in local coordinates for its definition. However, we have the following.

5.96 Proposition. *The second fundamental form is* intrinsic *on M, i.e., it does not depend on the parameterization of M.*

Proof. Let $\varphi : B(0,1) \to M$ and $\psi : \Delta \to M$ be two diffeomorphisms with image M. Denote by $u^1, u^2, \ldots, u^m$ the coordinates in $B(0,1)$ and by $v^1, v^2, \ldots, v^m$ the coordinates in Δ. We may and do asssume that $\varphi(0) = \psi(0) = 0$. From Proposition 5.8 $\psi = \varphi \circ h$ where $h : \Delta \to B(0,1)$, $h := \varphi^{-1} \circ \psi$ is a diffeomorphism between Δ and $B(0,1)$. We infer that $\mathbf{D}\psi(v) = \mathbf{D}\phi(h(v))\mathbf{D}h(v)$ $\forall v \in \Delta$, i.e., by writing $D_\alpha f$ for $\frac{\partial f}{\partial v^\alpha}$,

$$D_\alpha \psi^\ell(v) = \sum_{i=1}^m \frac{\partial \varphi^\ell}{\partial u^i}(h(v)) D_\alpha h^i(v),$$

hence

$$D_\beta D_\alpha \psi^\ell = \sum_{i,j=1}^m \frac{\partial^2 \varphi^\ell}{\partial u^i \partial u^j} D_\alpha h^i D_\beta h^j + \sum_{i=1}^m \frac{\partial \varphi^\ell}{\partial u^i} D_\beta D_\alpha h^i.$$

Now, if $a \in T_p M$, $a = \sum_{i=1}^n \xi^i \frac{\partial \varphi}{\partial u^i} = \sum_{\alpha=1}^m \eta^\alpha D_\alpha \psi$, or, equivalently,

$$\xi^i = \sum_{\alpha=1}^m \eta^\alpha D_\alpha h^i,$$

we get

$$\sum_{\alpha,\beta=1}^m \frac{\partial^2 \psi}{\partial v^\alpha \partial v^\beta} \eta^\alpha \eta^\beta = \sum_{i,j,\alpha,\beta=1}^m \frac{\partial^2 \varphi}{\partial u^i \partial u^j} D_\alpha h^i D_\beta h^j \eta^\alpha \eta^\beta$$

$$+ \sum_{\alpha,\beta,i=1}^m \frac{\partial \varphi}{\partial u^i} D_\beta D_\alpha h^i \eta^\alpha \eta^\beta$$

$$= \sum_{i,j=1}^m \frac{\partial^2 \varphi}{\partial u^i \partial u^j} \xi^i \xi^j + \sum_{i=1}^m \frac{\partial \varphi}{\partial u^i} \Big(\sum_{\alpha,\beta=1}^m D_\beta D_\alpha h^i \eta^\alpha \eta^\beta \Big).$$

Since the last term on the right-hand side is tangent to M, the vectors

$$\sum_{\alpha,\beta=1}^m \frac{\partial^2 \psi}{\partial v^\alpha \partial v^\beta} \eta^i \eta^j \qquad \text{and} \qquad \sum_{i,j=1}^m \frac{\partial^2 \varphi}{\partial u^i \partial u^j} \xi^i \xi^j$$

have the same normal component to $T_p M$. $\square$

5.97 Remark. For $u \in B$ and h small, the length of the vector $(\varphi(u + h) - \varphi(u))^N$ is the distance of $\varphi(u+h) - \varphi(u)$ from $T_p M$, $p = \varphi(u)$. Taylor's formula yields

$$(\varphi(u + h) - \varphi(u))^N = \frac{1}{2} \sum_{i,j=1}^m \Big(\frac{\partial^2 \varphi}{\partial u^i \partial u^j}(u) \Big)^N h^i h^j + o(|h|^2)$$

$$= \frac{1}{2} \mathbb{II}_p(h) + o(|h|^2) \qquad \text{as } h \to 0.$$

c. Curvature vector

5.98 Definition. *Let* $a \in T_p M$. *The curvature vector in the direction* a *of* M *at* p *is the normal vector* $\vec{k}(a) \in N_p M$ *defined by*

$$\vec{k}(a) := \frac{\mathbb{II}_p(a)}{I_p(a)} = \frac{1}{|a|^2} \mathbb{II}_p(a)$$

in particular $\vec{k}(a) = \mathbb{II}_p(a)$ *if* $|a| = 1$.

We have $\vec{k}(\lambda a) = \vec{k}(a)$ for all $\lambda \neq 0$. Moreover, it is easily seen that $\vec{k}(a)$ is the orthogonal projection into $N_p M$ of the curvature vector of a curve $\gamma(t)$ on M with $\gamma(0) = p$ and $\gamma'(0) = a$.

We may also regard the curvature as a *variation of the normal plane*. Let $a \in T_p M$ with components $a = (a^1, a^2, \ldots, a^n) \in \mathbb{R}^n$. For every vector field $X = (X^1, X^2, \ldots, X^n)$ of class $C^1(M)$, not necessarily tangent to M, we set

$$a \cdot D_a X := \sum_{i,j=1}^{n} a^i a^j \frac{\partial X^i}{\partial x^j}(p). \tag{5.45}$$

5.99 Proposition. *Let $\nu_1, \ldots, \nu_{n-m} : M \to \mathbb{R}^n$ be $n - m$ vector fields of class $C^1(M)$ that form an orthonormal basis at $p \in M$ of $N_p M$. Then for every $a \in T_p M$*

$$\vec{k}(a) := -\sum_{\alpha=1}^{n-m} (a \cdot D_a \nu_\alpha)\,\nu_a. \tag{5.46}$$

Proof. Let $a \in T_p M$ with $|a| = 1$ and let $\gamma :] - 1, 1[\to M$ be a curve of class C^2 on M parameterized by the arc length with $\gamma(0) = p$ and $\gamma'(0) = a$. Then $\vec{k}(a) = \mathrm{II}_p(a) = \gamma''(0)^N$. On the other hand, for every $\alpha = 1, \ldots, n - m$, $\nu_\alpha(\gamma(t))$ is orthogonal to $\gamma'(t)$ for every t, $(\gamma'(t)|\nu_\alpha(\gamma(t))) = 0$. Differentiating and evaluating at 0 we find

$$(\gamma''(0)|\nu_\alpha(p)) + \sum_{i=1}^{m} a^i a^j \frac{\partial \nu_\alpha^i}{\partial x^j}(p) = 0,$$

hence

$$(\vec{k}(a)|\nu_\alpha) = (\gamma''(0)|\nu_\alpha) = -a \cdot D_a \nu_\alpha$$

which proves the claim, since $\{\nu_a(p)\}$ is an orthonormal basis of $N_p M$.   □

d. Mean curvature vector

Let $(a_1, a_2, \ldots, a_m)$ be an orthonormal basis of $T_p M$. The normal vector

$$\vec{H} := \frac{1}{m} \sum_{i=1}^{m} \vec{k}(a_i) \tag{5.47}$$

is called the *mean curvature vector* of M at p.

By definition, the mean curvature vector is independent of the choice of the parameterization on M. Moreover, we can easily prove that it does not depend on the chosen basis we use in its definition. In fact, for every $i = 1, \ldots, m$ let X_i be such that $a_i := \mathbf{D}\varphi(u)X_i \in T_p M$, and let $\mathbf{X}$ be the $m \times m$-matrix $\mathbf{X} := [X_1|X_2|\ldots|X_m]$. We have $\mathbf{X}^T\mathbf{X} = G^{-1}$ and, consequently,

$$\vec{H} = \frac{1}{m} \sum_{i=1}^{m} X_i^T \Phi X_i = \frac{1}{m} \mathrm{tr}\,(\mathbf{X}^T \Phi \mathbf{X}) = \frac{1}{m} \mathrm{tr}\,(\Phi \mathbf{X}\mathbf{X}^T)$$
$$= \frac{1}{m} \mathrm{tr}\,(\Phi \mathbf{G}^{-1}) \tag{5.48}$$

where Φ is defined in (5.44).

Another useful expression for the mean curvature vector follows from (5.46):

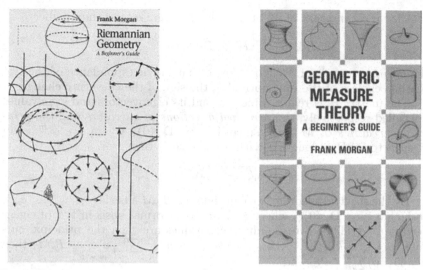

Figure 5.21. Two fascinating beginner's guides to differential geometry and geometric measure theory.

$$m\,\vec{H} := -\sum_{\alpha=1}^{n-m}\sum_{i=1}^{m}(a_i \cdot D_{a_i}\nu_\alpha)\nu_\alpha, \qquad (5.49)$$

where $(\nu_1, \nu_2, \ldots, \nu_n - m)$ are orthonormal vector fields at p that span N_pM and $(a_1, a_2, \ldots, a_m)$ are orthonormal vectors that span T_pM.

e. Curvature of surfaces of codimension one

Let M be an m-dimensional submanifold in $\mathbb{R}^{m+1}$ and, as before, let $\varphi : \Omega \to M$ be a diffeomorphism, $\Omega \subset \mathbb{R}^m$ open, $0 \in \Omega$ and $p = \varphi(0)$. Its normal space N_pM has dimension 1 and, if ν is a normal vector of norm one, the second fundamental form of M at p takes the form

$$\mathbb{I}_p(a) = (\,\mathbb{I}_p(a)\bullet\nu\,)\,\nu, \qquad a \in T_pM,$$

consequently,

$$\mathbb{I}_p(a)\bullet\nu := \sum_{\alpha,\beta=1}^{m} \mathbf{R}_{\alpha\beta}\xi^\alpha\xi^\beta = \xi^T\mathbf{R}\xi$$

where ξ is such that $a = \mathbf{D}\varphi(0)\xi$, $\xi = (\xi_1, \xi_2, \ldots, \xi_m)$, and $\mathbf{R}$ denotes the $m \times m$ matrix with real entries

$$\mathbf{R} = (\mathbf{R}_{\alpha\beta}), \qquad \mathbf{R}_{\alpha\beta} := \left(\frac{\partial^2\varphi}{\partial u^\alpha\partial u^\beta}(0)\bullet\nu\right).$$

The number

$$k(a) := \vec{k}(a)\bullet\nu = \frac{1}{|a|^2}\,\mathbb{I}_p(a)\bullet\nu$$

is called the *curvature* of M at p in the direction a, and the number

$$H := \vec{H} \bullet \nu$$

the *mean curvature* of M at p. Notice that in correspondence of the two possible choices of the unit normal ν, the sign of the curvature changes.

The matrix $\mathbf{R}$ is real symmetric, and its eigenvectors and eigenvalues are called respectively, the *principal directions of curvature* and the *principal curvatures* of M at p, since, if $a := \mathbf{D}\varphi(0)\xi$ where $\xi \in \mathbb{R}^m$ is an eigenvector of $\mathbf{R}$ of length 1 with eigenvalue λ, then

$$\lambda = \lambda|\xi|^2 = \xi^T \mathbf{R}\xi = k(a).$$

By the spectral theorem, we can choose in $T_p M$ a basis $(e_1, e_2, \ldots, e_m)$, such that $e_i = \mathbf{D}\varphi(0)\xi_i$ where ξ_i is an orthonormal basis in $\mathbb{R}^m$ of eigenvectors of $\mathbf{R}$. The corresponding eigenvalues are then the principal curvatures $k_1 := k(e_1), \ldots, k_m := k(e_m)$. If $a = \sum_{i=1}^m a^i e_i \in T_p M$, then $a = \mathbf{D}\varphi(0) \sum_{i=1}^m a^i \xi_i$ and

$$k(a) = \Big(\sum_{i=1}^m a^i \xi_i \Big)^T \mathbf{R} \Big(\sum_{i=1}^m a^i \xi \Big) = \sum_{i,j=1}^m a^i a^j \xi_j^T \mathbf{R}\xi_i$$

$$= \sum_{i=1}^m a^i a^j \delta_{ij} k(e_i) = \sum_{i=1}^m k_i (a^i)^2$$

known as the *Gauss formula* for the curvature.

Common standard choices for the sign of ν are the following:

o If M is given in parametric form, $\varphi : B \to M$, ν is chosen in such a way that $(D_1\varphi, D_2\varphi, \ldots, D_{n-1}\varphi, \nu)$ has positive determinant.

o If M is the boundary of an open set, ν is chosen as the interior normal in such a way that $\partial\Omega$ has nonnegative curvature if Ω is convex.

5.100 ¶. Let $\Omega \subset \mathbb{R}^2$ be an open set whose boundary $\partial\Omega$ is a 1-submanifold of $\mathbb{R}^2$. Prove that Ω is convex if and only if the curvature of $\partial\Omega$ is nonnegative.

5.101 ¶. Prove that the sphere of radius ρ in $\mathbb{R}^n$ has mean curvature $1/\rho$.

5.102 ¶. Let

$$T_2 := \Big\{ (x, y, z) \,\Big|\, (\sqrt{x^2 + y^2} - R)^2 + z^2 = \rho^2 \Big\}$$

be a two-dimensional torus in $\mathbb{R}^3$. Prove that its mean curvature is

$$H := \frac{R + 2\rho \cos\varphi}{2\rho(R + \rho \cos\varphi)}$$

where φ is the angle shown in Figure 5.22. Notice that $H \geq 0$ if $R > 2\rho$. This shows that the boundary of a nonconvex set in $\mathbb{R}^n$, $n \geq 3$, can have positive curvature. Finally, compute the principal curvatures of T_2.

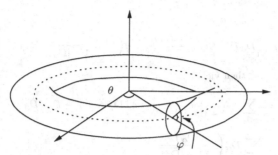

Figure 5.22. A two-dimensional torus.

5.103 Example (The graph of a function). Let M be the graph of the function $f : \Omega \rightarrow \mathbb{R}$, Ω open in $\mathbb{R}^n$. A parameterization of M is given by the map $\varphi(u) := (u, f(u))$, $u \in \Omega$, and, for $p := (u, f(u))$, we have:

(i) The vectors of $\mathbb{R}^{n+1}$

$$\frac{\partial \varphi}{\partial u^\alpha} = \Big(0, \dots, 0, 1, 0, \dots, 0, \frac{\partial f}{\partial u^\alpha}(u)\Big), \qquad \alpha = 1, \dots, n$$

where 1 is at the αth place, form a basis of $T_p M$.

(ii) The metric tensor of M is given by

$$\mathbf{G} = \mathrm{Id} + \mathbf{D}f^T \mathbf{D}f \qquad \text{or,} \qquad \mathbf{G} = (g_{\alpha\beta}), \ g_{\alpha\beta} = \delta_{\alpha\beta} + \frac{\partial f}{\partial u^\alpha}\frac{\partial f}{\partial u^\beta}.$$

(iii) We have $\mathbf{D}f\mathbf{D}f^T = |\mathbf{D}f|^2$.

(iv) The $n \times n$ matrix $\mathbf{D}f^T\mathbf{D}f$ has rank one, all its eigenvalues are zero except one that is $\mathrm{tr}\,(\mathbf{D}f^T\mathbf{D}f) = |\mathbf{D}f|^2$. Therefore, the eigenvalues of $\mathbf{G} = \mathrm{Id} + \mathbf{D}f^T\mathbf{D}f$ are

$$1 + |\mathbf{D}f|^2, 1, 1, \dots, 1,$$

and

$$\sqrt{g} = \sqrt{\det \mathbf{G}} = \sqrt{1 + |\mathbf{D}f|^2}.$$

(v) The inverse of the metric tensor is given by

$$\mathbf{G}^{-1} = \mathrm{Id} - \frac{1}{1 + |\mathbf{D}f|^2}\mathbf{D}f^T\mathbf{D}f$$

since for every $c \in \mathbb{R}$ we have

$$(\,\mathrm{Id} + \mathbf{D}f^T\mathbf{D}f)(\,\mathrm{Id} - c\,\mathbf{D}f^T\mathbf{D}f) = \mathrm{Id} + (1 - c)\mathbf{D}f^T\mathbf{D}f - c|\mathbf{D}f|^2\mathbf{D}f^T\mathbf{D}f$$
$$= \mathrm{Id} + (1 - c - c|\mathbf{D}f|^2)\mathbf{D}f^T\mathbf{D}f.$$

(vi) The unit vector

$$\nu := \frac{1}{\sqrt{1 + |\mathbf{D}f(u)|^2}}(-\mathbf{D}f(u), 1)$$

spans $T_p M$ and points upward; moreover, $\det[D_1\varphi \dots |D_n\varphi|\nu] = \sqrt{1 + |\mathbf{D}f|^2}$.

(vii) Since $\varphi(x) = (x, f(x))$, we have $D_\alpha D_\beta \varphi = (0, 0, \dots, 0, D_\alpha D_\beta f)$, hence

$$\Big[D_\alpha D_\beta \varphi\Big]^N = \frac{D_\alpha D_\beta f}{\sqrt{1 + |\mathbf{D}f|^2}}\,\nu.$$

(viii) For all $\alpha, \beta = 1, \ldots, n$ we have

$$g^{\alpha\beta}\left[D_\alpha D_\beta \varphi\right]^N = \frac{D_\alpha D_\beta f}{\sqrt{1 + |\mathbf{D}f|^2}}\left(\mathrm{Id} - \frac{1}{1 + |\mathbf{D}f|^2}D_\alpha f D_\beta f\right)\nu.$$

Equation (5.48) then yields

$$n\vec{H} = \sum_{\alpha,\beta=1}^{n}\frac{1}{(1 + |\mathbf{D}f|^2)^{3/2}}\left((1 + |\mathbf{D}f|^2)\delta_{\alpha\beta} - D_\alpha f D_\beta f D_\alpha D_\beta f\right)\nu$$

$$= \sum_{\alpha=1}^{n}D_\alpha\left(\frac{D_\alpha f}{\sqrt{1 + |\mathbf{D}f|^2}}\right)\nu.$$

(ix) Finally, the (scalar) mean curvature is given by

$$H = \frac{1}{n}\sum_{\alpha=1}^{n}D_\alpha\left(\frac{D_\alpha f}{\sqrt{1 + |\mathbf{D}f|^2}}\right).$$

f. Gradient and divergence on a surface

Let M be an embedded submanifold of $\mathbb{R}^n$, and, as usual, let $\varphi : \Omega \to M$, Ω open in $\mathbb{R}^m$, be a diffeomorphism, let $0 \in \Omega$ and let $p = \varphi(0)$. Let $f \in C^1(M)$, meaning that f is a function of class C^1 in an open neighborhood of the embedded submanifold M of $\mathbb{R}^n$. The orthogonal projection onto T_pM of the gradient ∇f of f at p in $\mathbb{R}^n$ is called the *tangential gradient* of f in M at p and is denoted by $\nabla_M f$,

$$\nabla_M f = \nabla f - (\nabla f)^N.$$

If $\mathbf{T} := \mathbf{D}\varphi(0)$, then $\nabla_M f = \mathbf{T}\xi$ for some $\xi \in \mathbb{R}^m$ and, since $\nabla f - \nabla_M f \in N_p(M)$, by the alternative theorem we have

$$\mathbf{T}^*(\nabla f - \mathbf{T}\xi) = 0.$$

This yields

$$\xi = (\mathbf{T}^*\mathbf{T})^{-1}\mathbf{T}^*\nabla f$$

i.e.,

$$\nabla_M f = \mathbf{T}(\mathbf{T}^*\mathbf{T})^{-1}\mathbf{T}^*\nabla f(p) = \mathbf{T}\mathbf{G}^{-1}\mathbf{T}^*\nabla f$$

or, more explicitly,

$$\nabla_M f = \sum_{\alpha=1}^{m}\xi^\alpha \frac{\partial\varphi}{\partial u^\alpha}(0), \qquad \xi^\alpha = \sum_{\beta=1}^{m}g^{\alpha\beta}(0)\frac{\partial(f \circ \varphi)}{\partial u^\beta}(0). \qquad (5.50)$$

5.104 Definition. *Let $X : M \to \mathbb{R}^n$ be a vector field of class $C^1(M)$, not necessarily tangent to M. The* divergence on M of X at p *is the number*

$$\mathrm{div}_M X(p) := \sum_{i=1}^{n}a_i \cdot D_{a_i}X(p)$$

see (5.45), where $(a_1, a_2, \ldots, a_n)$ is an orthonormal basis of T_pM.

5.105 Remark. Using the divergence operator, we rewrite (5.49) as

$$\vec{H} = -\frac{1}{m} \sum_{\alpha=1}^{n-m} (\operatorname{div}_M \nu_\alpha) \nu_\alpha.$$

5.106 Proposition. $\operatorname{div}_M X$ *does not depend on the chosen orthonormal basis in its definition. Moreover, we have*

(i) *If* $X : M \to T_p M$ *is a tangent vector field to* M, $X(\varphi(u)) = D\varphi(u)\xi(u)$, *where* $\xi = (\xi^1, \xi^2, \ldots, \xi^m) \in C^1(\Omega)$, *is its local representation, then*

$$\operatorname{div}_M X(\varphi(u)) = \operatorname{tr}\left(\mathbf{G}^{-1} D\varphi^T D X D\varphi\right)$$

$$= \sum_{\alpha,\beta=1}^{m} g^{\alpha\beta} \frac{\partial\varphi}{\partial u^\alpha} \cdot \frac{\partial(X \circ \varphi)}{\partial u^\beta} = \frac{1}{\sqrt{g}} \sum_{\alpha=1}^{m} \frac{\partial}{\partial u^\alpha}\left(\sqrt{g}\,\xi^\alpha\right).$$

(5.51)

(ii) *If* $X : M \to N_p M$ *is a normal vector field, then*

$$\operatorname{div}_M X(p) = -m\, X(p) \cdot \vec{H}(p)$$

(5.52)

where $\vec{H}$ *is the mean curvature vector of* M *at* p.

Proof. (i) If $(a_1, a_2, \ldots, a_m)$ is an orthonormal basis of $T_p M$, then $a_i = D\varphi(0)\xi_i$, $i = 1, \ldots m$ for some $(\xi_1, \xi_2, \ldots, \xi_m) \in \mathbb{R}^m$ such that the $n \times m$ matrix $\mathbf{B} = [\xi_1|\xi_2|\ldots|\xi_m]$ satisfies $\mathbf{BB}^T = \mathbf{G}^{-1}$. Therefore,

$$\operatorname{div}_M X(p) = \operatorname{tr}\left((D\varphi\mathbf{B})^T D X D\varphi\mathbf{B}\right) = \operatorname{tr}\left(D\varphi^T D X D\varphi\mathbf{BB}^T\right)$$
$$= \operatorname{tr}\left(D\varphi^T D X D\varphi\mathbf{G}^{-1}\right) = \operatorname{tr}\left(\mathbf{G}^{-1} D\varphi^T D X D\varphi\right).$$

This proves the independence of $\operatorname{div}_M X$ on the chosen orthonormal basis $a_1, \ldots, a_m$. Alternatively, we may compute more explicitly, shortening $\frac{\partial}{\partial u^\alpha}$ with D_α,

$$\operatorname{div}_M X = \sum_{i=1}^{m} a_i \cdot D_{a_i} X(p) = \sum_{i,\alpha,\beta=1}^{m} \sum_{\ell,k=1}^{n} \xi_i^\alpha \xi_i^\beta D_\alpha\varphi^k D_\beta\varphi^\ell \frac{\partial X^k}{\partial x^\ell}$$

$$= \sum_{\alpha,\beta=1}^{m} \sum_{\ell,k=1}^{n} g^{\alpha\beta} D_\alpha\varphi^k \frac{\partial X^k}{\partial x^\ell} D_\beta\varphi^\ell$$

$$= \sum_{k=1}^{n} \sum_{\alpha,\beta=1}^{m} g^{\alpha\beta} D_\alpha\varphi^k D_\beta(X^k \circ \varphi) = \sum_{\alpha,\beta=1}^{m} g^{\alpha\beta} D_\alpha\varphi \cdot D_\beta(X \circ \varphi)$$

recalling also that $\sum_{i=1}^{m} \xi_i^\alpha \xi_i^\alpha = g^{\alpha\beta}$.

Let us prove (5.51). With the convention that repeated indices are summed, we compute

$$\frac{1}{\sqrt{g}} D_\alpha\left(\sqrt{g}\,\xi^\alpha\right) = \frac{1}{2g}\xi^\alpha D_\alpha g + D_\alpha\xi^\alpha$$

and taking into account the formula of differentiation of determinants and the symmetry of $\mathbf{G}^{-1}$

$$\frac{1}{2g} D_\alpha g = \frac{1}{2}g^{\gamma\delta} D_\alpha g_{\gamma\delta} = \frac{1}{2}g^{\gamma\delta}\left(D_{\alpha\gamma}\varphi^\ell D_\delta\varphi^\ell + D_\gamma\varphi^\ell D_{\alpha\delta}\varphi^\ell\right)$$
$$= g^{\gamma\delta} D_\gamma\varphi^\ell D_{\alpha\delta}\varphi^\ell.$$

(5.53)

Therefore,

$$\frac{1}{\sqrt{g}}D_\alpha\left(\sqrt{g}\xi^\alpha\right) = D_\alpha\xi^\alpha + g^{\gamma\delta}D_\gamma\varphi^\ell D_{\alpha\delta}\varphi^\ell\xi^\alpha$$

$$= D_\alpha\xi^\alpha + g^{\gamma\delta}D_\gamma\varphi^\ell\left(D_\delta(\xi^\alpha D_\alpha\varphi^\ell) - D_\delta\xi^\alpha D_\alpha\varphi^\ell\right)$$

$$= D_\alpha\xi^\alpha + g^{\gamma\delta}D_\delta(X\circ\varphi)^\ell - g^{\gamma\delta}g_{\gamma\alpha}D_\delta\xi^\alpha$$

$$= g^{\gamma\delta}D_\gamma\varphi^\ell D_\delta(X\circ\varphi)^\ell.$$

i.e., (5.51).

(ii) From the definition of divergence, if $(\nu_1, \nu_2, \ldots, \nu_{n-m})$ are vector fields that form at p an orthonormal basis for N_pM, we have

$$m\,\vec{H}(p) = -\sum_{\alpha=1}^{n-m}(\operatorname{div}_M\nu_\alpha)\,\nu_\alpha. \tag{5.54}$$

On the other hand, since X is normal, we have $X = \sum_{\alpha=1}^{n-m}(X\bullet\nu_\alpha)\nu_\alpha$, hence

$$\operatorname{div}_M X = \sum_{\alpha=1}^{n-m}(\operatorname{div}_M\nu_\alpha)\,\nu_\alpha\bullet X.$$

By comparison with (5.54), we get $\operatorname{div}_M X = -m\,\vec{H}\bullet X$.   □

5.107 Corollary. *Let* $X : M \to \mathbb{R}^n$ *be a vector field of class* $C^1(M)$ *that vanishes near* $\varphi(\partial B) = \partial M$. *Then*

$$\int_M \operatorname{div}_M X\, d\mathcal{H}^m = -m\int_M X\bullet\vec{H}\,d\mathcal{H}^m.$$

Proof. We split X in its tangential and normal components

$$X = X^\perp + X^\top.$$

Since the operator div_M is linear, we have $\operatorname{div}_M X = \operatorname{div}_M X^\perp + \operatorname{div}_M X^\top$ and by (5.52)

$$\operatorname{div}_M X^\perp = -m\,X^\perp\bullet H = -m\,X\bullet\vec{H},$$

hence

$$\int_M \operatorname{div}_M X^\perp d\mathcal{H}^m = -m\int_M \vec{H}\bullet X\,d\mathcal{H}^m.$$

Finally, by writing $X^\top(\varphi(u)) := \sum_{\alpha=1}^m \xi^\alpha(u)\frac{\partial\varphi}{\partial u^\alpha}(u)$, using the area formula and Gauss–Green formulas in $B = B(0,1)$, we infer from (5.51)

$$\int_M \operatorname{div}_M X^\top d\mathcal{H}^m = \int_B \sum_{a=1}^m \frac{\partial}{\partial u^a}\left(\sqrt{g}\xi^\alpha\right)d\mathcal{L}^m = 0$$

since $\xi = (\xi^\alpha)$ vanishes near the boundary of B.   □

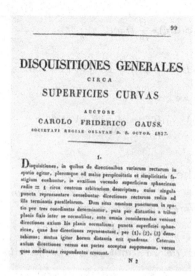

Figure 5.23. The first page of a paper by Carl Friedrich Gauss (1777–1855) where the "theorema egregium" appears, and the frontispiece of *Leçons sur la théorie générale des surfaces* by Gaston Darboux (1842–1917).

g. First variation of the area

The mean curvature and the divergence operator are tightly related to the *first variation* of the area of a surface.

5.108 Proposition (First variation of area). *Let* $\varphi : B(0,1) \subset \mathbb{R}^m \to \mathbb{R}^n$ *be a map of class* C^2, $M := \varphi(B(0,1))$, *and let* $\phi : [-1,1] \times \mathbb{R}^n \to \mathbb{R}^n$ *be a map of class* C^2 *that is the identity at* $\epsilon = 0$ *and does not move* $\varphi(\partial B(0,1))$,

$$\phi(\epsilon, x) = x, \qquad \forall x \in \varphi(\partial B(0,1)), \ \forall \epsilon.$$

Let M_ϵ *be the surface image of* $x \to \varphi_\epsilon(x) := \phi(\epsilon, x)$. *Then*

$$\frac{d\mathcal{H}^m(M_\epsilon)}{d\epsilon}\bigg|_{\epsilon=0} = \int_M \operatorname{div}_M V \, d\mathcal{H}^m = -m \int_M \vec{H} \bullet V \, d\mathcal{H}^m,$$

where $V(p) := \frac{\partial \phi}{\partial \epsilon}(0, p)$ *is the velocity of the flow* $\phi(\epsilon, x)$ *at* $\epsilon = 0$.

Proof. For all ϵ the area formula yields

$$\mathcal{H}^m(A_\epsilon) = \int_{B(0,1)} \sqrt{g_\epsilon(u)} \, d\mathcal{L}^m(u)$$

where $g_\epsilon(u) := \det \mathbf{G}_\epsilon(u)$ and $\mathbf{G}_\epsilon(u) := \mathbf{D}\varphi_\epsilon^T \mathbf{D}\varphi_\epsilon$. Differentiating the determinant,

$$\frac{\partial \sqrt{g_\epsilon}}{\partial \epsilon} = \frac{1}{2}\sqrt{g_\epsilon}\, g_\epsilon^{\alpha\beta}(u)\frac{\partial (g_\epsilon)_{\alpha\beta}}{\partial \epsilon}$$

and setting $g := \det \mathbf{G}_0$, $\mathbf{G}_0^{-1} := (g^{\alpha\beta})$, we compute at $\epsilon = 0$

$$\frac{\partial \sqrt{g_\epsilon}}{\partial \epsilon}\Big|_{\epsilon=0}(u) = \sqrt{g} \sum_{\alpha,\beta=1}^{m} g^{\alpha\beta}(u)\Big(D_\alpha\phi(0,u) \bullet \frac{\partial}{\partial \epsilon}D_\beta\phi(0,u)\Big)$$

$$= \sqrt{g} \sum_{\alpha,\beta=1}^{m} g^{\alpha\beta}\Big(D_\alpha\varphi \bullet D_\beta(V \circ \varphi)\Big) = \operatorname{div}_M V(u).$$

Therefore, differentiating under the integral sign we get

$$\frac{d\mathcal{H}^m(A_\epsilon)}{d\epsilon} = \int_B \frac{\partial \sqrt{g_\epsilon}}{\partial \epsilon}\,d\mathcal{L}^m == \int_B \operatorname{div}_M V\sqrt{g}\,d\mathcal{L}^m = \int_M \operatorname{div}_M V\,d\mathcal{H}^m.$$

and the result follows from Corollary 5.107. □

h. Laplace–Beltrami operator and the mean curvature

5.109 Definition. *As usual, let* $\varphi : \Omega \to \mathbb{R}^n$, $\Omega \subset \mathbb{R}^m$, *be a diffeomorphism,* $M = \varphi(\Omega)$ *and* $p = \varphi(u)$. *The differential operator on* M

$$\Delta_M f := \operatorname{div}_M(\nabla_M f), \qquad f \in C^2(M)$$

is called the Laplace–Beltrami operator *on* M, *and functions* f *with* $\Delta_M f = 0$ harmonic functions *on* M.

For $f \in C^2(M)$, $\Delta_M f$ is a continuous real valued function defined on M. Taking into account (5.50) and (5.51),

$$(\Delta_M f) \circ \varphi = \frac{1}{\sqrt{g}} \sum_{\alpha,\beta=1}^{m} \frac{\partial}{\partial u^\alpha}\Big(\sqrt{g}g^{\alpha\beta}\frac{\partial(f \circ \varphi)}{\partial u^\beta}\Big).$$

5.110 Proposition. *For* $i = 1, n$ *let* $f^i := \varphi^i \circ \varphi^{-1} : M \to \mathbb{R}$. *Then* $\Delta_M f(p) := (\Delta_M f^1(p), \ldots, \Delta_M f^n(p))$ *belongs to* $N_p M$ *and*

$$\Delta_M f(p) = mH(p).$$

Proof. In fact, we compute

$$(\Delta_M f^i) \circ \varphi = \frac{1}{\sqrt{g}} \sum_{\alpha,\beta=1}^{m} D_\alpha\Big(\sqrt{g}g^{\alpha\beta}D_\beta\varphi^i\Big)$$

and

$$\Delta_M\varphi \bullet D_\alpha\varphi = \frac{1}{\sqrt{g}}D_i\Big(\sqrt{g}g^{ij}D_j\varphi^\ell\Big)D_\alpha\varphi^\ell$$

$$= \frac{1}{2g}D_\alpha g - g^{ij}D_j\varphi^\ell D_i D_\alpha\varphi^\ell = 0$$

by recalling (5.53). The second claim follows, since

$$\Delta_M\varphi = \frac{1}{2g}D_i g g^{ij} D_j\varphi + D_i(g^{ij})D_j\varphi + g^{ij}D_i D_j\varphi$$

where the first two terms are tangential to M. Therefore, since we proved that $\Delta_M f$ is orthogonal to M,

$$\Delta_M f = \Big(\Delta_M f\Big)^N = (D_i D_j\varphi)^N g^{ij} = tr(\Phi \mathbf{G}^{-1}) = m\vec{H}(p),$$

see (5.48). □

Surfaces with vanishing mean curvature are called *minimal surfaces.* Proposition 5.110 then reads: *the coordinates of an embedded minimal surface are harmonic functions on the surface.*

i. Distance function

Let Ω be a bounded open domain, the boundary of which is an $(n-1)$-submanifold of class C^k, $k \geq 2$. For all $x \in \Omega$, we set $d(x) := \mathrm{dist}\,(x, \partial\Omega)$, $\Omega_\epsilon := \{x \in \Omega \,|\, d(x) < \epsilon\}$ and we denote by $\nu(\xi)$ the interior unit normal to $\partial\Omega$ at $\xi \in \partial\Omega$.

5.111 Theorem. *Let Ω be a bounded open domain in $\mathbb{R}^n$ with $\partial\Omega$ of class C^k, $k \geq 2$. Then there exists $\epsilon > 0$ such that the following holds.*

(i) *For all $x \in \Omega_\epsilon$ there is a unique point $\xi(x) \in \partial\Omega$ of least distance $d(x)$ from x; moreover, $x - \xi(x)$ is normal to $\partial\Omega$,*

$$x = \xi(x) + d(x)\,\nu(\xi).$$

(ii) *The functions $x \to \xi(x)$ is of class C^{k-1} and the function $x \to d(x)$ is of class C^k. Moreover, $d(x)$ solves the eikonal equation $|\mathbf{D}d| = 1$ in Ω_ϵ and*

$$\mathbf{D}d(x) = \nu(\xi(x)) \qquad \text{in } \overline{\Omega}_\epsilon.$$

(iii) *Consider for $0 \leq t < d(x)$ the t-level set of the distance function $M_t := \{y \in \Omega \,\big|\, d(y) = t\}$. Then M_t is of class C^k and*

$$-\Delta d(x) = (n-1)H_{M_t}(x) = \frac{1}{n-1}\sum_{i=1}^{n-1} \frac{k_i}{1-tk_i} \qquad (5.55)$$

where $k_1, k_2, \ldots, k_n - 1$ are the principal curvatures of $\partial\Omega$ evaluated at the least distance point $\xi(x)$. In particular,

$$H_{M_t}(x) \geq H_{\partial\Omega}(\xi(x)).$$

The coordinates (ξ, t) for $x = \xi + t\nu(\xi) \in \Omega_\epsilon$ are often called *Fermi's coordinates* of x.

Proof. Step 1. Let $B(z, \epsilon)$ be a ball centered at $z \in \partial\Omega$. Since the distance function is continuous, for every $x \in \Omega \cap B(z, \epsilon)$, there exists a point $y \in \partial\Omega$ of least distance from x. As we have seen, Fermat's principle implies that $x - y$ spans the normal line to $\partial\Omega$ through y. Thus, if $\nu(y)$ denotes the inward normal to $\partial\Omega$ at $y \in \partial\Omega$, and $d(x)$ is the least distance of x from $\partial\Omega$, we have $x = y + d(x)\nu(y)$.

Step 2. We prove that for every $z \in \partial\Omega$ there is a neighborhood $U(z)$ such that, for all $x \in U(z) \cap \Omega$ there is a unique $\xi(x) \in U(z) \cap \partial\Omega$ of least distance $d(x)$ from x, d is of class C^k and $\mathbf{D}d(x) = \nu(\pi(x)) \ \forall x \in U(z)$.

Since the claim is invariant by rigid motions, we can suppose that z is the origin, and that in an open neighborhood $U \subset \mathbb{R}^n$ of 0

(i) $\partial\Omega \cap U$ is the graph of a function h of class C^k defined on a ball $B(0, r)$ of $\mathbb{R}^{n-1}$ with $h(0) = 0$ and $\nabla h(0) = 0$,
(ii) the inward normal to $\partial\Omega$ at $0 \in \mathbb{R}^n$ is $(0, \ldots, 0, 1)$,
(iii) the axes of $\mathbb{R}^{n-1}$ are directed as the eigenvectors of $\mathbf{D}^2 h(0)$.

By our choice of the coordinate system, we have

$$\mathbf{D}^2 h(0) = \text{diag} \, (k_1, k_2, \ldots, k_{n-1}),$$

where $k_1, \ldots, k_{n-1}$ are the principal curvatures of $\partial \Omega$ at 0. Consider now the map $\phi : B(0, r) \times \mathbb{R} \to \mathbb{R}^n$ defined by

$$x = \phi(y, t) := (y, h(y)) + t \, \nu(y, h(y))$$

that is,

$$\begin{cases} x_i := y_i + t N_i(y) & \text{for } i = 1, \ldots, n-1, \\ x_n = h(y) + t N_n(y) \end{cases}$$

where $N(y) := \nu(y, h(y))$. We have

$$\begin{cases} N_i(y) = \dfrac{-D_i h(y)}{\sqrt{1 + |\nabla h(y)|^2}} & \text{for } i = 1, \ldots, n-1, \\ N_n(y) = \dfrac{1}{\sqrt{1 + |\nabla h(y)|^2}}. \end{cases}$$

and, since $\mathbf{D}h(0) = 0$, we infer

$$\begin{cases} D_j N_i(0) = -k_i \delta_{ij}, \\ D_n N_n(0) = 0, \end{cases} \tag{5.56}$$

hence

$$\mathbf{D}\phi(0, t) = \text{diag} \, (1 - tk_1, \ldots, 1 - tk_{n-1}, 1). \tag{5.57}$$

In particular det $\mathbf{D}\phi(0, 0) = 1$, hence ϕ is a locally invertible map of class C^{k-1} in a neighborhood of $0 \in \mathbb{R}^n$. Its inverse $\psi : B(0, \epsilon) \to \mathbb{R}^n$ defined on a ball centered at zero is of class C^{k-1}. We now set for $x \in B(0, \epsilon)$ $\pi(x) := (\psi^1(x), \ldots, \psi^{n-1}(x))$, $\xi(x) = (\pi(x), h(\pi(x)))$ and $t(x)) = \psi^n(x)$. Trivially $\xi(x)$ and $t(x)$ are of class C^{k-1}, $\xi(x) \in \partial \Omega$ and $\phi(\psi(x)) = x$ rewrites as

$$x = \xi(x) + t(x) \, \nu(\xi(x))) \qquad \forall x \in U(0), \tag{5.58}$$

from which we conclude that $\xi(x)$ is the unique point in $B(0, \epsilon)$ such that $x - \xi(x)$ is perpendicular to $\partial \Omega$, thus the least distance point by *Step 1*, and that $d(x) = |x - \xi(x)| = t(x)$. Consequently, $d(x)$ is of class C^{k-1}.

From (5.57) we easily get

$$\mathbf{D}\psi(x) = \text{diag} \, \Big(\frac{1}{1 - tk_1}, \ldots, \frac{1}{1 - tk_{n-1}}, 1 \Big) \tag{5.59}$$

where $t = d(x)$ provided $\pi(x) = 0$. In particular $\mathbf{D}d(x) = (0, 0, \ldots, 1) = \nu(\xi(x))$ when $\pi(x) = 0$.

The above construction can be repeated at each point $z \in \partial \Omega$, hence we conclude that

$$\mathbf{D}d(x) = \nu(\xi(x)) \qquad \forall x \in B(0, \epsilon). \tag{5.60}$$

In particular $\mathbf{D}d(x)$ is of class C^{k-1}. This concludes *Step 2*.

Step 3. For all $z \in \partial \Omega$ the results in *Step 2* hold in $B(z, \epsilon(z))$. Since $\partial \Omega$ is compact, covering it with finitely many balls $B_{z_1, \epsilon(z_1)}, \ldots, B_{z_k, \epsilon(z_k)}$ and choosing $\epsilon = \frac{1}{4} \min_{i=1,\ldots,k} \epsilon(z_i)$, we conclude that the results in *Step 2* hold in Ω_ϵ. This proves (i) and (ii).

Step 4. (5.49) and (5.60) yield for $x \in M_t$

$$(n - 1) H_{M_t}(x) = - \sum_{i=1}^{n} D_i \nu^i(\xi(x)) = -\text{tr} \, (\mathbf{D}\nu(\xi(x))) = -\text{tr} \, (\mathbf{D}^2 d(x)) = -\Delta d(x).$$

On the other hand, assuming that $\pi(x) = 0$ and $x \in M_t$, we get from (5.59) and (5.56)

$$\mathbf{D}^2 d(x) = \mathbf{D}(\nu(\xi(x))) = \mathbf{D}\nu(\pi(x))\mathbf{D}\xi(x)$$

$$= \begin{pmatrix} -k_1 & 0 & \cdots & 0 & 0 \\ 0 & -k_2 & 0 & 0 & 0 \\ \vdots & \vdots & \ddots & \vdots & \vdots \\ 0 & 0 & \cdots & k_{n-1} & 0 \\ 0 & 0 & \cdots & 0 & 0 \end{pmatrix} \operatorname{diag}\left(\frac{1}{1-tk_1}, \ldots, \frac{1}{1-tk_{n-1}}, 0\right)$$

$$= \operatorname{diag}\left(\frac{-k_1}{1-tk_1}, \ldots, \frac{-k_{n-1}}{1-tk_{n-1}}\right).$$

thus concluding that for $x \in \Omega_\epsilon$,

$$\Delta d(x) = -\sum_{i=1}^{n-1} \frac{k_i}{1 - tk_i}$$

where $t = d(x)$ and $k_1, k_2, \ldots, k_n - 1$ are evaluated at the least distance point $\xi(x)$. □

5.112 Remark. Notice that the assumption that $\partial\Omega$ is at least of class C^2 is truly necessary. In fact, let $0 < \alpha < 1$ and consider the open set

$$\Omega := \left\{ (x,y) \in \mathbb{R}^2 \,\middle|\, y = |x|^{2-\alpha}, \ y \le 1 \right\}$$

the boundary of which is of class $C^{1,1-\alpha}$ near $(0,0)$. It is easy to see that if $P = (0, y) \in \Omega$ is close to the curve $\{y = |x|^{2-\alpha}\}$, then P has two least distance points differing from $(0,0)$. Moreover one can show that the distance function is not differentiable at P, see Exercise 5.143.

5.5 Exercises

5.113 ¶. Study the transformations

$(x^2 - y^2, xy)$, $(\sqrt{x/y}, \sqrt{xy})$, $x, y > 0$,

$(e^x \cos y, e^x \sin y)$, $(x^2 - xy, y - x)$,

$(\sin(x + y), \cos(x + y))$.

5.114 ¶. Investigate the solvability in (x, y) of the system

$$\begin{cases} x + y + uv = 0, \\ uxy + v = 0 \end{cases}$$

when (u, v) is small, and of the system

$$\begin{cases} x + xyz = u, \\ y + xy = v, \\ z + 2x + 3z^2 = w \end{cases}$$

in (x, y, z) when (u, v, w) is small.

5.115 ¶. Prove that the relation

$$x^2 + \log(1 + zy) + xye^z = 0$$

defines a function $z = \varphi(x, y)$ in a neighborhood of the point $(0, 1, 0)$. Write its Taylor's polynomial of second degree with center (0.1) of $\varphi(x, y)$.

5.116 ¶. In thermodynamics one considers the equation $\phi(p, V, T) = \text{cost}$, where p, V, and T are respectively the pressure, the volume, and the temperature of a gas. In case we express one variable as function of the remaining two,

$$p = p(V, T), \qquad V = V(p, T), \qquad T = T(p, V),$$

with p, V, and T sufficiently regular, prove that

$$\frac{\partial p}{\partial T} \frac{\partial T}{\partial V} \frac{\partial V}{\partial p} = -1.$$

5.117 ¶. Prove that $M := \{(x, y, z) \mid z^2 = x^2 + y^2\}$ is not an r-submanifold of $\mathbb{R}^3$.

5.118 ¶. Prove that the maps defined on $\mathbb{R}^2$ by

$$(u, v) \rightarrow \left(\frac{2u}{u^2 + v^2 + 1}, \frac{2v}{u^2 + v^2 + 1}, \frac{u^2 + v^2 - 1}{u^2 + v^2 + 1} \right)$$

$$(u, v) \rightarrow \left(\frac{2u}{u^2 + v^2 + 1}, \frac{2v}{u^2 + v^2 + 1}, \frac{1 - u^2 - v^2}{u^2 + v^2 + 1} \right)$$

parameterize respectively $S^2 \setminus \{\text{North Pole}\}$ and $S^2 \setminus \{\text{South Pole}\}$ with $\mathbb{R}^2$.

5.119 ¶. Prove that the standard torus, obtained by rotating around the z-axis a circle of radius r around a point on the y axis at distance $R > r$, is a 2-submanifold of $\mathbb{R}^3$. Write it as a zero level set and find local parameterizations.

5.120 ¶. Let $\varphi : \mathbb{R} \rightarrow \mathbb{R}$ be a function of class C^1 with $|\varphi'(t)| \leq 1/2 \; \forall \, t$. Let $f : \mathbb{R}^2 \rightarrow \mathbb{R}^2$ be the map defined by $f(x, y) = (x + \varphi(y), y + \varphi(x))$. Prove that $f(\mathbb{R}^2) = \mathbb{R}^2$ and that f is globally invertible.

5.121 ¶. Visualize the sets $C = \{(x, y, u, v) \in \mathbb{R}^4 \mid x^2 + y^2 = 1, \; u^2 + v^2 = 1\}$ and $K := \{(x, y, u, v) \in \mathbb{R}^4 \mid x^2 + y^2 \leq 1, \; u^2 + v^2 \leq 1\}$ by describing their three-dimensional slices and prove that C and $K \setminus C$ are submanifolds of $\mathbb{R}^4$.

5.122 ¶. An ideal pointwise mass is constrained to move on the circle of center 0 and radius 1 and is connected to the point $(1, 0)$ by means of an ideal spring of elastic constant k. Find its equilibrium positions.

5.123 ¶. A particle is constrained to move on the ellipse $4x^2 + y^2 = 4$ and attracts another particle constrained to move on the line $3x + 2y = 25$. Find their equilibrim positions if they exist.

5.124 ¶. Find the maximum and minimum points of the function $\sum_{i=1}^{n} a_i x^i$ constrained to $\sum_{i=1}^{n} |x^i|^p = 1$, $p > 1$, and infer Hölder inequality.

5.125 ¶. Given $n + 1$ points $P_i = (x_i, y_i)$, $i = 0, 1, \ldots, n$, we denote with $[P_0 \ldots P_n]$ the closed polygonal line connecting successively $P_0, P_1, \ldots, P_n, P_0$. The length of this polygonal is

$$L := \sum_{i=0}^{n} |P_i - P_{i+i}| + |P_n - P_0|$$

and its enclosed *area* is

$$A := \left| \sum_{i=1}^{n} \text{oriented area of } [0P_{i-1}P_i] \right| = \frac{1}{2} \left| \sum_{i=1}^{n} \det \begin{pmatrix} x_{i-1} & y_{i-1} \\ x_i & y_i \end{pmatrix} \right|.$$

Show that this area is maximum among polygonals with n sides and given perimeter when the polygon is regular, in particular,

$$A \le \frac{1}{4\pi} \cot \left(\frac{\pi}{n} \right) L^2.$$

For $n \to \infty$ deduce that for any polygon with n sides we have the isoperimetric inequality

$$A \le \frac{1}{4\pi} L^2.$$

5.126 ¶ Simple roots. Prove that the simple zeros of a polynomial are C^∞ functions of the coefficients of the polynomial. [*Hint:* If x_0 is a simple root for P, then $P(x_0) = 0$ and $P'(x_0) \ne 0$.]

5.127 ¶. Prove that the simple eigenvalues of a matrix $\mathbf{A}$ are C^∞ functions of the entries of $\mathbf{A}$. Then infer the following.

Proposition. *Let $\mathbf{A}(t)$ be a differentiable curve in the space of $n \times n$ matrices. Suppose that λ_0 is a simple eigenvalue of $\mathbf{A}(0)$. Show that for t small $\mathbf{A}(t)$ has an eigenvalue $\lambda(t)$ that depends in a C^1-way from t and moreover $\lambda_0 = \lambda(0)$.*

[*Hint:* In order to prove the first claim, consider $f(x, \mu, \mathbf{A}) := (\mathbf{A}x - \mu x, |x|^2 - 1)$, prove

$$\det \frac{\partial f}{\partial(x, \lambda)}(x, \lambda, \mathbf{A}) = 2 \lim_{\mu \to \lambda} \frac{\det(\mathbf{A} - \mu \, \text{Id})}{\mu - \lambda}$$

and use the fact that λ is a simple zero of $\det(\mathbf{A} - \lambda \, \text{Id})$.]

5.128 ¶. The equations

$$\begin{cases} x^2 - yu = 0, \\ xy + uv = 0 \end{cases}$$

implicitly define (u, v) as functions of (x, y) in a neighborhood of (x_0, y_0, u_0, v_0) with $(y_0, u_0) \ne 0$. If $\varphi(x, y) := (u, v)$, compute $\det \mathbf{D}\varphi(x, y)$.

5.129 ¶. Prove that the system

$$\begin{cases} x_1 + x_2 + x_3 + x_4 = u_1, \\ x_2 + x_3 + x_4 = u_1 u_2, \\ x_3 + x_4 = u_1 u_2 u_3, \\ x_4 = u_1 u_2 u_3 u_4 \end{cases}$$

implicitly defines the x's as functions of the u's, and that

$$\det \frac{\partial(x_1, \ldots, x_4)}{\partial(u_1, \ldots, u_4)} = u_1^3 u_2^2 u_3.$$

5.130 ¶. Prove that the equations

$$\begin{cases} u = x + y + z, \\ v = u^2 + v^2 + z^2, \\ w = u(3v - u^2)/2 \end{cases}$$

define a 2-submanifold in $\mathbb{R}^3$.

5.131 ¶. Find the maximum and minimum points of $x^2 + y^2 + z^2$ in each of the following sets

$$\left\{ (x, y, z) \in \mathbb{R}^3 \,\middle|\, x + y + z = 3a \right\}, \qquad \left\{ (x, y, z) \in \mathbb{R}^3 \,\middle|\, xy + yz + xz = 3a^2 \right\},$$

$$\left\{ (x, y, z) \in \mathbb{R}^3 \,\middle|\, xyz = a^3 \right\}.$$

and of $x^2 + y^2 - 3x + 5y$ in the set

$$\left\{ (x, y, z) \in \mathbb{R}^3 \,\middle|\, (x + y)^2 = 4(x - y) \right\}.$$

5.132 ¶. Prove that the special group $SL(n, \mathbb{R})$ of $n \times n$ matrices with determinant 1 identified with points in $\mathbb{R}^{n^2}$ is a submanifold of $\mathbb{R}^{n^2}$ of dimension $n^2 - 1$.

5.133 ¶. Identify the group of symmetric $n \times n$ matrices $\mathrm{Sym}_n(\mathbb{R})$, with $\mathbb{R}^r$, $r = \binom{n+1}{2}$, and let $f : M_{n,n} \to \mathrm{Sym}_n(\mathbb{R})$ be the map $f(\mathbf{X}) := \mathbf{X}^T \mathbf{X}$. Then the orthogonal group is the counterimage of the identity, i.e., $O(n) = f^{-1}(\mathrm{Id})$. Prove that

$$df_{\mathbf{X}}(\mathbf{H}) = \mathbf{H}^T \mathbf{X} + \mathbf{X}^T \mathbf{H} \qquad \forall \mathbf{X} \in O(n),\ \forall \mathbf{H} \in M_{n,n},$$

and infer that $O(n)$ is a submanifold of $\mathbb{R}^{n^2}$ of dimension $\binom{n}{2} = \frac{1}{2}n(n-1)$.

5.134 ¶. Let $\mathbf{A}, \mathbf{B}$ be two self-adjoint matrices in $M_{n,n}(\mathbb{R})$. Find the critical values of $\mathbf{A}x \bullet x$ constrained to $\mathbf{B}x \bullet x = 1$.

5.135 ¶. Prove that a graph over $[0, 1]$ has zero mean curvature if and only if it is a straight line. Prove also that a graph over $[0, 1]$ has constant mean curvature $k > 0$ provided $k \leq 2$ and, in this case, it is a piece of a circle of radius larger than or equal to $1/2$.

5.136 ¶. As we have seen, there exists a unique (up to rigid motions) planar curve $\gamma(s)$ parameterized by the arc length with given *positive* scalar curvature $k(s)$. Prove that the same result holds if $k(s) \geq 0$ provided k is analytic. Finally, show examples of nonuniqueness if $k(s)$ vanishes and is not analytic. [*Hint:* Compare scalar and oriented curvatures.]

5.137 ¶ Envelopes. Let Ω be an open set of $\mathbb{R}^2$ and let Γ be an interval around 0. Let $f : \Omega \times \Gamma \to \mathbb{R}$ $f = f(x, y, c)$, be a map of class C^1, with $f(x, y, c) = 0$ at some point and $\nabla f(x, y, c) \neq 0$ in $\Omega \times \Gamma$. Consider the 1-parameter family of curves

$$M_c := \left\{ (x, y) \in \Omega \,\middle|\, f(x, y, c) = 0 \right\}, \qquad c \in \Gamma.$$

(i) A curve $c \to \varphi(c) := (\xi(c), \eta(c))$, $c \in \Gamma$, is such that

$$f(\xi(c), \eta(c), c) = 0 \quad \text{and} \quad \nabla f(\xi(c), \eta(c), c) \perp \varphi'(c)$$

for all $c \in \Gamma$ if and only if

$$f(\xi(c), \eta(c), c) = 0 \quad \text{and} \quad \frac{\partial}{\partial c} f(\xi(c), \eta(c), c) = 0$$

for all $c \in \Gamma$. In this case, the curve $\varphi(c)$ is called the *envelope* of the family $\{M_c\}$.

(ii) Prove that, if $f(x_0, y_0, c_0) = 0$ and $f_{cc}(x_0.y_0, c_0) \neq 0$, then locally, i.e., for small $|c - c_0|$, the family $\{M_c\}$ has an envelope.

(iii) Finally, show that, if moreover,

$$f_x f_{cy} - f_y f_{cx} \neq 0 \quad \text{in } (x_0, y_0, c_0)$$

then the envelope curve $\varphi(c)$ is regular, i.e., $\varphi'(c) \neq 0$.

5.138 ¶ Evolute. Let $\gamma : I \to \mathbb{R}^2$ be a curve with $k(t) \neq 0$ for all t. The curve

$$\sigma(t) := \gamma(t) + \frac{1}{k(t)} n(t)$$

is called the *evolute* of γ. Prove that the tangent to $\sigma(t)$ is the normal to γ at t.

5.139 ¶. Prove that if all normal lines to a planar curve meet at the same point, then the trajectory of the curve is a circle.

5.140 ¶. Let $\gamma(s) : I \to \mathbb{R}^3$ be a curve parameterized by the arc length. Suppose that $\tau(s) \neq 0$ and $k'(s) \neq 0$. Prove that the trajectory of γ lies in a sphere if and only if

$$R^2(s) + \left(\frac{R'(s)}{t(s)}\right)^2 = cost, \quad R(s) := \frac{1}{k(s)}.$$

5.141 ¶. Show that a curve in polar coordinates $\rho = \rho(\theta)$ has curvature given by

$$k(\theta) := \frac{2\rho'^2 - \rho\rho'' + \rho^2}{(\rho'^2 + \rho^2)^{3/2}}.$$

5.142 ¶ Evolute. Let $X \in C^k(I, \mathbb{R}^2)$, $k \geq 3$, be a simple regular curve with $k(t)$ as curvature, $\rho(t)$ as radius of curvature, and $n(t)$ as normal at $X(t)$. The *evolute* of $X(t)$ is the curve $Y(t) := X(t) + \rho(t)n(t)$ If $X(t) = (x(t), y(t))$ and $Y(t) = (\xi(t), \eta(t))$, prove that

$$\xi(t) = x - y' \frac{x'^2 + y'^2}{x'y'' - y'x''}, \quad \eta(t) = y + x' \frac{x'^2 + y'^2}{x'y'' - y'x''}.$$

Prove that the evolute of a cycloid $x(t) = R(t + \sin t)$, $y(t) = R(1 - \cos t)$, is again a cycloid (Christiaan Huygens (1629–1695)) and the evolute of the parabola $y = x^2/2$ is Neile's parabola $8(y - 1)^3 = 27x^2$ (William Neile (1637–1670)).

5.143 ¶. For $1 < \alpha \leq 2$, let $M_\alpha \subset \mathbb{R}^2$ be the graph of $f_\alpha(x) = x^\alpha$, $x \in \mathbb{R}$. Prove that $f_\alpha \in C^1(\mathbb{R})$ and that $f_\alpha \in C^2(\mathbb{R})$ if and only if $\alpha = 2$. If

$$M_{\alpha,\epsilon} := \left\{(x, y) \,\middle|\, y > x^\alpha, \text{dist}\,((x, y), M_\alpha) = \epsilon\right\},$$

prove that

(i) $M_{\alpha,\epsilon}$ is a submanifold for all $\epsilon > 0$ if $\alpha = 2$,

(ii) $M_{\alpha,\epsilon}$ is singular for all $\epsilon > 0$ at the point $(0, y) \in M_{\alpha,\epsilon}$ if $0 < \alpha < 2$.

5.144 ¶. Let $f : \mathbb{R}^n \to \mathbb{R}$ be a function of class C^1 and $\Omega := \{x \mid f(x) < 0\}$. Suppose that $\partial\Omega = \{x \mid f(x) = 0\}$ and that $\mathbf{D}f(x) \neq 0 \; \forall x \in \partial\Omega$. Prove that $\partial\Omega$ is a $(n-1)$-dimensional submanifold of $\mathbb{R}^n$ with exterior normal $\nu(x) = \nabla f(x)/|\nabla f(x)|$ at $x \in \partial\Omega$, and that

$$\int_\Omega \Delta f \, dx = \int_{\partial\Omega} |\nabla f| \, d\mathcal{H}^{n-1}.$$

5.145 ¶. Let Ω be a bounded open domain of $\mathbb{R}^n$ with boundary of class C^3, let $d(x)$ be the distance function between x and $\partial\Omega$ and let

$$\Omega_\epsilon := \overline{\Omega} \cup \left\{ x \in \Omega^c \; \middle| \; d(x) < \epsilon \right\}.$$

Prove that for ϵ small, we have

$$\mathcal{H}^{n-1}(\partial\Omega_\epsilon) - \mathcal{H}^{n-1}(\partial\Omega) = \int_{\Omega_\epsilon \setminus \Omega} \Delta d \, dx.$$

6. Systems of Ordinary Differential Equations

The system of ordinary differential equations

$$\begin{cases} x_1' = f_1(t, x_1, x_2, \ldots, x_n), \\ x_2' = f_2(t, x_1, x_2, \ldots, x_n), \\ \ldots \\ x_n' = f_n(t, x_1, x_2, \ldots, x_n) \end{cases}$$

or, in short,

$$x' = f(t, x), \qquad t \in I \subset \mathbb{R}, \; x = x(t) : I \to \mathbb{R}^n,$$

where f is a map from a domain $\Omega \subset \mathbb{R} \times \mathbb{R}^n$ into $\mathbb{R}^n$, needs not have solutions defined on the entire interval I, even in the case $\Omega = \mathbb{R} \times \mathbb{R}^n$ and f smooth. For instance, see [GM1], for $n = 1$ all solutions of $x' = x^2$ are of the form $x(t) = 1/(c - t)$, $c \in \mathbb{R}$, thus defined either for $t < c$ or $t > c$. If $f(t, x) : \Omega \subset \mathbb{R}^n \to \mathbb{R}^n$ is continuous in t and locally Lipschitz in Ω, then the Picard–Lindelöf theorem says, see [GM3], that the Cauchy problem

$$\begin{cases} x' = f(t, x), \\ x(t_0) = x_0 \end{cases}$$

has a unique local solution, indeed on a maximal interval containing t_0 in the sense that the trajectory $(t, x(t))$ reaches the boundary of Ω. Actually, local solvability holds for continuous functions f, but uniqueness does not in general. Finally, under the assumption of the Picard–Lindelöf theorem the solution depends continuously on the initial datum, see [GM3], and, as we saw in Chapter 5 Section 5.3, f depends in a C^k way on the initial datum if f is of class C^k.

In this chapter we discuss a selection of classic results from the basic theory of ODE with the partial motivations of illustrating structures and techniques we have introduced. Of course, we refrain from any attempt of completeness and systematicity both for reasons of space and because this would lead into the *theory of ODE* and the *theory of dynamical systems* that have their autonomous development.

M. Giaquinta and G. Modica, *Mathematical Analysis: An Introduction to Functions of Several Variables*, DOI: 10.1007/978-0-8176-4612-7_6,
© Birkhäuser Boston, a part of Springer Science + Business Media, LLC 2010

6.1 Linear Systems

a. Linear systems of first-order ODEs

If for $f : I \times \mathbb{R}^n \to \mathbb{R}^n$ we have

$$|f(t,x)| \le a(t)|x| + b(t)$$

where $a(t)$ and $b(t)$ are bounded and continuous functions in I, then every local solution of $x' = f(t,x)$ extends as a solution in the whole interval I, see [GM3]. In particular, for every $(t_0, x_0) \in I \times \mathbb{R}^n$, the system of linear differential equations in *normal form*

$$x' = \mathbf{A}(t)x + b(t), \qquad \mathbf{A} \in C^0(I, M_{n,n}(\mathbb{R})), \ b \in C^0(I, \mathbb{R}^n),$$

has a unique solution $x(t) = x(t; t_0, x_0)$ defined on I, of class C^1, and such that $x(t_0) = x_0$. In other words, if we denote by $\mathcal{S} \subset C^1(I, \mathbb{R}^n)$ the set of solutions of $x' = f(t,x)$,

$$\mathcal{S} := \Big\{ y \in C^1(I, \mathbb{R}^n) \,\Big|\, y \text{ is a solution of } x' = f(t,x) \Big\}$$

and by $\mathcal{F} : \mathbb{R}^n \to \mathcal{S}$ the map that associates to the initial data x_0 at time t_0 the unique solution $x(t; t_0, x_0)$, we infer that $\mathcal{F}$ is well defined, injective, and onto.

In the *linear homogeneous case*,

$$x' = \mathbf{A}(t)x, \qquad \mathbf{A} \in C^0(I, M_{n,n}(\mathbb{R})), \tag{6.1}$$

$\mathcal{F}$ is also *linear*, hence we can state the following.

6.1 Proposition. *The space $\mathcal{S}$ of all solutions of the linear system* (6.1) *is a real vector space of dimension n.*

6.2 Definition. *We say that a map $t \to \mathbf{Z}(t) \in M_{n,n}(\mathbb{R})$, $t \in I$ is a* fundamental system of solutions *of* (6.1) *if $\mathbf{Z}(t)$ has as columns n solutions of* (6.1) *that form a basis for the space $\mathcal{S}$ of all solutions of* (6.1).

Again from the linearity of $\mathcal{F}$, we infer at once the following.

6.3 Proposition. *A map $t \to \mathbf{Z}(t) \in M_{n,n}(\mathbb{R})$, $t \in I$, is a fundamental system of solutions of* (6.1) *if and only $\mathbf{Z}(t)$ has as columns n solutions of* (6.1) *and* $\det \mathbf{Z}(s) \ne 0$ *at some $t_0 \in I$ (and therefore $\det \mathbf{Z}(s) \ne 0$ at every $s \in I$); in matrix notation*

$$\begin{cases} \mathbf{Z}'(t) = \mathbf{A}(t)\mathbf{Z}(t) & \forall t \in I, \\ \det \mathbf{Z}(s) \ne 0 & \forall s \in I. \end{cases}$$

In the linear case, for every $t, t_0 \in I$ also the map $x_0 \rightarrow x(t, ; t_0, x_0)$ is injective and linear from $\mathbb{R}^n$ onto itself; therefore there is a nonsingular matrix $\mathbf{W}(t, t_0)$ such that

$$x(t; t_0, x_0) = \mathbf{W}(t, t_0)x_0 = \mathbf{W}(t, t_0)x(t_0; t_0, x_0), \qquad (6.2)$$

called the *transition matrix* (from the value of x_0 at "time" t_0 to the value of $x = \mathbf{W}(t, t_0)x_0$ at "time" t). By definition

$$\mathbf{W}(s, s) = \mathrm{Id}$$

and for every $j = 1, \ldots, n$, the j-column $w^j(t)$ of $\mathbf{W}(t, s)$, $j = 1, \ldots, n$ solves the Cauchy problem

$$\begin{cases} w_t^j(t, s) = \mathbf{A}(t)w^j(t, s), \\ w^j(s) = e_j, \end{cases}$$

$(e_1, e_2, \ldots, e_n)$ being the canonical basis of $\mathbb{R}^n$. Therefore we infer that *for fixed $s \in I$, $t \rightarrow \mathbf{W}(t, s)$ is the fundamental system of solutions of* (6.1) *for which $\mathbf{W}(s, s) = \mathrm{Id}$*,

$$\begin{cases} \mathbf{W}_t(t, s) = \mathbf{A}(t)\mathbf{W}(t), \\ \mathbf{W}(s, s) = \mathrm{Id}. \end{cases} \qquad (6.3)$$

According to (6.2) the map $t \rightarrow \mathbf{Z}(t) : I \rightarrow M_{n,n}$ is a fundamental system of (6.1) if and only if

$$\mathbf{Z}(t) = \mathbf{W}(t, s)\mathbf{Z}(s).$$

In particular, we may compute $\mathbf{W}(t, s)$ from a given fundamental system of solutions $t \rightarrow \mathbf{Z}(t)$ as

$$\mathbf{W}(t, s) = \mathbf{Z}(t)\mathbf{Z}(s)^{-1} \qquad \forall t, s \in I. \qquad (6.4)$$

since in this case $\mathbf{Z}(s)$ is invertible.

6.4 Proposition. *Let $\mathbf{W}(t, s)$ be the transition matrix associated to $\mathbf{A}(t)$. Then we have:*

(i) $\mathbf{W}(t, t) = \mathrm{Id}$ *for all $t \in I$.*
(ii) $\mathbf{W}(t, s)\mathbf{W}(s, r) = \mathbf{W}(t, r)$.
(iii) $\mathbf{W}(t, s)^{-1} = \mathbf{W}(s, t)$.
(iv) *We have*

$$\frac{\partial \mathbf{W}(t, s)}{\partial t} = \mathbf{A}(t)\mathbf{W}(t, s), \qquad \frac{\partial \mathbf{W}(t, s)}{\partial s} = -\mathbf{W}(t, s)\mathbf{A}(s).$$

(v) Liouville's equation holds

$$\frac{\partial}{\partial t} \det \mathbf{W}(t,s) = \operatorname{tr}(\mathbf{A}(t)) \det \mathbf{W}(t,s),$$

in particular Abel's formula holds

$$\det \mathbf{W}(t,s) = \exp\left(\int_s^t \operatorname{tr}\mathbf{A}(\tau)\,d\tau\right).$$

Proof. We leave to the reader the proofs of (i), (ii),..., and (iv), and we prove (v). Since

$$\mathbf{W}(t+\epsilon,s) = \mathbf{W}(t,s) + \frac{\partial \mathbf{W}}{\partial t}(t,s)\,\epsilon + o(\epsilon)$$
$$= \mathbf{W}(t,s) + \epsilon\,\mathbf{A}(t)\mathbf{W}(t,s) + o(\epsilon) == (\operatorname{Id} + \epsilon\mathbf{A}(t))\mathbf{W}(t,s) + o(\epsilon)$$

we have

$$\det \mathbf{W}(t+\epsilon,s) = \det(\operatorname{Id} + \epsilon\mathbf{A}(t))\det\mathbf{W}(t,s) + o(\epsilon)$$
$$= (1 + \epsilon\operatorname{tr}\mathbf{A}(t) + o(\epsilon))\det\mathbf{W}(t,s) + o(\epsilon)$$
$$= \det W(t,s) + \epsilon\operatorname{tr}\mathbf{A}(t)\det\mathbf{W}(t,s) + o(\epsilon)$$

hence (v). □

6.5 ¶. Noticing that $\mathbf{W}(t,s) = \mathbf{W}(t,t_0)\mathbf{W}(t_0,s)$, prove that $\mathbf{W}(t,s)$ is of class C^1 in (t,s).

Either by a direct check or by the method of *variations of constants*, that is, looking for a solution of the type $u(t) := \mathbf{W}(t,s)c(t)$, $c: I \to\in \mathbb{R}^n$, we easily get the following.

6.6 Theorem. *The unique solution of the Cauchy problem*

$$\begin{cases} x' = \mathbf{A}(t)x + f(t), \\ x(t_0) = x_0 \end{cases}$$

is given by

$$x(t) = \mathbf{W}(t,t_0)x_0 + \int_{t_0}^t \mathbf{W}(t,\tau)f(\tau)\,d\tau$$

where $\mathbf{W}(t,s)$ *is the transition matrix associated to* $\mathbf{A}(t)$.

Now, define by induction

$$\begin{cases} \mathbf{W}_0(t,s) := \operatorname{Id}, \\ \mathbf{W}_{k+1}(t,s) := \int_s^t \mathbf{A}(\tau)\mathbf{W}_k(\tau,s)\,d\tau, \quad k \geq 0, \end{cases} \tag{6.5}$$

i.e.,

$$\begin{cases} \mathbf{W}_0(t,s) = \mathrm{Id}, \\ \mathbf{W}_1(t,s) = \int_s^t \mathbf{A}(\tau)\,d\tau, \\ \mathbf{W}_k(t,s) = \int_s^t \int_s^{\tau_k} \cdots \int_s^{\tau_2} \mathbf{A}(\tau_k) \cdots \mathbf{A}(\tau_1)\,d\tau_1 \cdots d\tau_k, \quad k \ge 2. \end{cases}$$

By applying the contraction theorem as for proving existence, see [GM3], and using the linearity we easily infer the following.

6.7 Proposition. *For every interval* $J \subset\subset I$, *we have*

$$|\mathbf{W}_k(t,s)| \le \|\mathbf{A}\|_{\infty,J}^k \frac{|t-s|^k}{k!} \qquad \forall t,s \in J,$$

where $\|\mathbf{A}\|_{\infty,J} := \sup_{t\in J} \|\mathbf{A}(t)\|$. *Therefore, the series* $\sum_{k=0}^{\infty} \mathbf{W}_k(t,s)$ *converges uniformly on the compact sets of* $I \times I$ *to the transition matrix,*

$$\mathbf{W}(t,s) = \sum_{k=0}^{\infty} \mathbf{W}_k(t,s),$$

and the following estimate holds

$$\|\mathbf{W}(t,s)\| \le e^{|t-s|\,\|\mathbf{A}\|_{\infty,J}} \qquad \forall t,s \in J \subset\subset I.$$

b. Linear systems with constant coefficients

Suppose that $\mathbf{A}(t)$ commutes with

$$\mathbf{B}(t,s) := \int_s^t \mathbf{A}(\tau)\,d\tau$$

for every $t \in \mathbb{R}$ and fixed s. For instance, this happens if $\mathbf{A}(t)$ and $\mathbf{A}(\tau)$ commute for all t and τ,

$$[\mathbf{A}(t), \mathbf{A}(\tau)] := \mathbf{A}(t)\mathbf{A}(\tau) - \mathbf{A}(\tau)\mathbf{A}(t) = 0,$$

in particular when $\mathbf{A}(t) := \mathbf{A}$ is a constant matrix.

If $\mathbf{A}(t)$ and $\mathbf{B}(t,s)$ commute for every t and fixed s, then

$$\frac{\partial}{\partial t}\mathbf{B}^k(t,s) = k\mathbf{B}^{k-1}(t,s)\frac{\partial \mathbf{B}}{\partial t}(t,s) = k\mathbf{B}^{k-1}(t,s)\mathbf{A}(t)$$

and we infer by induction from (6.5) that $\mathbf{W}_k(t,s) = \frac{1}{k!}\mathbf{B}^k(t,s)\ \forall k \ge 0$, hence

$$\begin{aligned} \mathbf{W}(t,s) &= \sum_{k=0}^{\infty} \frac{1}{k!}\mathbf{B}^k(t,s) \\ &= \sum_{k=0}^{\infty} \frac{1}{k!}\left(\int_s^t \mathbf{A}(\tau)\,d\tau\right)^k = \exp\left(\int_s^t \mathbf{A}(\tau)\,d\tau\right). \end{aligned} \qquad (6.6)$$

The exponential of a matrix:

(i) $e^{\mathbf{A}} = \sum_{k=0}^{\infty} \frac{\mathbf{A}^k}{k!} = \lim_{n \to \infty} \left(\mathrm{Id} + \frac{1}{n} \mathbf{A} \right)^n$.

(ii) $\left\| e^{\mathbf{A}} \right\| \leq e^{\|\mathbf{A}\|}$.

(iii) If $\mathbf{AB} = \mathbf{BA}$, then $e^{\mathbf{A}} \mathbf{B} = \mathbf{B} e^{\mathbf{A}}$, $e^{\mathbf{A}+\mathbf{B}} = e^{\mathbf{A}} e^{\mathbf{B}}$, $\left(e^{\mathbf{A}} \right)^{-1} = e^{-\mathbf{A}}$, $\frac{d}{dt} e^{t\mathbf{A}} = \mathbf{A} e^{t\mathbf{A}}$.

(iv) If $\mathbf{B}$ is invertible, then $e^{\mathbf{BAB}^{-1}} = \mathbf{B} e^{\mathbf{A}} \mathbf{B}^{-1}$.

(v) $\det e^{\mathbf{A}} = e^{\operatorname{tr} \mathbf{A}}$.

(vi) $\partial_{\mathbf{A}} \exp(\mathbf{A})(\mathbf{H}) = \sum_{h,k \geq 0} \frac{\mathbf{A}^h \mathbf{H} \mathbf{A}^k}{(k+h+1)!}$ ($= \mathbf{H} e^{\mathbf{A}}$ if $\mathbf{AH} = \mathbf{HA}$).

Figure 6.1. Some properties of the exponential of a matrix, see Chapter 1 and [GM3].

Therefore for systems with constant coefficients, $\mathbf{A}(t) = \mathbf{A}$, we conclude

$$\mathbf{W}(t, s) = e^{(t-s)\mathbf{A}}$$

Notice that this implies that

$$\mathbf{W}(t, s) = \mathbf{W}(t - s, 0)$$

and by (6.4)

$$\mathbf{Z}(t)\mathbf{Z}(s)^{-1} = \mathbf{Z}(t - s)\mathbf{Z}(0)^{-1}$$

for every fundamental matrix $\mathbf{Z}(t)$ of $x' = \mathbf{A}x$. These formulas can also be proved by direct computation. As a consequence of Theorem 6.6 we get the following.

6.8 Corollary. *The unique solution of the Cauchy problem*

$$\begin{cases} x'(t) = \mathbf{A}x(t) + f(t), \\ x(t_0) = x_0 \end{cases}$$

is given by

$$x(t) = e^{(t-t_0)\mathbf{A}} x_0 + \int_{t_0}^{t} e^{(t-s)\mathbf{A}} f(s)\, ds.$$

c. More about linear systems

Consider a linear system with constant coefficient $x' = \mathbf{A}x$ where $\mathbf{A} \in M_{n,n}(\mathbb{R})$. By a change of variables $x = \mathbf{P}y$, it tranforms into the equivalent linear system $y' = \mathbf{P}^{-1}\mathbf{A}\mathbf{P}y$ with coefficient matrix $\mathbf{B} := \mathbf{P}^{-1}\mathbf{A}\mathbf{P}$ that is similar to $\mathbf{A}$. Of course, if we are able to find the solutions of $y' = \mathbf{B}y$, then the solutions of the original system are given by $x(t) = \mathbf{P}y(t)$, that is

$$e^{t\mathbf{A}} = \mathbf{P} e^{t\mathbf{B}} \mathbf{P}^{-1}.$$

For instance, if $(u_1, u_2, \ldots, u_n)$ is a basis of $\mathbb{R}^n$ made by eigenvectors of $\mathbf{A}$, then

$$\mathrm{diag}\,(\lambda_1, \ldots, \lambda_n) = \mathbf{P}^{-1}\mathbf{A}\mathbf{P}$$

with $\mathbf{P} = \left[u_1|u_2|\ldots|u_n\right]$. Therefore, we find

$$
\begin{aligned}
e^{t\mathbf{A}} &= \exp\left(t\mathbf{P}\mathrm{diag}\,(\lambda_1, \lambda_2, \ldots, \lambda_n)\mathbf{P}^{-1}\right) \\
&= \mathbf{P}\exp\left(t\mathrm{diag}\,(\lambda_1, \lambda_2, \ldots, \lambda_n)\right)\mathbf{P}^{-1} \\
&= \mathbf{P}\,\mathrm{diag}\,(e^{\lambda_1 t}, e^{\lambda_2 t}, \ldots, e^{\lambda_n t})\mathbf{P}^{-1} \\
&= \left[u_1 e^{\lambda_1 t}|u_2 e^{\lambda_2 t}|\ldots|u_n e^{\lambda_n t}\right]\mathbf{P}^{-1}.
\end{aligned}
$$

6.9 Example. We can find the same result by noticing that $x(t) = e^{\lambda t}u$ solves $x' = \mathbf{A}x$ with initial data $x(0) = u$ if u is an eigenvector of $\mathbf{A}$ with associated eigenvalue λ. Therefore, if $(u_1, u_2, \ldots, u_n)$ is a basis of eigenvectors of $\mathbf{A}$ with corresponding eigenvalues $\lambda_1, \lambda_2, \ldots, \lambda_n$, then the matrix

$$\mathbf{Z}(t) = \left[e^{\lambda_1 t}u_1 \,\middle|\, e^{\lambda_2 t}u_2 \,\middle|\, \ldots \,\middle|\, e^{\lambda_n t}u_n\right]$$

is a fundamental matrix for $x' = \mathbf{A}x$ with $\mathbf{Z}(0) = \left[u_1 \,|\, u_2 \,|\, \ldots \,|\, u_n\right]$, hence

$$e^{t\mathbf{A}} = \mathbf{W}(t, 0) = \mathbf{Z}(t)\mathbf{Z}(0)^{-1} = \mathbf{Z}(t)\mathbf{P}^{-1}.$$

In the general case, one considers $\mathbf{A}$ as a complex valued matrix and uses one of its Jordan canonical forms.

Let $\lambda_1, \lambda_2, \ldots, \lambda_k$ be the distinct eigenvalues of $\mathbf{A}$ with relative algebraic multiplicities $m_1, m_2, \ldots, m_k$. For every i, $1 \leq i \leq k$, let p_i be the dimension of the eigenspace relative to λ_i. Then, see [GM3], there exists a linear change of basis $\mathbf{P} \in M_{n,n}(\mathbb{C})$ such that $\mathbf{J} := \mathbf{P}^{-1}\mathbf{A}\mathbf{P}$ has the form

$$
\mathbf{J} = \begin{pmatrix}
\mathbf{J}_{1,1} & 0 & 0 & \cdots & 0 \\
0 & \mathbf{J}_{1,2} & 0 & \cdots & 0 \\
\cdots & & & & \\
0 & 0 & 0 & \cdots & \mathbf{J}_{k,p_k}
\end{pmatrix}
$$

where $i = 1, \ldots, k$, $j = 1, \ldots, p_i$ and

$$\mathbf{J}_{i,j} = \begin{cases} \lambda_i & \text{if } \mathbf{J}_{i,j} \text{ has dimension } \ell_{i,j} = 1, \\[2em] \begin{pmatrix} \lambda_i & 1 & 0 & 0 & \cdots & 0 \\ 0 & \lambda_i & 1 & 0 & \cdots & 0 \\ \cdots & & & & & \\ \cdots & & & & & \\ 0 & 0 & 0 & \cdots & \lambda_i & 1 \\ 0 & 0 & 0 & \cdots & 0 & \lambda_i \end{pmatrix} & \text{if } \ell_{i,j} = \dim \mathbf{J}_{i,j} > 1. \end{cases}$$

If $\mathbf{J}' = \mathbf{J}_{i,j} = (\lambda)$ has dimension 1, then $e^{t\mathbf{J}'} = e^{\lambda t}$. Instead, if $\mathbf{J}' = \mathbf{J}_{i,j}$ is one of the blocks of dimension $\ell \geq 2$,

$$\mathbf{J}' = \begin{pmatrix} \lambda & 1 & 0 & \cdots & 0 \\ 0 & \lambda & 1 & \cdots & 0 \\ \cdots & & & & \\ 0 & \cdots & 0 & \lambda & 1 \\ 0 & 0 & \cdots & 0 & \lambda \end{pmatrix},$$

then

$$\mathbf{J}' = \lambda \operatorname{Id} + \mathbf{N}, \qquad \mathbf{N}_{ij} = \delta_{i+1,j}.$$

Since $\mathbf{N}$ and Id commute, we have

$$e^{t\mathbf{J}'} = e^{t\lambda \operatorname{Id}} e^{t\mathbf{N}} = e^{\lambda t} e^{t\mathbf{N}}.$$

On the other hand, since

$$(\mathbf{N}^k)_{ij} = \begin{cases} \delta_{i+k,j} & \text{if } k < \ell, \\ 0 & \text{if } k \geq \ell \end{cases}$$

we have

$$e^{t\mathbf{N}} = \sum_{k=0}^{\infty} \frac{1}{k!} \mathbf{N}^k = \sum_{k=0}^{\ell} \frac{1}{k!} \mathbf{N}^k,$$

hence

$$\exp(t\mathbf{J}') = e^{\lambda t} e^{t\mathbf{N}} = e^{\lambda t} \begin{pmatrix} 1 & t & \dfrac{t^2}{2} & \cdots & \dfrac{t^{\ell-1}}{(\ell-1)!} \\[1em] 0 & 1 & t & \cdots & \dfrac{t^{\ell-2}}{(\ell-2)!} \\[1em] \cdots & & & & \\[1em] 0 & 0 & \cdots & 1 & t \\[1em] 0 & 0 & \cdots & 0 & 1 \end{pmatrix}.$$

In conclusion, we have

$$\exp\left(t\mathbf{A}\right) = \mathbf{P}e^{t\mathbf{J}}\mathbf{P}^{-1} \tag{6.7}$$

with

$$
e^{t\mathbf{J}} := \begin{pmatrix} e^{t\mathbf{J}_{1,1}} & 0 & \cdots & 0 \\ 0 & e^{t\mathbf{J}_{1,2}} & \cdots & 0 \\ & \cdots & & \\ 0 & 0 & \cdots & e^{t\mathbf{J}_{k,p_k}} \end{pmatrix}.
$$

We observe that every entry of the matrix $e^{t\mathbf{A}}$ has the form

$$\sum_{j=1}^{k} p_j(t)\exp\left(\lambda_j t\right)$$

where $p_j(t)$ is a polynomial of degree at most $p_j - 1$ and $\lambda_1, \lambda_2, \ldots, \lambda_n$ are the eigenvalues of $\mathbf{A}$. It follows that for every $\rho > \max(\Re\lambda_1, \Re\lambda_2, \ldots, \Re\lambda_n)$ there is a constant C_ρ such that

$$\left|\exp\left(t\mathbf{A}\right)\right| \leq C_\rho e^{t\rho}, \qquad 0 \leq t < \infty.$$

In particular, we infer the following *stability result.*

6.10 Theorem. *Suppose that all eigenvalues of $\mathbf{A}$ have negative real part. Then every solution of $x' = \mathbf{A}x$ converges to zero when $t \to +\infty$. Indeed, if $\max_{i=1,n}(\Re\lambda_i) < -\sigma < 0$, then there exists a constant C_σ such that*

$$|x(t)| \leq C_\sigma e^{-\sigma t}, \qquad t \geq 0.$$

6.11 ¶. Let $x(t)$ solve $x' = \mathbf{A}x + f$. Prove that $x(t)$ does not grow more than exponentially at $+\infty$ if f does not grow more than exponentially at $+\infty$.

6.12 ¶. Let $\mathbf{Z}(t)$ be a fundamental system of solutions of the $n \times n$ first-order system of ODE $v' = \mathbf{A}v$, $\mathbf{A} \in M_{n,n}(\mathbb{R})$. Prove that

$$\det \mathbf{Z}(t) = \det \mathbf{Z}(t_0)\exp\left(\int_{t_0}^{t} \operatorname{tr}\mathbf{A}(\tau)\,d\tau\right)$$

for all $t, t_0 \in I$. [*Hint:* Use Abel's formula.]

6.1.1 Higher-order equations

a. Higher-order equations and first-order systems

A linear differential equation of order n in normal form,

$$u^{(n)} + a_{n-1}(t)u^{(n-1)} + \cdots + a_1(t)u' + a_0(t)u = f(t), \qquad (6.8)$$

is equivalent to a system of linear differential equations of first-order in the unknown $v := (u, u', \ldots, u^{(n-1)})$. Indeed, if u solves (6.8) and we set

$$v = (v_1, v_2, \ldots, v_n) := (u, u', \ldots, u^{(n-1)}), \qquad (6.9)$$

then v solves the system

$$\begin{cases} v_0' = v_1, \\ v_1' = v_2, \\ \ldots, \\ v_n' = -a_0(t)v_0 - a_1(t)v_1 + \cdots - a_{n-1}(t)v_{n-1} + b(t) \end{cases}$$

that has the vector form

$$v' = \mathbf{A}(t)v + f(t) \qquad (6.10)$$

with

$$\mathbf{A}(t) := \begin{pmatrix} 0 & 1 & 0 & \cdots & 0 \\ 0 & 0 & 1 & \cdots & 0 \\ \cdots & & & & \\ 0 & 0 & 0 & \cdots & 1 \\ -a_0(t) & -a_1(t) & -a_2(t) & \cdots & -a_{n-1}(t) \end{pmatrix} \qquad (6.11)$$

and

$$f(t) = (0, 0, \ldots, 0, b(t))^T.$$

Conversely, if $v : I \to \mathbb{R}^n$ solves (6.10) with $\mathbf{A}$ as in (6.11) and $f(t) = f(t) = (0, 0, \ldots, 0, b(t))^T$, then the first component $u := v_0$ of v solves (6.8) and $v = (u, u', \ldots, u^{(n-1)})$.

Therefore we can apply the theory of systems of linear first-order equations to represent the solutions of an equation of order n.

6.13 Proposition. *Let $\mathbf{W}(t, s)$ be the transition matrix of the system (6.10), then the solutions of (6.8) are given by*

$$u(t) = \sum_{j=1}^{n} \mathbf{W}_{1j}(t, t_0)c_j + \int_{t_0}^{t} \mathbf{W}_{1n}(t, \tau)f(\tau) \, d\tau, \qquad c = (c_1, \ldots, c_n) \in \mathbb{R}^n.$$

$$(6.12)$$

In particular, if $f(t) = 0 \; \forall t$, then the space of solutions is an n-dimensional vector space.

If $u_1, u_2, \ldots, u_n$ are n solutions of the homogeneous equation

$$u^{(n)} + a_{n-1}(t)u^{(n-1)} + \cdots + a_1(t)u' + a_0(t)u = 0 \qquad (6.13)$$

associated to (6.8), then the columns of the $n \times n$-matrix

$$\mathbf{Z}(t) := \begin{pmatrix} u_1 & u_2 & \cdots & u_n \\ u_1' & u_2' & \cdots & u_n' \\ u_1'' & u_2'' & \cdots & u_n'' \\ & \cdots & & \\ u_1^{(n-1)} & u_2^{(n-1)} & \cdots & u_n^{(n-1)} \end{pmatrix} \qquad (6.14)$$

are solutions of the system $v' = \mathbf{A}(t)v$. Therefore the following claims are equivalent

(i) $u_1, u_2, \ldots, u_n$ are linearly independent functions that solve (6.13),
(ii) the columns of $\mathbf{Z}(t)$ are linearly independent $\forall t$,
(iii) $\mathbf{Z}(t)$ is a fundamental system of solutions of $v' = \mathbf{A}v$,
(iv) $\det \mathbf{Z}(t_0) \neq 0$ for some $t_0 \in I$.

In this case $u_1, u_2, \ldots, u_n$ form a basis for the vector space of the solutions of (6.13). The function $w(t) := \det \mathbf{Z}(t)$ is called the *Wronskian* of the solutions $u_1, u_2, \ldots, u_n$.

6.14 ¶. Let $u_1, u_2, \ldots, u_n$ be n solutions of the homogeneous equation (6.13) and let $w(t) := \det \mathbf{Z}(t)$ as in (6.14). Prove that $w'(t) = -a_{n-1}(t)w$, hence

$$w(t) = w(t_0)\exp\left(-\int_{t_0}^{t} a_{n-1}(\tau)\, d\tau\right).$$

b. Homogeneous linear equations with constant coefficients

When the coefficients $a_0, \ldots, a_{n-1}$ are constant, we can compute a basis of solutions of the homogeneous equation

$$u^{(n)} + a_{n-1}u^{(n-1)} + \cdots + a_1u' + a_0u = 0 \qquad (6.15)$$

in terms of the roots of the *characteristic polynomial*

$$p(\lambda) := \sum_{k=0}^{n} a_k \lambda^k.$$

But, it is more convenient to work with *complex-valued solutions* of (6.15).

6.15 Theorem. *Let $\lambda_1, \lambda_2, \ldots, \lambda_k$ be the distinct roots of the characteristic polynomial $p(\lambda)$ of (6.15) with multiplicity, respectively, $r_1, r_2, \ldots, r_k$, so that $\sum_{i=1}^{k} r_i = n$. Then the functions*

$$t^h e^{\lambda_k t}, \qquad 1 \leq h \leq r_i,\ 1 \leq i \leq k,$$

form a basis of solutions for the homogeneous equation (6.15). In particular the vector space of solutions of (6.15) has dimension n.

Proof. Let $x' = \mathbf{A}x$ be the $n \times n$ first-order system associated to (6.15). One shows, proceeding for instance by induction on the dimension n, that $p(\lambda) = \det(\lambda \operatorname{Id} - \mathbf{A})$, so that the roots of $p(\lambda)$ are precisely the eigenvalues of $\mathbf{A}$. Denote by $u_{i,1}, \ldots u_{i,p_i}$ a basis of the eigenspace associated to λ_i and, for $j = 1, \ldots, p_i$ by $\ell_{i,j}$ the dimension of the Jordan block associated to $u_{i,j}$. Then $\mathbf{A} = \mathbf{P}\mathbf{J}\mathbf{P}^{-1}$ and from (6.7) we infer that

$$\mathbf{P}e^{t\mathbf{J}}$$

is a fundamental system of solutions of $x' = \mathbf{A}x$. Recall that

$$e^{t\mathbf{J}} := \begin{pmatrix} e^{t\mathbf{J}_{1,1}} & 0 & \cdots & 0 \\ 0 & e^{t\mathbf{J}_{1,2}} & \cdots & 0 \\ \cdots & & & \\ 0 & 0 & \cdots & e^{t\mathbf{J}_{k,p_k}} \end{pmatrix}$$

and

$$\exp(t\mathbf{J}_{i,j}) = e^{\lambda_i t} \begin{pmatrix} 1 & t & \dfrac{t^2}{2} & \cdots & \dfrac{t^{\ell_{i,j}-1}}{(\ell_{i,j}-1)!} \\ 0 & 1 & t & \cdots & \dfrac{t^{\ell_{i,j}-2}}{(\ell_{i,j}-2)!} \\ \cdots & & & & \\ 0 & 0 & \cdots & 1 & t \\ 0 & 0 & \cdots & 0 & 1 \end{pmatrix}.$$

Therefore the n functions in the first row of $\mathbf{P}e^{t\mathbf{J}}$,

$$u_i(t) := \left(\mathbf{P}e^{t\mathbf{J}}\right)^1_i, \qquad i = 1, \ldots, n \tag{6.16}$$

form a basis of solutions for (6.15). Since $e^{t\mathbf{J}}$ is a block-triangular matrix, (6.16) rewrites as

$$\begin{pmatrix} u_1(t) \\ \vdots \\ u_n(t) \end{pmatrix} = \mathbf{C}y(t)$$

where $y(t)$ denotes the vector made by the functions in the first rows of the Jordan blocks

$$y(t) := \Big(e^{\lambda_1 t}, te^{\lambda_1 t}, \ldots, \frac{t^{\ell_{1,1}}}{\ell_{1,1}!}e^{\lambda_1 t}, \ldots, \ldots, e^{\lambda_1 t}, te^{\lambda_1 t}, \ldots, \frac{t^{\ell_{1,p_1}}}{\ell_{1,p_1}!}e^{\lambda_1 t},$$

$$e^{\lambda_2 t}, te^{\lambda_2 t}, \ldots, \frac{t^{\ell_{2,1}}}{\ell_{2,1}!}e^{\lambda_2 t}, \ldots, \ldots, e^{\lambda_2 t}, te^{\lambda_2 t}, \ldots, \frac{t^{\ell_{2,p_2}}}{\ell_{2,p_2}!}e^{\lambda_2 t}$$

$$\cdots$$

$$e^{\lambda_2 t}, te^{\lambda_2 t}, \ldots, \frac{t^{\ell_{k,p_1}}}{\ell_{k,p_1}!}e^{\lambda_2 t} \ldots, \ldots, e^{\lambda_k t}, te^{\lambda_k t}, \ldots, \frac{t^{\ell_{k,p_k}}}{\ell_{k,p_k}!}e^{\lambda_k t} \Big)^T$$

and

$$\mathbf{C} := \begin{pmatrix} C_{1,1} & 0 & \cdots & 0 \\ 0 & C_{1,2} & \cdots & 0 \\ & \cdots & & \\ 0 & 0 & \cdots & C_{k,p_k} \end{pmatrix}$$

where each block is triangular with the same constant in its diagonal.

Since u_1, $u_2, \ldots, u_n$ are linearly independent, $\mathbf{C}$ is nonsingular and the components of $y(t)$ are linearly independent, too. In particular, for every $i = 1, \ldots, k$ there is a unique Jordan block corresponding to λ_i, hence $p_i = 1$, $\ell_{i,1} = r_i$ and the functions

$$t^h e^{\lambda_k t}, \qquad 1 \le h \le r_i, \ 1 \le i \le k, \tag{6.17}$$

are the components of y, hence are linearly independent. Finally, since $\mathbf{C}$ is nonsingular, the functions in (6.17) are solutions of (6.15).   □

c. Nonhomogeneous linear ODEs

Now let us consider the nonhomogeneous equation

$$u^{(n)} + a_{n-1} u^{(n-1)} + \cdots + a_1 u' + a_0 u = b(t) \tag{6.18}$$

As we have already solved the corresponding homogeneous equation (6.15), it suffices to find a *particular solution*.

6.16 Theorem (Duhamel's formula). *Let $v(t)$ be the solution of the homogeneous equation (6.15) with initial data*

$$u(0) = 0, \ u'(0) = 0, \ \ldots, \ u^{(n-1)}(0) = 1.$$

Then the function

$$u(t) = \int_0^t k(t - \tau) b(\tau) \, d\tau$$

solves (6.18) with initial data

$$u(0) = 0, \ u'(0) = 0, \ \ldots, \ u^{(n-1)}(0) = 0.$$

Proof. Let $x' = \mathbf{A}x + f(t)$ be the linear system of first-order associated to (6.18). The last column of $\mathbf{W}(t, 0) := e^{t\mathbf{A}}$ solves $x' = \mathbf{A}x$ with initial value $x(0) = (0, 0, \ldots, 0, 1)^T$, and the first component of x, $x^1(t) = W_{1n}(t)$, solves the homogeneous equation (6.15) with initial data

$$u(0) = 0, \ u'(0) = 0, \ \ldots, \ u^{(n-1)}(0) = 1,$$

so that $v(t) = W_{1n}(t) \ \forall t$. Since $W(t, s) = W(t - s, 0) = e^{(t-s)\mathbf{A}}$, the result follows from (6.12).   □

6.2 Stability

In this section we consider *autonomous* systems of ordinary differential equations, i.e., systems of the type

$$x' = f(x) \tag{6.19}$$

where $f : \mathbb{R}^n \to \mathbb{R}^n$ is a smooth vector field, possibly defined in an open domain $\Omega \subset \mathbb{R}^n$. Clearly, those systems are *invariant* by time-translation, and every solution describes in t an *orbit* or *trajectory* that can be interpreted as the path of a particle that moves with velocity $f(x)$ at the point x; finally, for every $x \in \mathbb{R}^n$ there is an orbit going through it. The whole family of orbits of (6.19) is sometimes called the *flux* generated by f.

We are interested in *local* and *global* behavior of the whole family of orbits more than on each orbit, i.e., as one says, we want to look at (6.19) as a *dynamical system*. Then, we should expect in general, see [GM2], sensitive dependence from the initial conditions, strange attractors, chaotic behavior, etc., and even an introductive study would not be possible both for space reasons and as it would force us to deviate too much from our path: we refer the interested reader to any of the many monographs about ODE's and dynamical systems.

Here and in the next section, we confine ourselves to illustrating some classical results relative to 2×2-systems, or, as one says in physics, with one degree of freedom, with the goal of showing absence, in this case, of chaotic behavior. Such a chaotic behavior appears instead for 3×3-systems, but we shall not dwell on this.

6.2.1 Critical points and linearization

As we saw in Chapter 5, in a neighborhood of a point x_0 for which $f(x_0) \neq 0$, the orbits of the system (6.19) are trivial, meaning that they are diffeomorphic to a bundle of parallel straight lines.

6.17 Definition. *A point $x_0 \in \mathbb{R}^n$ for which $f(x_0) = 0$ is called a* critical *or* equilibrium point *for* (6.19).

If x_0 is a critical point, then the constant vector $x(t) := x_0$ is a solution with orbit the point x_0. In physics, critical points correspond to equilibrium states: For the pendulum,

$$\theta'' + \frac{g}{L} \sin \theta = 0,$$

that is equivalent to the system

$$\begin{cases} x_1' = x_2, \\ x_2' = -\frac{g}{L} \sin x_1, \end{cases}$$

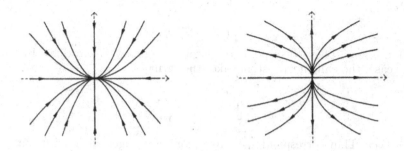

Figure 6.2. On the left: (a) A nodal point asymptotically stable with $\lambda_1 < \lambda_2 < 0$, and, on the right (b) a nodal point unstable with $0 < \lambda_1 < \lambda_2$.

the equilibrium points are $(n\pi, 0)$, $n = 0, \pm 1, \pm 2, \ldots$.

We begin by classifying the behavior of the orbits of the *linearization* of a 2×2-system in a neighborhood of a critical point. Such a linearization in general has the form

$$\begin{cases} x' = ax + by, \\ y' = cx + dy \end{cases} \tag{6.20}$$

where $ad - bc \neq 0$. The behavior of its orbits near zero is classified by the eigenvalues λ_1, λ_2 of the matrix $\mathbf{A} := \begin{pmatrix} a & b \\ c & d \end{pmatrix}$. In fact, a linear isomorphism of $\mathbb{R}^2$ transforms the system (6.20) in one of the following canonical forms in the variables (ξ, η).

(i) *Nodal points and saddle points.* They correspond respectively to real, distinct, and nonzero eigenvalues of the same sign and to real, distinct, and nonzero eigenvalues of opposite signs, $\lambda_1 < 0 < \lambda_2$. In both cases the canonical form is

$$\begin{cases} \xi' = \lambda_1 \xi, \\ \eta' = \lambda_2 \eta. \end{cases}$$

(ii) *Degenerate nodal points.* They correspond to a double eigenvalue. There are two possibilities:

(a) Rank $\begin{pmatrix} a - \lambda & b \\ c & d - \lambda \end{pmatrix} = 0$. In this case, the canonical system is

$$\begin{cases} \xi' = \lambda \xi, \\ \eta' = \lambda \eta. \end{cases}$$

(b) Rank $\begin{pmatrix} a - \lambda & b \\ c & d - \lambda \end{pmatrix} = 1$. In this case, the canonical system takes the form

$$\begin{cases} \xi' = \frac{a+b}{2}\xi + \eta, \\ \eta' = \frac{a+d}{2}\eta. \end{cases}$$

(iii) *Centers.* They correspond to purely imaginary eigenvalues. In this case, the canonical system takes the form

$$\begin{cases} \xi' = \alpha\xi - \beta\eta, \\ \eta' = \beta\xi + \alpha\eta. \end{cases}$$

(iv) *Foci.* They correspond to complex conjugate eigenvalues with nonzero real part.

6.18 Definition. *We say that a critical point x_0 for the system $x' = f(x)$ is* stable *if for every open neighborhood $U(x_0)$ of x_0 there exists another open neighborhood $V(x_0) \subset U(x_0)$ of x_0 such that every orbit passing through $V(x_0)$ at t_0 remains in $U(x_0)$ for all $t \geq t_0$.*

We say that x_0 is asymptotically stable *if it is stable and there is an open neighborhood $W(x_0)$ of x_0 such that every orbit through $W(x_0)$ converges to x_0 when $t \to \infty$.*

An isolated critical point that is not stable is said to be unstable.

Now, consider the nonlinear 2×2-system

$$\begin{cases} x' = P(x,y), \\ y' = Q(x,y) \end{cases} \tag{6.21}$$

and assume that $(0,0)$ is a critical point, and actually that

$$P(x,y) = ax + by + o\left(\sqrt{x^2 + y^2}\right),$$

$$Q(x,y) = cx + dy + o\left(\sqrt{x^2 + y^2}\right)$$

with $ad - bc \neq 0$. Then the following result, which we state without proof, holds.

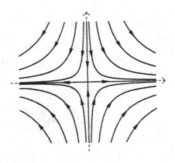

Figure 6.3. A saddle point (always unstable) $\lambda_1 < 0 < \lambda_2$.

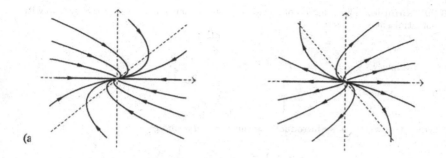

Figure 6.4. On the left: (a) A asymptoticallly stable with $\lambda < 0$, and on the right: (b) a degenerate nodal point unstable with $\lambda > 0$.

6.19 Theorem (Linearization theorem). *Suppose that* $(0,0)$ *is a focus, or a nodal or a saddle point for the linearized system. Then the behavior of the orbits of* (6.20) *and* (6.21) *is similar, meaning that the orbits are stable (respectively, asymptotically stable or unstable) for both systems.*

Notice that nothing is stated for centers of degenerate nodal points. In fact, in both cases a small perturbation of the coefficients of the linearized system moves the eigenvalues out of the imaginary axes or, respectively, separates double eigenvalues, modifying the nature of the critical point. In this case, the higher-order terms are decisive to establish stability.

6.2.2 Lyapunov's method

A different approach to determine the stability of a nonlinear suystem is due to Aleksandr Lyapunov (1857–1918); the idea behind it being the following classic result of Lagrange: The equilibrium position of a conservative mechanical system is stable if and only if its potential energy has a local minimum at this point.

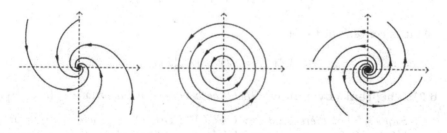

Figure 6.5. The case of conjugate complex eigenvalues: respectively, from the left, (a) $\Re\lambda < 0$, (b) $\Re\lambda = 0$, and (c) $\Re\lambda > 0$.

6.20 Example. The equation of motion of a one-dimensional conservative system with a potential energy U is

$$mx'' = -\frac{dU}{dx},$$

or, equivalently,

$$\begin{cases} x_1' = x_2, \\ mx_2' = -\dfrac{dU}{dx_1}(x_1) \end{cases}$$

where $x(t) = x_1(t)$. If we introduce the function *total energy*

$$H(x_1, x_2) := \frac{1}{m}\left(U(x_1) + \frac{1}{2}x_2^2\right),$$

the equation of motion of a one-dimensional conservative system with a potential energy U takes its *Hamiltonian form*, see [GM5],

$$\begin{cases} x_1' = \dfrac{\partial H}{\partial x_2}(x_1, x_2), \\[2mm] x_2' = -\dfrac{\partial H}{\partial x_1}(x_1, x_2), \end{cases}$$

and one easily finds that the orbits that lie on the level sets of H are closed lines that retract to zero if U has a minimum at 0. Therefore, 0 is a stable equilibrium point.

We begin with a few simple remarks coming from the existence and unicity theorems for the Cauchy problem for ODE, see [GM3]. We leave the details to the reader.

(i) The orbits of the system $x' = f(x)$ through a noncritical point cannot reach in finite time a critical point, i.e., if $x(t)$ is a solution and $x(t) \to x_0$ when $t \to t_0$, then x_0 is not a critical point.

(ii) An orbit of $x' = f(x)$ through a noncritical point has no self-intersections, except when it is a simple and closed curve, corresponding to a periodic solution.

Let $x(t)$ be a solution of $x' = f(x)$ and let $V : \mathbb{R}^n \to \mathbb{R}$ be a scalar function. We have

$$\frac{d}{dt}V(x(t)) = \nabla V(x(t)) \bullet f(x(t)),$$

and it is convenient to set

$$V^*(x) := \nabla V(x) \bullet f(x).$$

6.21 Theorem (Lyapunov). *Let 0 be a critical point for $x' = f(x)$.*

(i) *Suppose that there exist $r > 0$ and $V : B(0,r) \to \mathbb{R}$ with $V(0) = 0$, $V(x) > 0$ for $x \neq 0$ for which $V^*(x) \leq 0$ $\forall x \in B(0,r)$. Then 0 is a stable critical point. Moreover, if $V^*(0) = 0$ and $V^*(x) < 0$ for $x \neq 0$, then 0 is asymptotically stable.*

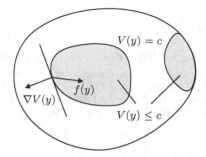

Figure 6.6. Illustration of the proof of Lyapunov's theorem.

(ii) *If there exists $V : B(0,r) \to \mathbb{R}$ with $V(0) = 0$ and either $V(x) > 0$ for all $x \neq 0$ or $V(x) < 0$ for all $x \neq 0$ and if, moreover, in every neighborhood of 0 there exists $\bar{x} \neq 0$ such that $\operatorname{sgn} V(\bar{x}) = \operatorname{sgn} V^*(\bar{x})$, then 0 is an unstable critical point.*

Theorem 6.21 applies to many interesting situations that we are not going to specify. But, of course, it also has limitations. It does not tell us how to construct the function V or find V, often called a *Lyapunov function* of the system; it provides us with only sufficient conditions and, finally, it gives no estimate of the region of asymptotic stability. Further inquiries are needed and could be done, but we will not dwell on this topic.

Proof. The idea of the proof is contained in Figure 6.6. Formally, we proceed as follows:

(i) By assumption there is $r > 0$ such that

$$V(x) > 0 \ \forall x \in B(0,r) \setminus \{0\} \qquad \text{and} \qquad V^*(x) \leq 0 \ \forall x \in B(0,r).$$

For $0 < \epsilon < r$ we set

$$\mu_\epsilon := \min_{\epsilon \leq |\xi| < r} V(x).$$

Trivially $\mu_\epsilon > 0$ and, by the continuity of V, there exists $\delta > 0$ such that $V(x) < \mu_\epsilon$ for all $|x| < \delta$. Now we shall show that for $|x_0| < \delta$ the solution $x(t; x_0)$ of $x' = f(x)$ with $x(0; x_0) = x_0$ is well defined for every $t \geq 0$ and $|x(t)| \leq \epsilon \ \forall t$.

Let $[0, t_1[$ be the maximal interval of existence of $x(t; x_0)$. Since $\frac{d}{dt} V(x(t)) = V^*(x(t)) \leq 0$, we have

$$0 < V(x(t)) \leq V(x_0) \leq \mu_\epsilon \qquad \text{for } 0 < t < t_1$$

hence $|x(t)| \leq \epsilon \ \forall t \in [0, t_1[$, because of the definition of μ_ϵ. Now we claim that $t_1 = +\infty$. Otherwise, there is t_2 such that $|x(t_2; x_0)| = \epsilon$ hence

$$\mu_\epsilon \leq V(x(t_2; x_0)) \leq V(x_0) < \mu_\epsilon,$$

a contradiction.

To prove the second part, it suffices to show that $x(t; x_0) \to 0$ as $t \to +\infty$, or, equivalently, $V(x(t; x_0)) \to 0$ as $t \to +\infty$. If not, since $t \to V(x(t; x_0))$ is monotone, there is $\eta > 0$ such that $V(x(t; x_0)) > \eta > 0$ for all $t \geq 0$. If $0 < \delta_1 < r$ is such that $V(x) < \eta$ for $|x| < \delta_1$, then $|x(t; x_0)| \geq \delta_1 \ \forall t$. We set

$$\mu := \min_{\delta_1 \leq |x| \leq \rho} (-V^*(x)),$$

then $\mu > 0$ and

$$-\frac{d}{dt}V(x(t)) = -V^*(x(t)) \geq \mu \qquad \text{for } t > 0.$$

By integration we finally get $V(x(t)) \leq V(x_0) - \mu t$, which is absurd for t large.

(ii) Suppose $V^*(x) > 0$ on $B(0, r)$ and, without loss of generality, $|V(x)| \leq M$ for some $M > 0$ on $B(0, r)$. If $V(\overline{x}) > 0$, we choose $\delta > 0$ so that $V(x) < V(\overline{x})$ if $|x| < \delta$, and we denote by $x(t; \overline{x})$ the solution of $x' = f(x)$ with $x(0; \overline{x}) = \overline{x}$ defined in its maximal interval $[0, t_1[$. Since $\frac{d}{dt}V(x(t; \overline{x})) = V^*(x(t; \overline{x})) \geq 0$, we have

$$V(x(t; \overline{x})) \geq V(\overline{x})$$

from which $x(t, \overline{x}) \geq \delta \ \forall t \in [0, t_1[$. Therefore, if we set

$$\mu^* := \min_{\delta \leq |x| \leq r} V^*(y),$$

we have $\mu^* > 0$ and

$$V(x(t; \overline{x})) \geq V(\overline{x}) + \mu^* t.$$

On the other hand, t_1 cannot be $+\infty$ as, otherwise, we would have $V(x(t, \overline{x})) \to +\infty$, which contradicts $|V(x)| \leq M$. Consequently, there exists t_2 such that $|x(t_2, \overline{x})| = r$. In this way we find a sequence of points that converges to zero whose orbits leave $B(0, r)$ in finite time, i.e., 0 is an unstable critical point. □

6.22 ¶ Nonlinear damped oscillator. Let $p : \mathbb{R}^2 \to \mathbb{R}$ and $q : \mathbb{R} \to \mathbb{R}$ be continuous functions such that $p(u, s) \geq 0$, $u\, q(u) \geq 0$ for all $(u, s) \in \mathbb{R}^2$ and

$$\int_0^y q(u)\, dx \to +\infty \qquad \text{as } |y| \to \infty.$$

Prove the following.

(i) the solutions of the homogeneous equation

$$x'' + p(x, x')x' + q(x) = 0$$

remain bounded in time, i.e., there exists $K = K(x(0), x'(0))$ such that $|x(t)| + |x'(t)| \leq K \ \forall t \geq 0$.

(ii) For all x_0, $x_1 \in \mathbb{R}$, the Cauchy problem

$$\begin{cases} x'' + p(x, x')x' + q(x) = 0, \\ x(0) = x_0, \ x'(0) = x_1 \end{cases}$$

has a unique solution.

[*Hint:* Consider the Lyapunov function

$$V(t) := \frac{x'^2(t)}{2} + \int_0^{x(t)} p(u)\, du.]$$

6.3 Poincaré–Bendixson Theorem

In this section we deal with the behavior of orbits of first-order differential systems in the plane; in particular, we shall see that no complications such as infinitely many periodic orbits, invariant Cantor type sets and many of the phenomena common to map iterations or to difference systems, or differential systems in 3 or more dimensions can occur for differential systems in the plane. The reason for that is in the Poincaré–Bendixson theorem, which states that, outside the stationary points, the most complicated orbit that a planar system can have is an orbit converging when $t \to +\infty$ and $t \to -\infty$ to a closed orbit.

After a few preliminaries in Section 6.3.1, we prove the Poincaré–Bendixson theorem in Section 6.3.2, and finally, in Section 6.3.3 we shall discuss the behavior of the orbits of a differential system on a torus without critical points.

6.3.1 Limit sets and invariant sets

From now on, we shall assume that for every $p \in \mathbb{R}^2$ the solution $\xi(t; p)$ of

$$\begin{cases} \xi' = f(\xi), \\ \xi(0) = p \end{cases}$$

exists for all $t \in \mathbb{R}$. We denote by $\gamma(p)$ the orbit of p,

$$\gamma(p) := \Big\{ x \,\Big|\, x = \xi(t; p), -\infty < t < +\infty \Big\},$$

and with $\gamma^+(p)$ and $\gamma^-(p)$ respectively the positive and negative semiorbit through p,

$$\gamma^+(p) := \Big\{ x \,\Big|\, x = \xi(t; p), 0 \le t < +\infty \Big\},$$

$$\gamma^-(p) := \Big\{ x \,\Big|\, x = \xi(t; p), -\infty < t \le 0 \Big\}.$$

We say that a point q belongs to the *ω-limit* or *positive limit set* of $\omega(\gamma)$ of γ if $\xi(t_k; p) \to q$ for some sequence $\{t_k\}$ such that $t_k \to +\infty$; similarly, q belongs to the *α-limit* or *negative limit* $\alpha(\gamma)$ of γ if $\gamma(t_k; p) \to q$ for some sequence $\{t_k\}$ such that $t_k \to -\infty$.

It is easily seen that the ω-limit and the α-limit of γ are given respectively by

$$\omega(p) := \bigcap_{\tau \in \mathbb{R}} \overline{\bigcup_{t \ge \tau} \xi(t; p)} \qquad \alpha(p) := \bigcap_{\tau \in \mathbb{R}} \overline{\bigcup_{t \le \tau} \xi(t; p)}.$$

Finally, we say that $M \subset \mathbb{R}^2$ is an *invariant* (respectively *positively invariant*) set for the system $\xi' = f(\xi)$ if $\gamma(p) \in M$ (respectively $\gamma^+(p) \in M$) for all $p \in M$. An orbit is invariant by definition.

6.23 ¶. Prove the following.

(i) For the system

$$\begin{cases} t' = 1, \\ x' = x, \end{cases}$$

or, equivalently, for the equation $x' = x$, the ω-limit is empty, and the α-limit is $\{0\}$.

(ii) For the system

$$\begin{pmatrix} x \\ y \end{pmatrix}' = \begin{pmatrix} y \\ -x \end{pmatrix} + (1 - x^2 - y^2) \begin{pmatrix} x \\ y \end{pmatrix},$$

the ω-limit of all trajectories, except the 0-trajectory, is the unit circle.

6.24 Proposition. *We have:*

(i) $\omega(p)$ *and* $\alpha(p)$ *are closed and invariant. For all* $q \in \omega(p)$ *we have* $\gamma(q) \subset \omega(p)$; *in particular,* $\omega(q) \subset \omega(p)$ *and* $\alpha(q) \subset \omega(p)$.
(ii) *If* $\gamma^+(p)$ *(respectively,* $\gamma^-(p)$*) is bounded, then* $\omega(p)$ *(respectively,* $\alpha(p)$*) is nonempty, compact, and connected, and* dist $(\xi(t; p), \omega(p)) \to 0$ *as* $t \to +\infty$ *(respectively* dist $(\xi(t; p), \alpha(p)) \to 0$ *as* $t \to -\infty$*).*

Proof. (i) The closure is trivial. Let $q \in \omega(p)$ and $t \in \mathbb{R}$. There exists $\{t_k\}$, $t_k \to +\infty$, such that $\xi(t_k; p) \to q$. It follows that $\xi(t + t_k; p) = \xi(t; \xi(t_k, p)) \to \xi(t; q)$ because of the continuous dependence on the initial data, hence the orbit through q is contained in $\omega(p)$.

(ii) Let us prove the claim for the ω-limit. If $\gamma^+(p)$ is bounded, clearly $\omega(p)$ is bounded, hence compact. In particular, since $\omega(p)$ is made of limit points of $\gamma^+(p)$, it is nonempty. Finally, we trivially have

$$\text{dist}\,(\xi(t, p), \omega(p)) \to 0 \qquad \text{as } t \to +\infty.$$

It remains to prove that $\omega(p))$ is connected. If it is not connected, we can find two compact sets K_1 and K_2 such that $\omega(p) = K_1 \cup K_2$ and $K_1 \cap K_2 = \emptyset$, and, consequently, two disjoint open sets U_1 and U_2 such that $K_1 \subset U_1$ and $K_2 \subset U_2$. Since

$$\text{dist}\,(\xi(t, p), \omega(p)) \to 0 \qquad \text{as } t \to +\infty,$$

$\gamma(t; p) \in U_1 \cup U_2$ for t large. Since $\gamma(t; p) \in U_1$ and $\gamma(s; p) \in U_2$ for some t, s, it follows that the set $\{x = \gamma(t; p) \mid t \text{ large}\}$ is not connected, and this is absurd. $\qquad \square$

By using Zorn's lemma one can show the following.

6.25 Proposition. *Every compact and invariant set contains a minimal invariant set.*

6.26 Example. The circle $\{r = 1\}$ is an invariant set for the system in (ii) Exercise 6.23 that in polar coordinates reads as

$$\begin{cases} \theta' = \sin^2 \theta - (1 - r)^3, \\ r' = r(1 - r), \end{cases}$$

and the minimal invariant sets in $\{r = 1\}$ are $\theta = 0$ and $\theta < \pi$.

Finally, we prove the following.

6.27 Proposition. *If K is a positively invariant set and K is homeomorphic to the unit ball, then K contains at least a critical point.*

Proof. For any $\tau_1 > 0$, consider the map $K \to K$ taking $p \in K$ to $\xi(\tau_1; p)$. From Brouwer's fixed point theorem there is $p_1 \in K$ such that $\xi(\tau_1; p_1) = p_1$, hence a periodic orbit of period τ_1. Similarly, for $\tau_m > 0$, $\tau_m \to 0$ as $m \to +\infty$, we find p_m with $\xi(\tau_m; p_m) = p_m$, and we may assume that $p_m \to p^*$, by taking a subsequence. For all t and all integers m, there exists an integer $k_m(t)$ such that $k_m(t)\tau_m \le t < k_m(t)\tau_m + \tau_m$ and $\xi(k_m(t)\tau_m; p_m) = p_m$ for all t, since $\xi(t, p_m)$ is periodic of period τ_m in t. We therefore find

$$|\xi(t; p^*) - p^*| \le |\xi(t; p^*) - \xi(t; p_m)| + |\xi(t; p_m) - p_m| + |p_m - p^*|$$
$$= |\xi(t; p^*) - \xi(t; p_m)| + |\xi(t - k_m(t)\tau_m; p_m) - p_m| + |p_m - p^*|,$$

and, the right-hand side being infinitesimal as $m \to \infty$, we conclude $\xi(t; p^*) = p^*$, i.e., p^* is a critical point. □

6.3.2 Poincaré–Bendixson theorem

6.28 Theorem (Poincaré–Bendixson). *Let p be a point in $\mathbb{R}^2$ such that the solution $\xi(t; p)$ of $\xi' = f(\xi)$ is defined for all $t > 0$ and is bounded. The following holds.*

(i) *Either $\omega(p)$ is a closed orbit, or for all $q \in \omega(p)$, f vanishes on $\omega(q)$. Consequently, if $\omega(p)$ does not contain any zero of f, $\omega(p)$ is a closed orbit; moreover,*
   (a) either $\gamma^+(p) = \omega(p)$,
   (b) or $\omega(p) = \overline{\gamma^+(p)} \setminus \gamma^+(p)$, i.e., $\gamma^+(p)$ spirals $\omega(p)$.
(ii) *Similarly, either $\alpha(p)$ is a closed orbit, or for all $q \in \alpha(p)$, f vanishes on $\alpha(q)$. Consequently, if $\alpha(p)$ does not contain any zero of f, $\alpha(p)$ is a closed orbit; moreover,*
   (a) either $\gamma^-(p) = \alpha(p)$,
   (b) or $\alpha(p) = \overline{\gamma^-(p)} \setminus \gamma^-(p)$, i.e., $\gamma^-(p)$ spirals from $\alpha(p)$.

The following lemmas are useful to prove Theorem 6.28.

6.29 Lemma (Monotonicity). *Let $I \subset \mathbb{R}^2$ be a transversal segment to f. If $\xi(t)$ is a solution of $\xi' = f(\xi)$ that meets I in three points, $A_i = \xi(t_i)$ with $t_1 < t_2 < t_3$, then A_2 is in between A_1 and A_3 in I.*

Proof. The set $\xi([t_1, t_2])$ union the segment of extreme points A_1 and A_2 forms a closed curve in $\mathbb{R}^2$. From Jordan's theorem, this curve bounds a region U. The vector field f along I either enters or exits the region; by possibly changing sign to f, we can assume that it enters. Then, $\xi([t_2, \infty[)$ lies in U and the part of $I \setminus A_2$ containing A_1 is on the boundary of U or outside U. Therefore, A_3 is on the connected component of $I \setminus A_2$ that does not contain A_1. □

6.30 Lemma. *Let I be a transversal segment to f and let $A \in I$. For all $\epsilon > 0$ there exists $r > 0$ such that every orbit that at time $t = 0$ is in $\partial B(A, r)$ goes through I in a time t_0 with $|t_0| < \epsilon$.*

Proof. We may assume that A is the origin and I is along the x-axis. If $\xi(t;(x_0,y_0)) = (x(t,x_0,y_0), y(t,x_0,y_0))$ is the trajectory through $p := (x_0,y_0)$, then by assumption $\frac{\partial y}{\partial t}(0,0,0) \neq 0$, since the x-axis is transversal to the trajectory at $y(0,0,0) = 0$. The implicit function theorem yields the result. $\qquad\square$

6.31 Lemma. *Let $p \in \mathbb{R}^n$ and let I be an open segment that is transversal to f at p. Then $I \cap \omega(p)$ contains at most a point.*

Proof. Since $z \in \omega(p)$, there exists a sequence $\{t_j\}$, $t_j \to \infty$ such that $\xi(t_j;p) \to z$. On the other hand, since $z \in I \cap \omega(p)$, for any $\epsilon > 0$, we can find a neighborhood V_ϵ of I such that every point in V_ϵ flows to I in a time less than ϵ, by Lemma 6.30. Therefore, we find $\{t'_j\}$ with $t'_j \to +\infty$ such that $\xi(t'_j;p) \in I$ and $\xi(t'_j;p) \to z$. But by Lemma 6.29, $\xi(t'_j;p) \to z$ monotonically. Thus $I \cap \omega(x)$ cannot contain more than one point. $\qquad\square$

Proof of Theorem 6.28. Let p be as in the claim, $q \in \omega(p)$ and $z \in \omega(q)$ with $f(z) \neq 0$. We can find a segment I_z transverse to f. We take a sequence $\{t_j\} \to \infty$ such that $\xi(t_j;q) \to z$, and moving along the flow, we find a sequence $\{t'_j\}$ with $t'_j \to +\infty$, $\xi(t_j;q) \in I_z$, and $\xi(t'_j;q) \to q$. Thus, $\gamma^+(q)$ intersects I_z, in particular

$$\emptyset \neq \gamma^+(q) \cap I_z \subset \omega(p) \cap I_z.$$

Since $I_z \cap \omega(p)$ contains at most one point by Lemma 6.31, $\gamma^+(q)$ intersects I_z at a unique point that must be z since $z \in \omega(q) \cap I_z \subset \omega(p) \cap I_z$. Hence, there exists $t, s \in \mathbb{R}$, $t > s$, such that $\xi(t;q) = \xi(s;q) = z$. If we set $\tau := t - s$, we have

$$z = \xi(\tau + s;q) = \xi(\tau;\xi(s;q)) = \xi(\tau;z),$$

i.e., $\gamma_+(q)$ is periodic of period τ.

It remains to show that $\omega(p) = \gamma_+(q)$. We notice that $\gamma_+(q)$ is closed, since it is periodic; hence $\omega(p) \setminus \gamma^+(q)$ is open in $\omega(p)$. If $\omega(p) \neq \gamma_+(q)$, we find a sequence of points $\{z_n\} \subset \omega(p) \setminus \gamma^+(q)$ such that $z_n \to z \in \gamma^+(q)$. Since $\omega(p)$ is invariant, the orbits through z_n are all contained in $\omega(p)$; moreover, moving along the flow, we find $\{y_n\}$ with $y_n \in \omega(p) \cap I_z$ and $y_n \to z$. But Lemma 6.31 yields $y_n = z$ for large n, hence $z_n \in \gamma^+(q)$ for large n, a contradiction. $\qquad\square$

The Poincaré–Bendixson theorem can be used to prove the existence of periodic solutions of an autonomous system $x' = f(x)$ in the plane, provided one is able to find a domain $\Omega \subset \mathbb{R}^2$ possibly invariant and without critical points.

6.3.3 Systems on a torus

In this section we deal with the trajectories of a first-order system

$$\begin{cases} \varphi' = \Phi(\varphi,\theta), \\ \theta' = \Theta(\varphi,\theta) \end{cases} \tag{6.22}$$

where Φ and Θ are doubly periodic given functions with

$$\Phi(\varphi+1,\theta) = \Phi(\varphi,\theta) = \Phi(\varphi,\theta+1),$$
$$\Theta(\varphi+1,\theta) = \Theta(\varphi,\theta) = \Theta(\varphi,\theta+1),$$

V. ARNOLD

ÉQUATIONS
DIFFÉRENTIELLES
ORDINAIRES

CHAMPS DE VECTEURS
GROUPES À UN PARAMÈTRE
DIFFÉOMORPHISMES
FLOTS
SYSTÈMES LINÉAIRES
STABILITÉS DES POSITIONS
D'ÉQUILIBRE
THÉORIE DES OSCILLATIONS
ÉQUATIONS DIFFÉRENTIELLES
SUR LES VARIÉTÉS

ÉDITIONS MIR·MOSCOU

Figure 6.7. Jurgen Moser (1928–1999) and a book by Vladimir Arnold (1937–) on ODEs.

$\Phi(x, y)$ and $\Theta(x, y)$ are continuous in $\mathbb{R}^2$ and never zero, hence bounded and bounded away from zero. Moreover, we shall assume that for every given initial condition, (6.22) has a unique solution. Since Φ never vanishes, $t \to \varphi(t)$ is strictly monotone, hence invertible. Therefore, the trajectory of every solution $t \to (\varphi(t), \theta(t))$ of (6.22) is the graph of a function that we denote by $\theta = \theta(\varphi)$ and solves the first-order equation

$$\frac{d\theta}{d\varphi} = A(\varphi, \theta), \qquad A(\varphi, \theta) := \frac{\Theta(\varphi, \theta)}{\Phi(\varphi, \theta)}. \tag{6.23}$$

Of course,

$$A(\varphi + 1, \theta) = A(\varphi, \theta + 1) = A(\varphi, \theta) \tag{6.24}$$

and $A(\theta, \varphi)$ is bounded and bounded away from zero. In particular the orbits $\theta = \theta(\varphi)$ are defined for all $\varphi \in \mathbb{R}$.

By identifying the opposite sides of a square of size 1 in the plane (φ, θ), we may interpret the system (6.22) as a system of differential equations on a flat torus.

6.32 ¶. Suppose $\Phi = 1$, $\Theta = \omega$ constant, hence $A(\varphi, \theta) = \omega$. Prove that the orbits are closed, equivalently, the solutions of (6.23) are periodic, if ω is rational. Prove that the orbits are dense in the torus if ω is irrational.

6.33 ¶. Suppose $A(\varphi, \theta) := \sin 2\pi\theta$. Prove that $\theta = 0$ and $\theta = 1/2$ are two closed orbits and that every orbit through θ_0 with $0 < \theta_0 < 1$ has $\theta = 1/2$ as ω–limit and $\theta = 0$ as α–limit.

Every orbit goes through the meridian $C := \{(\varphi, \theta) \,|\, \varphi = 0\}$, therefore, it suffices to restrict to the initial values $(0, \xi)$, $\xi \in \mathbb{R}$. Let $\theta(\varphi, \xi)$ be the solution through $(0, \xi)$. On account of uniqueness $\theta(\varphi, \xi)$ is increasing in ξ for every fixed φ and

$$\theta(\varphi, \xi + m) = \theta(\varphi, \xi) + m, \qquad \forall \varphi \in I,$$
$$\theta(m, \theta(n, \xi)) = \theta(n, \theta(m, \xi)) = \theta(m + n, \xi) \qquad \forall n, m \in \mathbb{Z}, \ \forall \xi \in \mathbb{R}.$$

In particular, the map $T := \mathbb{R} \to \mathbb{R}$, given by $\xi \to \theta(1, \xi)$, is continuous and increasing, hence a homeomorphism of $\mathbb{R}$, and

$$T^0(\xi) = \xi, \qquad T^n(\xi) = \xi(n, \xi), \qquad T^{n+m}(\xi) = T^n(T^m(\xi)),$$

for all $n, m = 0, \pm 1, \pm 2, \ldots$.

6.34 Theorem. *The limit, called the rotation number of* (6.22),

$$\rho := \lim_{n \to \infty} \frac{\theta(n; \xi)}{n} = \lim_{n \to \infty} \frac{T^n(\xi)}{n},$$

exists and is independent of ξ. Moreover, ρ is rational if and only if a power of T has a fixed point, i.e., if and only if (6.22) *has a periodic orbit.*

Proof. For $0 \leq \xi, \overline{\xi} \leq 1$, we have
$$\theta(\varphi; \overline{\xi} - 1) = \theta(\varphi, \overline{\xi} - 1) \leq \theta(\varphi, \xi) \leq \theta(\varphi, \overline{\xi} + 1) = \theta(\varphi, \overline{\xi}) + 1;$$
this implies that ρ is independent of ξ. If $0 \leq \xi - m \leq 1$, $m \in \mathbb{N}$, we have
$$\theta(\varphi, 0) \leq \theta(\varphi, \xi) - m \leq \theta(\varphi, 0) + 1,$$
$$\theta(\varphi, 0) - 1 \leq \theta(\varphi, \xi) - \xi \leq \theta(\varphi, 0) + 1,$$
in particular,
$$\theta(m, 0) - 1 \leq \theta(m, \xi) - \xi \leq \theta(m, 0) + 1,$$
$$n\theta(m, 0) - n \leq \theta(nm, 0) \leq n\theta(m, 0) + n,$$
$$n\theta(-m, 0) - n \leq \theta(-nm, 0) \leq n\theta(-m, 0) + n,$$
from which we get
$$\left| \frac{\theta(nm, 0)}{nm} - \frac{\theta(m, 0)}{m} \right| \leq \frac{1}{|m|},$$
$$\left| \frac{\theta(nm, 0)}{nm} - \frac{\theta(n, 0)}{n} \right| \leq \frac{1}{|n|}$$
and, finally,
$$\left| \frac{\theta(n, 0)}{n} - \frac{\theta(m, 0)}{m} \right| \leq \frac{1}{|n|} + \frac{1}{|m|}.$$
The existence of the limit ρ and the estimate
$$\left| \rho - \frac{\theta(m, 0)}{m} \right| \leq \frac{1}{|m|}$$

now follow at once.

If $T^m \xi = \xi$, then there exists an integer k such that $\theta(m, \xi) = \xi + k$ and
$$\rho = \lim_{|n| \to \infty} \frac{\theta(nm, \xi)}{nm} = \lim_{|n| \to \infty} \frac{\xi + nk}{nm} = \frac{k}{m},$$

i.e., ρ is rational. Conversely, suppose $\rho = k/m$ and that T^m has no fixed point, in particular, $\theta(m, \xi) \neq \xi + k$. Suppose $\theta(m, \xi) > \xi + k \ \forall \xi \in [0, 1[$, equivalently, let $a > 0$ be such that $\theta(m, \xi) - \xi - k \geq a > 0 \ \forall \xi \in [0, 1[$. For all $\zeta \in \mathbb{R}$ we deduce $\theta(m, \zeta) - \zeta - k \geq a$ and, iterating, $\theta(rm, \xi) - \xi \geq r(k + a) \ \forall r$. Dividing by rm, and letting $r \to \infty$, we conclude $\rho \geq \frac{k}{m} + \frac{a}{m}$, a contradiction. $\quad\square$

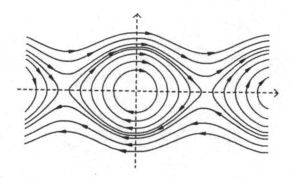

Figure 6.8. The orbits of the pendulum equation.

6.35 Theorem. *If the rotation number ρ of the system* (6.22) *is rational, then every orbit of* (6.22) *on the torus is either periodic or converges to a closed curve.*

Proof. Since ρ is rational, there exists a closed trajectory γ that intersects every meridian of the torus T. Therefore, $T \setminus \gamma$ is equivalent to an annulus Γ and the differential system is equivalent to a differential system on Γ without critical points. The result then follows from the Poincaré–Bendixson theorem.   □

One can carry on the analysis to cover the case of irrational rotation numbers. For every $\xi \in \mathbb{R}$, i.e., for every point $P = (0, \xi)$ in the meridian $\{(\varphi, \theta) \,|\, \varphi = 0\}$, we set

$$D(\xi) := \left\{ T^n \xi \,\middle|\, n \in \mathbb{Z} \right\} \subset \mathbb{R},$$

and denote by $D(\xi)'$ the set of limit points $D(\xi)$. We state the following without proof.

6.36 Theorem. *Suppose the rotation number ρ is irrational. Then the set $F := D(\xi)'$ is independent of $\xi \in \mathbb{R}$ and is invariant under T, $T(F) = F$. Moreover, only one of the following two situations can occur*

 (i) (ERGODIC CASE) $F = \mathbb{R}$,
 (ii) *F is a Cantor type set, i.e., F has no isolated points and its closure has no interior point.*

Finally, in the case that T has continuous first derivative, $T' > 0$ and T has bounded variation, then $F = \mathbb{R}$.

For further information the reader is referred to one of the monographs in the final bibliographical remarks.

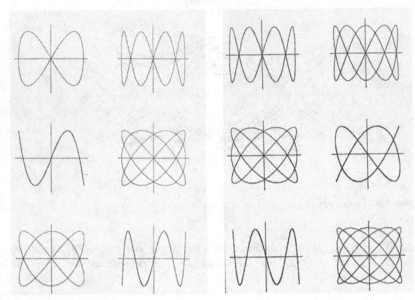

Figure 6.9. The trajectories of $z(t) = (\alpha \sin(\gamma t + \varphi), \beta \sin t)$ with $\alpha = 9$, $\beta = 8$ and from the top-left (a) $\varphi = 0$, $\gamma = 1/2$, $1/3$, $2/3$, $1/4$, $3/4$, $1/5$. (b) $\varphi = \pi$, $\gamma = 1/4$, $3/4$, $1/5$, $2/5$, $3/5$, $4/5$.

6.4 Exercises

6.37 ¶. Find the general integral of the equation

$$y' = \frac{x - y}{x + 2y}.$$

6.38 ¶. The solutions of

$$y' = x + \frac{x^2}{1 + x^2}$$

are globally defined in $\mathbb{R}$. Find their asymptotic development when $x \to +\infty$ modulus $o(1/x)$ terms.

6.39 ¶. Let $f : \mathbb{R}^n \to \mathbb{R}$ be of class $C^2(\mathbb{R})$. Prove that, if $\nabla f(0) = 0$ and $\mathbf{H}f(0) < 0$, then 0 is a point of stable equilibrium for the system $x'' = \nabla f(x)$.

6.40 ¶. Let $\mathbf{A} \in M_{n,n}(\mathbb{R})$ be a nonnegative symmetric matrix. Prove that the behavior of every solution of $x'' = \mathbf{A}x$ in a suitable orthonormal basis is a simple harmonic motion. When $n = 2$, the trajectories of the solutions of $x'' = \mathbf{A}x$ form the so-called *Lissajous figures*, see Figure 6.9.

6.41 ¶. Consider the equation

$$x' = x^3 - x + \lambda \sin t, \qquad |\lambda| < \frac{2}{3\sqrt{3}}.$$

Prove the following.

(i) If $\alpha(t)$ is a solution in $[0,T]$ with $|x(0)| < 1/\sqrt{3}$, then $|x(t)| \le 1/\sqrt{3}$ $\forall t \in [0,T]$.

(ii) For all x_0 with $|x_0| < 1/\sqrt{3}$, there is a solution of the Cauchy problem in $[0, +\infty[$ with initial condition $x(0) = x_0$ in $[0,\infty[$.

(iii) There is a periodic solution with period 2π.

6.42 ¶. Consider the differential system

$$\begin{cases} x' = -y + (x^2 + y^2)x, \\ y' = x + (x^2 + y^2)y \end{cases} \qquad (6.25)$$

and its linearization

$$\begin{cases} x' = -y, \\ y' = x \end{cases}$$

that has the periodic solutions $z(t) = x(t) + iy(t) = iAe^{it}$, $A \in \mathbb{C}$. Prove that (6.25) has no periodic solution near the origin. [*Hint:* Notice that $\frac{d}{dt}\frac{1}{2}(x^2 + y^2) = (x^2 + y^2)^4$.]

A. Mathematicians and Other Scientists

George Airy (1801–1892)
Stefan Banach (1892–1945)
Eugenio Beltrami (1835–1899)
Ivar Otto Bendixson (1861–1935)
Johann Bernoulli (1667–1748)
Jacques Binet (1786–1856)
Ludwig Boltzmann (1844–1906)
L. E. Brouwer (1881–1966)
Constantin Carathéodory (1873–1950)
Felice Casorati (1835–1890)
Augustin-Louis Cauchy (1789–1857)
Bonaventura Cavalieri (1598–1647)
Pafnuty Chebyshev (1821–1894)
Richard Courant (1888–1972)
Jean d'Alembert (1717–1783)
Gaston Darboux (1842–1917)
Ulisse Dini (1845–1918)
Diocles (240BC–180BC)
Lejeune Dirichlet (1805–1859)
Jean–Marie Duhamel (1797–1872)
Leonhard Euler (1707–1783)
Pierre Fatou (1878–1929)
Pierre de Fermat (1601–1665)
Enrico Fermi (1901–1954)
Richard Feynman (1918–1988)
Leonardo Pisano (1170–1250), called
    Fibonacci
Joseph Fourier (1768–1830)
Maurice Fréchet (1878–1973)
Ivar Fredholm (1866–1927)
Jean Frenet (1816–1900)
Augustin Fresnel (1788–1827)
Guido Fubini (1879–1943)
René Gateaux (1889–1914)
Carl Friedrich Gauss (1777–1855)
Edouard Goursat (1858–1936)
George Green (1793–1841)
Thomas Grönwall (1877–1932)
Paul Guldin (1577–1643)
Jacques Hadamard (1865–1963)
Hans Hahn (1879–1934)
Felix Hausdorff (1869–1942)
Charles Hermite (1822–1901)
Otto Hesse (1811–1874)

David Hilbert (1862–1943)
Otto Hölder (1859–1937)
Robert Hooke (1635–1703)
Adolf Hurwitz (1859–1919)
Christiaan Huygens (1629–1695)
Carl Jacobi (1804–1851)
Johan Jensen (1859–1925)
Camille Jordan (1838–1922)
Paul Koebe (1882–1945)
Olga Ladyzhenskaya (1922–2004)
Joseph-Louis Lagrange (1736–1813)
Pierre-Simon Laplace (1749–1827)
Pierre Laurent (1813–1854)
Henri Lebesgue (1875–1941)
Jean Leray (1906–1998)
Beppo Levi (1875–1962)
Sophus Lie (1842–1899)
Ernst Lindelöf (1870–1946)
Joseph Liouville (1809–1882)
Rudolf Lipschitz (1832–1903)
Jules Lissajous (1822–1880)
Nikolai Lusin (1883–1950)
Aleksandr Lyapunov (1857–1918)
Jozef Marcinkiewicz (1910–1940)
Lorenzo Mascheroni (1750–1800)
Hjalmar Mellin (1854–1933)
John Milnor (1931–)
Hermann Minkowski (1864–1909)
Gösta Mittag-Leffler (1846–1927)
August Möbius (1790–1868)
Paul Montel (1876–1975)
Giacinto Morera (1856–1909)
Harald Marston Morse (1892–1977)
Jurgen Moser (1928–1999)
Claude Navier (1785–1836)
William Neile (1637–1670)
Sir Isaac Newton (1643–1727)
Giuseppe Peano (1858–1932)
Émile Picard (1856–1941)
Max Planck (1858–1947)
J. Henri Poincaré (1854–1912)
Siméon Poisson (1781–1840)
Hans Rademacher (1892–1969)
G. F. Bernhard Riemann (1826–1866)

Frigyes Riesz (1880–1956)
Eugène Rouché (1832–1910)
Arthur Sard (–)
Erhard Schmidt (1876–1959)
Erwin Schrödinger (1887–1961)
Hermann Schwarz (1843–1921)
Joseph Serret (1819–1885)
Jakob Steiner (1796–1863)
James Stirling (1692–1770)
George Gabriel Stokes (1819–1903)
James Joseph Sylvester (1814–1897)
Brook Taylor (1685–1731)

Leonida Tonelli (1885–1946)
Stanislaw Ulam (1909–1984)
Giuseppe Vitali (1875–1932)
Vincenzo Viviani (1622–1703)
John von Neumann (1903–1957)
Karl Weierstrass (1815–1897)
Hassler Whitney (1907–1989)
Hoëné Wronski (1778–1853)
William Young (1863–1942)
Max Zorn (1906–1993)
Antoni Zygmund (1900–1992)

There exist many web sites dedicated to the history of mathematics; we mention, e.g.,
http://www-history.mcs.st-and.ac.uk/~history.

B. Bibliographical Notes

We collect here a few suggestions for the readers interested in delving deeper into some of the topics treated in this volume. Of course, the list could be either longer or shorter. It reflects the taste and the knowledge of the authors.

General references:

o H. Cartan, *Calcul Différentiel*, Hermann, Paris, 1977.
o E. Di Benedetto, *Real Analysis*, Birkhäuser, Basel, 2002.
o J. Dieudonné, *Élements d'Analyse*, Gauthiers–Villars, Paris, 1968.
o J. J. Duistermatt, J. A. C. Kolk, *Multidimensional Real Analysis*, vols. 2, Cambridge University Press, Cambridge, 2004.
o W. H. Fleming, *Functions of Several Variables*, Addison-Wesley, Reading, MA, 1966.
o S. Hildebrandt, *Analysis 2*, Springer-Verlag, Berlin, 2003.
o J. Jost, *Postmodern Analysis*, Springer-Verlag, Berlin, 1998.
o J. T. Schwartz, *Nonlinear Functional Analysis*, Gordon and Breach, New York, 1969.
o W. Rudin, *Principles of Mathematical Analysis*, Madison, 1953.

About the *theory of holomorphic functions*:

o L. V. Ahlfors, *Complex Analysis*, McGraw-Hill, New York, 1979.
o H. Cartan, *Théorie des Fonctions Analytiques*, Hermann, Paris, 1961.
o J. B. Conway, *Functions of one Complex Variable*, vols. 2, Springer-Verlag, New York, 1978.
o E. Hille, *Analytic Function Theory*, vols. 2, Ginn and Co., Boston, 1959.
o M. Lavrentiev, B. Chabat, *Méthodes de la Théorie des Fonctions d'une Variable Complexe*, Éditions Mir, Moscow, 1972.
o G. Sansone, J. Gerretsen, *Lectures on the Theory of Functions of a Complex Variable*, vols. 2, P. Noordhoff, Gröningen, 1960.
o E. M. Stein, R. Shakarchi, *Complex Analysis*, Princeton University Press, Princeton, NJ, 2003.

About the *theory of surfaces*:

o M. do Carmo, *Differential Geometry of Curves and Surfaces*, Prentice Hall, Englewood Cliffs, NJ, 1976.
o B. A. Dubrovin, A. T. Fomenko, S. P. Novikov, *Modern Geometry, Methods and Applications*, vols. 3, Springer-Verlag, Berlin, 1984.
o J. Milnor, *Topology from the Differential Viewpoint*, Princeton University Press, Princeton, NJ, 1977.
o M. Spivak, *Differential Geometry*, vols. 5, Publish or Perish, Berkeley, 1979.

About *differential forms*:

o M. do Carmo, *Differential Forms and Applications*, Springer-Verlag, New York, 1994.
o H. F. Flanders, *Differential Forms with Applications to the Physical Sciences*, Dover, New York, 1989.
o M. Spivak, *Calculus on Manifolds*, Benjamin, New York, 1965.

About Fourier series:

o B. Burke Hubbard, *Ondes et Ondelettes*, Pour la Science, Paris, 1985.
o J. P. Kahane, P. G. Lemarié–Rieusset, *Série de Fourier et Ondelettes*, Cassini, Paris, 1998.
o T. W. Körner, *Fourier Analysis*, Cambridge University Press, Cambridge, 1996.
o E. M. Stein, R. Shakarchi, *Fourier Analysis*, Princeton University Press, Princeton, NJ, 2003.

About *ODEs*:

o H. Amann, *Ordinary Differential Equations, An Introduction to Nonlinear Analysis*, W. de Gruyter, Berlin, 1990.
o V. Arnold, *Équations Différentielles Ordinaires*, Éditions Mir, Moscow, 1978.
o V. Arnold, *Chapitres Supplémentaires sur la Théorie des Équations Différentielles Ordinaires*, Éditions Mir, Moscow, 1980.
o F. Brauer, J. A. Nohel, *The Qualitative Theory of Ordinary Differential Equations, An Introduction*, W. A. Benjamin Inc., New York, 1969, Dover, New York, 1989.
o J. Hale, *Ordinary Differential Equations*, R. E. Krieger, Malabar, 1980.
o P. Hartmann, *Ordinary Differential Equations*, Wiley, New York, 1964.
o J. H. Hubbard, B. H. West, *Differential Equations: A Dynamical System Approach*, Springer-Verlag, New York, 1995.
o W. Hurewicz, *Lectures on Ordinary Differential Equations*, MIT Press, Cambridge, MA, 1958, Dover, New York, 1990.
o A. Katok, B. Hasselblatt, *Introduction to the Modern Theory of Dynamical Systems*, Ency. Mat. Appl. 54, Cambridge University Press, Cambridge, 1995.
o G. Sansone, R. Conti, *Nonlinear Differential Equations*, Pergamon Press, New York, 1964.

C. Index